Current Topics in Behavioral Neurosciences

For further volumes:
http://www.springer.com/series/7854

About this Series

Current Topics in Behavioral Neurosciences provides critical and comprehensive discussions of the most significant areas of behavioral neuroscience research, written by leading international authorities. Each volume offers an informative and contemporary account of its subject, making it an unrivalled reference source. Titles in this series are available in both print and electronic formats.

With the development of new methodologies for brain imaging, genetic and genomic analyses, molecular engineering of mutant animals, novel routes for drug delivery, and sophisticated cross-species behavioral assessments, it is now possible to study behavior relevant to psychiatric and neurological diseases and disorders on the physiological level. The *Behavioral Neurosciences* series focuses on "translational medicine" and cutting-edge technologies. Preclinical and clinical trials for the development of new diagostics and therapeutics as well as prevention efforts are covered whenever possible.

Marie-Christine Pardon
Mark W. Bondi
Editors

Behavioral Neurobiology of Aging

Dr. Marie-Christine Pardon
School of Biomedical Sciences
Queen's Medical Centre
University of Nottingham Medical School
Nottingham NG7 2UH
UK

Dr. Mark W. Bondi
Division of Geriatric Psychiatry
University of California, San Diego
Gilman Drive 9500
La Jolla CA 92093-0603
USA

ISSN 1866-3370 e-ISSN 1866-3389
ISBN 978-3-642-23874-1 e-ISBN 978-3-642-23875-8
DOI 10.1007/978-3-642-23875-8
Springer Heidelberg New York Dordrecht London

Library of Congress Control Number: 2012932741

Printed on acid-free paper

Springer is part of Springer Science+Business Media (www.springer.com)

Preface

Modern societies face the substantial economic and social challenge of an unprecedented increase in life expectancy, given the rise in chronic medical and systemic conditions associated with aging. In addition, the baby boom generation has begun to reach retirement age. The result of this expansion of the demographic make-up of our societies will be that, by 2025, one third of the population of developed countries will be aged over 60 years (10% worldwide)[1]. This surge will be associated with a particularly rapid increase in the number of older adults aged 80 years and older and accompanied by increases in the prevalence of age-related disorders as well. For example, the prevalence of Alzheimer's disease worldwide was about 26 million in 2006 and is expected to quadruple to more than 106 million by 2050.

Public health perspectives aside, people do not want to simply live longer; they want to age successfully and remain physically and mentally active in their later years. Improving the quality of life in our later years must start by understanding when and how functional declines of the central nervous system occur. Major advances in our understanding of brain aging and, in particular, the distinction between normal and pathological aging are therefore required before suitable preventive and curative strategies can be developed. In this volume we present the current state of research findings related to healthy brain and cognitive aging by integrating contributions from leading authorities on human clinical studies and translational research in animal models. The goals of such cross-disciplinary coverage are to lessen compartmentalization within one's own discipline, encourage communication across basic and clinical science areas, generate seed-beds of hypothesis generation, and ultimately maximize the potential for seamless translation of discoveries to clinical application.

In the opening chapter of this volume, Drs. Hayden and Welsh-Bohmer provide an overview of the determinants of cognitive aging and dementia. Their epidemiologic study and analysis is followed by several chapters describing age-related

[1] World Health Organization, http://www.who.int/whr/1998/media_centre/50facts/en/.

changes in cognition and emotion. Dr. Depp and colleagues review the definition and biological, psychological, and environmental determinants of successful cognitive aging while Drs. Kaszniak and Menchola provide a comprehensive overview of the behavioral neuroscience of emotion and creativity in human aging. Dr. Marighetto and colleagues critically discuss the preclinical studies examining the impact of aging on memory systems and how they can be translated to humans.

Three subsequent chapters then explore changes in the brain that accompany normal aging. Dr. Guidotti-Breting and colleagues give a thorough overview of advances in our understanding of normal aging achieved through the use of functional neuroimaging as well as important avenues for future research while Drs. Woodard and Sugarman offer insights into how such functional neuroimaging techniques can allow for the differentiation between normal aging and dementia and help predict cognitive decline. Then, Drs. Juraska and Lowry provide a detailed analysis of the neuroanatomical changes in the brain associated with age-related cognitive decline at the level of neuronal loss, white matter and synaptic changes, by integrating data from neuroimaging and stereological studies in human, nonhuman primates and rodents. They conclude by examining whether the course of neuroanatomical aging can be altered by hormone replacement in females. In the next chapter, Dr. Boulware and colleagues review studies in the same species that examine the effects of reproductive aging and hormone replacement on cognitive functions mediated by the hippocampus and prefrontal cortex.

The next series of chapters cover medical and psychiatric conditions that can negatively impact cognition in late life. This section starts with Dr. Salmon's examination of Mild Cognitive Impairment, a clinical condition characterized by significant cognitive impairment in the absence of dementia, but which frequently progresses to dementia. Then, Dr. Seidel and colleagues and Dr. Wijeratne, and colleagues highlight how functional consequences of cerebrovascular changes or psychiatric conditions in older adults exacerbate cognitive decline, respectively.

The closing chapters of this volume are devoted to an exploration of strategies to diminish and delay age-related cognitive declines, both pharmacologically and non-pharmacologically. Dr. Jak provides a critical summary of the ever-growing body of research focusing on participation in physical and cognitive activities among older adults and their impact on cognition, the brain, and cognitive aging outcomes. Drs. Redolat and Mesa-Gresa critically discuss preclinical work addressing the potential impact of physical exercise on cognition in aged rodents. Then, Dr. Kinsley and colleagues provide an overview of how reproductive experience delays the aging process in rats. Finally, Dr. Corey-Bloom presents an overview of the clinical trials for mild cognitive impairment, their limitations as well as the potential strategies for overcoming the identified problems in future trials.

This volume provides topics that will be useful to researchers, clinicians and students interested in the current knowledge and research challenges in neuro-biological perspectives in aging as well as future research directions in aging research.

Reference

Brookmeyer R, Johnson E, Ziegler-Graham K, Arrighi HM (2007) Forecasting the global burden of Alzheimer's disease. Alzheimer's Dementia 3(3):186–91

Nottingham, La Jolla

Marie-Christine Pardon
Mark W. Bondi

Contents

Part I Epidemiologic Perspective in Aging

Epidemiology of Cognitive Aging and Alzheimer's Disease: Contributions of the Cache County Utah Study of Memory, Health and Aging 3
Kathleen M. Hayden and Kathleen A. Welsh-Bohmer

Part II Cognitive and Emotional Perspectives in Aging

Successful Cognitive Aging 35
Colin A. Depp, Alexandria Harmell and Ipsit V. Vahia

Behavioral Neuroscience of Emotion in Aging 51
Alfred W. Kaszniak and Marisa Menchola

Studying the Impact of Aging on Memory Systems: Contribution of Two Behavioral Models in the Mouse 67
Aline Marighetto, Laurent Brayda-Bruno and Nicole Etchamendy

Functional Neuroimaging Studies in Normal Aging 91
Leslie M. Guidotti Breting, Elizabeth R. Tuminello and S. Duke Han

Functional Magnetic Resonance Imaging in Aging and Dementia: Detection of Age-Related Cognitive Changes and Prediction of Cognitive Decline 113
John L. Woodard and Michael A. Sugarman

Neuroanatomical Changes Associated with Cognitive Aging 137
Janice M. Juraska and Nioka C. Lowry

Part III Reproductive Aging

The Impact of Age-Related Ovarian Hormone Loss on Cognitive and Neural Function 165
Marissa I. Boulware, Brianne A. Kent and Karyn M. Frick

Part IV Medical and Psychiatric Factors in Aging

Neuropsychological Features of Mild Cognitive Impairment and Preclinical Alzheimer's Disease 187
David P. Salmon

Cerebrovascular Disease and Cognition in Older Adults 213
Gregory A. Seidel, Tania Giovannetti and David J. Libon

Psychiatric Disorders in Ageing 243
C. Wijeratne, S. Reutens, B. Draper and P. Sachdev

Part V Modifiers of Brain Aging

The Impact of Physical and Mental Activity on Cognitive Aging 273
Amy J. Jak

Potential Benefits and Limitations of Enriched Environments and Cognitive Activity on Age-Related Behavioural Decline 293
Rosa Redolat and Patricia Mesa-Gresa

Reproductive Experience may Positively Adjust the Trajectory of Senescence 317
Craig Howard Kinsley, R. Adam Franssen and Elizabeth Amory Meyer

Treatment Trials in Aging and Mild Cognitive Impairment 347
Jody Corey-Bloom

Erratum E1

Index 357

Contributors

Marissa I. Boulware Department of Psychology, University of Wisconsin-Milwaukee, 2441 E. Hartford Ave, Milwaukee, WI 53211, USA

Laurent Brayda-Bruno Neurocentre Magendie-Inserm U862, 146 Rue Leo Saignat, Bordeaux-Cedex 33077, France

Jody Corey-Bloom Shiley-Marcos Alzheimer Disease Research Center, University of California, San Diego, 8950 Villa La Jolla Drive (Suite C129), La Jolla, CA 92037, USA, e-mail: jcoreybloom@ucsd.edu

Colin A. Depp Sam and Rose Stein Institute for Research on Aging, San Diego, CA, USA; Department of Psychiatry, University of California, San Diego, CA, USA, e-mail: cdepp@ucsd.edu

B. Draper School of Psychiatry, University of New South Wales, Sydney, NSW, Australia

S. Duke Han Department of Behavioral Sciences, Rush University Medical Center, 1653 W. Congress Parkway, Chicago, IL 60612-3244, USA, e-mail: Duke_Han@rush.edu

Nicole Etchamendy NutriNeuro Bordeaux University, avenue des facultés, 33406 Talence, France

R. Adam Franssen Department of Biology, Longwood University, Farmville, VA, USA

Karyn M. Frick Department of Psychology, University of Wisconsin-Milwaukee, 2441 E. Hartford Ave, Milwaukee, WI 53211, USA, e-mail: frickk@uwm.edu

Tania Giovannetti Department of Psychology, Temple University, Philadelphia PA, USA, e-mail: tgio@temple.edu

Leslie M. Guidotti Breting Department of Psychiarty and Behavioral Sciences, NorthShore University HealthSystem, Evanston, IL, USA

Alexandria Harmell Joint Doctoral Program in Clinical Psychology, San Diego State University/University of California, San Diego, CA, USA

Kathleen M. Hayden Department of Psychiatry, Joseph and Kathleen Bryan Alzheimer's Disease Research; Center—Division of Neurology, 2200 W. Main Street, Suite A200, Durham, NC 27705, USA

Amy J. Jak Department of Psychology Service, Veteran's Affairs San Diego Healthcare System, San Diego, CA, USA; Department of Psychiatry School of Medicine, University of California, San Diego, CA, USA, e-mail: ajak@ucsd.edu

Janice M. Juraska Department of Psychology and Program in Neuroscience, University of Illinois, 603 E Daniel, Champaign, IL 61820, USA, e-mail: jjuraska@illinois.edu

Alfred W. Kaszniak Department of Psychology, University of Arizona, 1503 E. University, Tucson, AZ 85721, USA, e-mail: kaszniak@email.arizona.edu

Brianne A. Kent Department of Psychology, Yale University, New Haven, CT 06520, USA

Craig Howard Kinsley Department of Psychology, Center for Neuroscience, Gottwald Science Center and 116 Richmond Hall, University of Richmond, B-326/328, 28 Westhampton Way, Richmond, VA 23173, USA, e-mail: ckinsley@richmond.edu

David J. Libon Department of Neurology, Drexel University College of Medicine, Philadelphia, PA, USA, e-mail: David.Libon@drexelmed.edu

Nioka C. Lowry Department of Psychology, University of Illinois, 603 E Daniel, Champaign, IL 61820, USA, e-mail: nchisho2@illinois.edu

Aline Marighetto Neurocentre Magendie-Inserm U862, 146 Rue Leo Saignat, Bordeaux-Cedex 33077, France, e-mail: a.marighetto@cnic.u-bordeaux1.fr

Marisa Menchola Department of Psychology, University of Arizona, 1503 E. University, Tucson, AZ 85721, USA; Department of Family and Community Medicine, University of Arizona, 1450 N Cherry, Tucson, AZ 85719, USA

Patricia Mesa-Gresa Departamento de Psicobiología, Universitat de València, Blasco Ibáñez 21, 46010 Valencia, Spain, e-mail: Patricia.mesa@uv.es

Elizabeth Amory Meyer Department of Psychology, Center for Neuroscience, Gottwald Science Center and 116 Richmond Hall, University of Richmond, B-326/328, 28 Westhampton Way, Richmond, VA 23173, USA

Rosa Redolat Departamento de Psicobiología, Universitat de València, Blasco Ibáñez 21, 46010 Valencia, Spain, e-mail: Rosa.redolat@uv.es

S. Reutens School of Psychiatry, University of New South Wales, Sydney, NSW, Australia

P. Sachdev School of Psychiatry, University of New South Wales, Sydney, NSW, Australia, e-mail: p.sachdev@unsw.edu.au

David P. Salmon Department of Neurosciences (0948), University of California, 9500 Gilman Drive, La Jolla, CA 92093-0948, USA, e-mail: dsalmon@ucsd.edu

Gregory A. Seidel Department of Psychology, Temple University, Philadelphia PA, USA, e-mail: gregory.seidel@temple.edu

Michael A. Sugarman Department of Psychology, Wayne State University, 5057 Woodward Ave., 7th Floor, Detroit, MI 48202, USA, e-mail: msugarman5@gmail.com

Elizabeth R. Tuminello Department of Psychology, Loyola University Chicago, Chicago, IL, USA

Ipsit V. Vahia Sam and Rose Stein Institute for Research on Aging, San Diego, CA, USA; Department of Psychiatry, University of California, San Diego, CA, USA

Kathleen A. Welsh-Bohmer Department of Psychiatry, Joseph and Kathleen Bryan Alzheimer's Disease Research; Center—Division of Neurology, 2200 W. Main Street, Suite A200, Durham, NC 27705, USA, e-mail: Kathleen.WelshBohmer@duke.edu

C. Wijeratne School of Psychiatry, University of New South Wales, Sydney, NSW, Australia

John L. Woodard Department of Psychology, Wayne State University, 5057 Woodward Ave., 7th Floor, Detroit, MI 48202, USA, e-mail: john.woodard@wayne.edu

Part I
Epidemiologic Perspective in Aging

Epidemiology of Cognitive Aging and Alzheimer's Disease: Contributions of the Cache County Utah Study of Memory, Health and Aging

Kathleen M. Hayden and Kathleen A. Welsh-Bohmer

Abstract Epidemiological studies of Alzheimer's disease (AD) provide insights into changing public health trends and their contribution to disease incidence. The current chapter considers how the population-based approach has contributed to our understanding of lifetime exposures that contribute to later disease risk and may act to modify onset of symptoms. We focus on the findings from a recent survey of an exceptionally long-lived population, the Cache County Utah Study of Memory, Health, and Aging. This study is confined to a single geographic population has allowed estimation of the genetic and environmental influences on AD expression *across the expected human lifespan* of 95+ years. Given the emphasis of this text on the behavioral neurosciences of aging, we highlight within the current chapter the particular contributions of this population-based study to the neuropsychology of aging and AD. We also discuss hypotheses generated from this survey with respect to factors that may either accelerate or delay symptom onset in AD and the conditions that appear to be associated with successful cognitive aging.

Keywords Epidemiology · Alzheimer's disease · Population-based study · Prodromal AD · Mildcognitive impairment

---

K. M. Hayden · K. A. Welsh-Bohmer (✉)
Department of Psychiatry,
Joseph and Kathleen Bryan Alzheimer's Disease Research Center—Division of Neurology, 2200 W. Main Street, Suite A200, Durham, NC 27705, USA
e-mail: Kathleen.WelshBohmer@duke.edu

Curr Topics Behav Neurosci (2012) 10: 3–31
DOI: 10.1007/7854_2011_152

Published Online: 2 August 2011

Contents

1 Introduction 4
1.1 The Cache County Utah Study of Memory, Health, and Aging 5
1.2 Population Characteristics of Cache County Utah 6
1.3 Sequential Population Screening for Detection of AD and MCI 7
1.4 Cache County Contributions: Defining AD Expression and Prevention Targets 9
2 Part I: Defining Disease Outcomes for Prevention Studies of AD 10
2.1 Defining Normal Cognitive Aging and AD in the Population 10
2.2 Neuropsychological Features of Normative Aging 11
2.3 Neuropsychological Characteristics of Prodromal VaD and prAD 12
2.4 Refining the Clinical Diagnosis of Prodromal AD 14
3 Part II: Factors Modifying Cognitive Decline and AD Progress 18
3.1 Medical Conditions Influencing Cognitive Decline 18
3.2 Medications Delaying Symptom Expression and Cognitive Decline 20
3.3 Lifestyle Factors and Cognitive Decline 23
4 Conclusions 25
4.1 Preventing AD and Successful Cognitive Aging 25
References 26

1 Introduction

Alzheimer's disease (AD) is a growing public health problem globally (Brookmeyer et al. 2007). Current estimates of the disease prevalence in the United States indicates that at least 2.5 million Americans are affected (Plassman et al. 2007), with as many as 5.3 million affected when milder forms of the disease and mixed forms of dementia are included in the estimates (Hebert et al. 2003). The resulting human burden and economic consequences are enormous. Each year the US now spends over $90 billion on AD (almost 10% of all health care costs), making it the country's third most costly medical condition after cancer and heart disease. Without effective means of prevention, the problem of AD will only become more extreme. AD incidence doubles with each five years of added age (Jorm et al. 1987; Breteler et al. 1992; Jorm and Jolley 1998), implying ever-larger numbers of new cases as populations survive longer.

Effective means of preventing AD dementia are imperative. Toward this end, converging scientific approaches have focused on identifying early diagnostic features of the disease and potential contributing factors underlying its pathogenesis. Epidemiological studies of AD have provided important information regarding the disease incidence with changing population trends and have highlighted risk factors associated with disease onset and the rate of disease progression (Launer and Brock 2004; Lindsay et al. 2004; Brayne et al. 2011). To exemplify the contributions of this approach toward advancing understanding of AD, this chapter focuses on the findings from one of the recent national

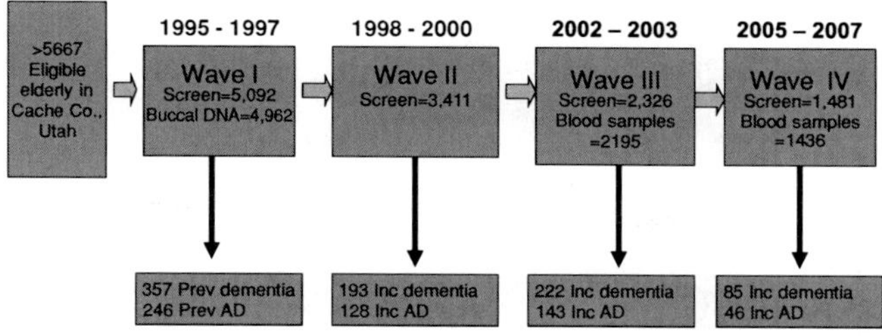

Fig. 1 Cache County study design

population surveys: the Cache County Utah Study of Memory, Health, and Aging. Confined to a single population, unique for its exceptional longevity, the Cache Study has allowed estimation of the genetic and environmental influences on AD expression *across the expected human lifespan* of 95+ years. Given the emphasis of this text on the behavioral neurosciences of aging, we highlight within the current chapter the particular contributions of the Cache County investigation to the neuropsychology of aging and AD. We also discuss hypotheses generated from this survey with respect to factors that may either accelerate or delay symptom onset in AD and the conditions associated with successful cognitive aging.

1.1 The Cache County Utah Study of Memory, Health, and Aging

The Cache County Utah Study of Memory, Health, and Aging (hereafter referred to as the *Cache County Study*: CCS) is an epidemiological investigation of dementia and cognitive decline conducted within a single county in northern Utah. The study attempted to survey and follow all members of the county who were aged 65 or older as of January 1st, 1995. Over 14 years of detailed observation until the close of the investigation in 2008, the CCS reported the prevalence and incidence of cognitive disorders and dementia across advanced aging, and has explored a broad array of genetic and environmental antecedents that influence AD symptom onset and dementia progression.

The starting sample was comprised of nearly all members of the permanent resident population of the Cache County, Utah, who were aged 65 or older at the start of the study (n = 5,092 of 5,667 eligible). Ongoing population surveillance for dementia was achieved by regular cognitive screening, spaced approximately every three years with a staged clinical assessment design. The approach involved the administration of a cognitive screener which was followed by more detailed assessments for subgroups suspected of having a cognitive disorder and in "designated controls," individuals matched to cases based on age, gender, and apolipoprotein E (*APOE*) genotype. The overall study design and the corresponding sample sizes and number of cases of dementia identified at each phase of study are summarized in Fig. 1.

1.2 Population Characteristics of Cache County Utah

The CCS endeavors to study the incidence, prevalence, and risk factors of AD and mild cognitive disorders into late-old age. With an emphasis on "late-old age" that is after age 85, the population in Cache County Utah has proven highly advantageous both for the study of AD in the very old but also for investigation of the converse situation, factors that contribute to robust cognitive aging. Among the most long-lived populations in the US, there are four population characteristics that have made this cohort ideal for longitudinal studies of aging:

First, as already noted, the population is exceptionally long-lived. The conditional life expectancy for males at age 65 among the highest in the United States (Murray et al. 1988), exceeding national norms by almost 10 years (Manton and Tolley 1991). The low-population mortality probably reflects a healthy lifestyle including the low use of alcohol and tobacco—risk factors for such common killers such as hypertensive or atherosclerotic cardiovascular disease (CVD) and several prevalent cancers (McGinnis and Foege 1993). Because of the exceptional longevity, there are a large number of survivors 85 years of age and older. The ability to identify many members of a closed population surviving to late-old age makes this population an ideal one for the examination of a number of key issues, including: (a) the incidence of AD in the very old, (b) the clinical expression of AD in late-old age, and finally (c) the cognitive characteristics and normative values for normal brain aging across the lifespan, relatively un-confounded by chronic illnesses.

A second key feature of the population is its healthy nature. Physically active, the community of 50,000 permanent residents embraces a high quality of life. The area is known for its outdoor recreational activities and the local economy is based largely on education, manufacturing, and agriculture industries. In recent statistics from the Wisconsin's Population Health Institute, in conjunction with the Robert Wood Johnson Foundation and the Centers for Disease Control (http://www.countyhealthrankings.org/), Cache County was rated as the second highest county in terms of overall health within the state of Utah, and first with respect to longevity. There is a low prevalence of chronic diseases in the population (e.g. diabetes, hypertension). This factor reduces some forms of variance in neuropsychological testing and it also greatly simplifies the differential diagnosis of dementia, especially in late-old age.

A third strength of the population for studies of aging is the cultural attitude toward volunteerism and research. Utah State University is located in Logan Utah, the seat of the county. The local population is highly educated, with the majority (70%) possessing a high-school degree or higher (US Census Bureau 2010). Perhaps due to the large number of faculty members from the local universities residing in the community, the population is very supportive of research as evidenced by high-participation rates (90% at Wave 1, +75% at later waves) in the CCS (Norton et al. 1994). Reliable estimation of prevalence and incidence hinges on effective enumeration of the population and careful surveillance and case ascertainment. In population surveys of AD, participation rates of 90% are unique and as nearly complete as one can hope to achieve. Commonly, medically frail individuals and those experiencing cognitive impairments are less likely to participate in longitudinal research studies of AD. This problem of differential attrition in the group at highest risk of disease can complicate estimates of dementia prevalence and incidence. As a consequence, the low-refusal rate in the CCS mitigates some of these methodological concerns and facilitates the reliable projections of disease burden and population trends.

Finally, tracking individuals over extended periods of observation is fairly simplified due to a highly close-knit community structure and a tendency for residents to maintain permanent roots in the valley. The CCS participants come from fairly large families and a very close-knit community. Approximately 90% of the population cohort are members of the Church of Latter Day Saints. Strong social ties and community networks allow ease in tracking individuals who change residence over time or transition from their home to either an assisted care or nursing home facility. The Cache valley's geographic location, nestled in the Wasatch front, also has contributed to fairly low rates of in- and out-migration. With many people remaining in the local area, it becomes a fairly straight-forward task to locate participants over the extended years of observation. This helps to ensure nearly complete ascertainment of the population over time, a characteristic of key importance when estimating dementia incidence and secular trends that may influence dementia onset and progression.

1.3 Sequential Population Screening for Detection of AD and MCI

As previously noted, the goals of the CCS were to determine the prevalence (Wave 1: baseline wave) and incidence (Wave 2) of AD and other dementias, as well as the genetic and environmental factors associated with risk of disease. Follow-up waves of study, Waves 3 and 4, were designed, respectively, for the detection of prevalent mild cognitive impairment (MCI) and its incidence over three years, for purposes of empirically defining prodromal AD in the very old and the factors affecting disease expression in advanced aging. Because it was not practical to perform detailed clinical examinations in all 5,000 participants, the study employed a unique *staged screening* approach of the population and had built in within it, a nested- case–control design. The methods employed are well described in previous publications (Breitner et al. 1999; Miech et al. 2002). Briefly, the sequential screening stages for each wave were as follows:

(1) The *initial population screening stage* was comprised of mental-status screening of the participant (modified Mini Mental State examination [3MS (Tschanz et al. 2002)] or administration of a proxy interview using the IQCODE (Jorm and Jacomb 1989) in those for whom self-report was not possible. Usually this occurred in medically frail individuals, in those with extreme hearing impairments, or in those with significant dementia. The screening examination also included detailed risk factor ascertainment, comprehensive medication review, and collection of a DNA sample.
(2) Detailed *informant interviews* were conducted using the Dementia Questionnaire, a guided interview for dementia and health history that was administered over the telephone (Kawas et al. 1994) in the subset of individuals who had low-cognitive screening scores.
(3) Full *clinical assessments* for dementing disorders were then completed within the participant home by a clinical team (nurse, psychometrist) for all individuals whose screening examinations were rated by two senior clinicians as either suggesting: (1) definite dementia, (2) likely dementia, or (3) possible mild dementia or other cognitive disorder. Additionally, all individuals over the age of 90, regardless of cognitive status were clinically examined, were 19% of the population who were selected as "designated controls." The latter group was comprised of individuals selected at random from the population to match suspected cases in a 2:1 fashion, based on age, gender, and *APOE* genotype.

The clinical assessments involved a neuropsychological evaluation administered by a psychometrist and took approximately 90 min to complete. The battery included measures tapping key domains affected in neurological disorders and degenerative dementias (i.e. orientation, expressive and receptive language, abstraction, attention and processing speed, executive control and working memory, verbal and non-verbal learning and memory, visuospatial and constructional praxis) (Tschanz et al. 2000). Additionally, the participant was examined by a research nurse who performed a complete neurological examination, physical evaluation, obtained vital signs, and a detailed health and medicine inventory. At the completion of the clinical assessment, all available clinical information was reviewed by a geropsychiatrist and a neuropsychologist at which time a preliminary diagnosis was assigned using DSM-IIIR criteria for dementia (American Psychiatric Association 1987) at Waves 1 and 2 and DSM-IV criteria for dementia and mild cognitive disorders (American Psychiatric Association 1994) at the later assessment waves (Waves 3 and 4).

(4) *Physician assessments* were conducted in the home to verify diagnoses for all individuals who at the case assessment had either a dementia diagnosis or were classified as having cognitive impairment, falling short of dementia, a condition suggesting an early-stage dementing disorder. Detailed laboratory testing was ordered at this point, which included standard laboratory blood panels and a head MRI scan to assist the differential diagnosis. After these evaluations, a panel of expert clinicians reviewed all available data and

assigned to each individual a *consensus diagnosis* of AD, vascular dementia (VaD), or other diagnosis using standard criteria (McKhann et al. 1984; Roman et al. 1993).

(5) Follow-up evaluations at 18 months after the clinical assessment were scheduled for all individuals diagnosed with dementing disorders and MCIs suspected to be early prodromal AD. These assessments used identical clinical procedures as already described.

Each subsequent wave of study in the population were conducted within the surviving individuals from the previous wave who were nondemented, including those with mild cognitive disorders who remained dementia free at the 18 month follow-up. Clinical assessment methods were similar across waves with the exception of that the 3MS cut-scores were adjusted at each wave to be more stringent, accommodated for expected test–retest effects common in cognitively normal populations (Miech et al. 2002). The designated control panel was longitudinally examined at follow-up waves, regardless of their screening scores. Replenishment of the panel occurred in 2002 (Wave 3) and we also examined all surviving individuals, age 85 and older at Waves 3 and 4, to permit full ascertainment of mild cognitive disorders in the very old.

It should be noted that the unique methods of the CCS, specifically the inclusion of a designated control panel, allowed for a nested case–control study in the population setting. Importantly, this design inclusion permitted an opportunity to establish the efficacy of our population screening methods for the detection of dementia and to use this information to adjust our population estimates of AD and MCI prevalence and incidence. Additionally, detailed, longitudinal information in the population setting permitted the establishment of normative information for common neuropsychological tools used in the very old as well as opportunities to determine empirically the expression of prodromal AD and other dementias over 3-, 6-, and 10-year intervals prior to the diagnosis of a dementing disorder. By systematically evaluating all individuals in the population over age 90 regardless of their screening examination results (Waves 1 and 2) and broadening this age window at Waves 3 and 4 to include all individuals over age 85, the CCS has one of the largest, completely clinically ascertained population-based cohorts of very old individuals, thereby permitting investigations not only of AD in the very old, but also the predictors of successful cognitive aging into very old age (see Breitner et al. 1999). Some of the most important neuropsychological contributions from the CCS are in the sections that follow.

1.4 Cache County Contributions: Defining AD Expression and Prevention Targets

There are currently over 70 publications that have come from the CCS. Among these are important findings regarding the neuropsychological expression of AD and other common dementias of late life, along with the clinical distinctions between these pathological conditions and normal cognitive aging. The results which are summarized in the sections to follow provide a population perspective, an approach which is complimentary to other observational studies involving clinical cohorts and convenience samples. Although the CCS is not representative of the US population as a whole with respect to educational and cultural diversity, the study allows the estimation of genetic and environmental influences into late-old age and provides information that can be tested in more nationally representative samples, such the Aging and Demographics of Memory Study (ADAMS; Langa et al. 2005; Plassman et al. 2008, 2011). Ultimately, clues from epidemiological studies such as CCS suggest potential protective agents that can be tested in randomized clinical trials or in animal models to determine a causal relationship between exposure and protection against cognitive decline. The sections that follow are divided into two parts: *Part I* considers the lessons from the CCS epidemiological study in terms of defining important disease outcomes, such as prodromal AD and normal cognitive aging in the very old. *Part II* focuses on exposures revealed in the epidemiological context that appear to have a beneficial relationship on cognitive decline and point to potential therapeutic approaches that can be tested in clinical trials.

2 Part I: Defining Disease Outcomes for Prevention Studies of AD

2.1 Defining Normal Cognitive Aging and AD in the Population

Population-based studies of AD are conducted primarily for purposes of estimating prevalence and incidence of the disease and monitoring trends. Case definition is critical in this context, and is particularly challenging in very old populations where frailty and sensory loss are common confounds. Variation in defining dementia and distinguishing AD reliably from more benign effects of brain aging can result in quite different estimates of incidence across studies (Kryscio et al. 2004), making it challenging to evaluate the presence of true population trends. Unlike the clinical setting, population ascertainment does not permit exhaustive medical testing due to costs. Epidemiological methods typically must rely on multi-stage assessment methods, which reserve expensive clinical evaluations for identified subsets drawn from the population. The resulting projections made regarding disease prevalence and incidence then rest on assumptions made in the methods and are influenced by the efficiency of screening in identifying disease, the sensitivity and specificity of the methods, and sampling decisions. If neuropsychological tests form the basis of dementia determinations, the cut-scores applied can dramatically affect the estimates (Launer 2011). In studies where mild forms of cognitive disorder are included, high-AD incidence is reported compared to studies where AD is defined by significant impairment in functional capabilities (Brookmeyer et al. 2011).

The CCS used a combination of broad screening and deep sampling of identified cases and controls in the population, as already described. Because the study intentionally oversampled the oldest age groups, neuropsychological information is available in the very old, permitting normative studies of successful cognitive aging (Ostbye et al. 2006; Welsh-Bohmer et al. 2009) and the opportunity to empirically define the clinical characteristics of AD and other dementias across a reasonable human lifespan of 100 years.

2.2 *Neuropsychological Features of Normative Aging*

Arriving at firm distinctions between normal aging, mild cognitive disorders, and emerging AD dementia requires the availability of normative standards, ascertained ideally in clinically well-characterized groups observed over time, permitting the retrospective removal of cases later discovered to have indolent dementia. Such "robust" norms provide ideal standards, unconfounded by early, difficult to detect cognitive disorders (Sliwinski et al. 1997). Sampling in community populations adds important socio-cultural dimensions to normative standards (e.g. socioeconomic status, regional differences in quality of education), which may otherwise be under-represented when using paid volunteers or convenience samples (Manly et al. 2005). The CCS, with its over-representation of very old individuals (aged 85 years+), offers a unique opportunity to determine normal test performance in advanced aging on many common neuropsychological tests employed in practice. Because the very old individuals in the CCS population are all clinically well-characterized and followed longitudinally, contamination of the sample with undiagnosed early cases of dementia is relatively minimized.

Recent work conducted in the Cache County population examined neuropsychological performance of all individuals over the age of 65 who were clinically examined and determined to be cognitively and functionally normal. The study examined the effects of common demographic modifiers (age, education, gender) on test performance as well as the influence of variations in apolipoprotein E genotype (Welsh-Bohmer et al. 2009). A standard neuropsychological battery (see Tschanz et al. 2000) has been employed in this population study as well as in other national surveys (Plassman et al. 2007). Included are measures of mental status, episodic memory, expressive language, attention and concentration, executive control, and speeded motor performance, as summarized in Table 1.

In the analysis of normative performance across the life span, there were over 500 individuals who had detailed clinical evaluations. Of these, there were 330 individuals over the age of 75 and a total of 112 subjects who were over age 85. Although neuropsychological test performance was influenced significantly by advancing age, as expected, the effects of age were not uniformly observed across all domains.

Table 1 Neuropsychological tests used in Cache County population study

Domain	Tests commonly used	References
Orientation/global mental status	Mini mental state examination	Folstein et al. (1975)
Premorbid estimation of intellect	Shipley vocabulary test	Shipley (1967)
Language	Controlled oral word association test (COWA) of multilingual aphasia examination	Benton and Hamsher (1983)
	Category fluency (animals)	Strauss et al. (2006)
	Boston naming	Kaplan et al. (1978)
Memory	Logical memory I,II of WMS- R	Wechsler (1997)
	CERAD word list learning test	Welsh-Bohmer and Mohs (1997)
	Benton visual retention test	Benton et al. (1992)
	CERAD delayed visual recall test	
Attention/ concentration	Trail making test-Part A	Reitan (1958)
Executive function	Trail making Test-B	Reitan (1958)
	Symbol digit modalities Test	Smith (1991)
Visuoperception	CERAD constructional praxis	Welsh-Bohmer and Mohs (1997)
Personality and mood	Beck depression inventory-II	Beck et al. (1996)
	Neuropsychiatric inventory	Cummings et al. (1994)

Learning, memory, and executive measures, particularly those tests requiring a speeded motor component, were particularly influenced by age. Intelligence, assessed through vocabulary measures, remained highly robust across advanced age, consistent with other reports (Park and Reuter-Lorenz 2009). Education had powerful effects across nearly all measures, but gender and *APOE* genotype had negligible and inconsistent effects. Only isolated aspects of expressive speech (animal fluency) and verbal memory (delayed CERAD word list memory) were modified by these variables. Women outperformed men on these measures, and individuals with an *APOE* ε4 allele performed less well; however, all these effects largely disappeared once age and education were controlled. The findings suggest that some cognitive functions remain quite robust into late-old age. With care to remove cases of indolent dementia from the normative sample, *APOE* does not appear to exert any major effect on cognition across the normal life span.

2.3 Neuropsychological Characteristics of Prodromal VaD and prAD

With the benefit of longitudinal observation in this well-characterized population it has been possible to define the early cognitive signatures of brain diseases common in aging, such as AD and VaD, well before the full dementia syndrome

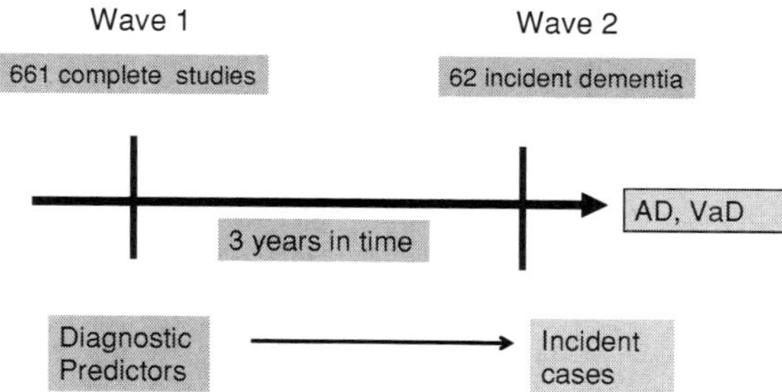

Fig. 2 Cache County study design—detection of prodromal VaD and AD

has expressed. Of particular interest is mild forms of vascular cognitive impairment (VCI), a broad group of disorders that includes VaD. Risk factors for VCI include hypertension, diabetes, elevated circulating cholesterol, and obesity. Because many of these are treatable conditions, early identification of VCI has been identified as a potential treatment target (Rockwood et al. 1997).

Using data from the CCS it has been possible to explore whether there is an identifiable prodromal stage to VaD and the neuropsychological characteristics distinguishing these mild forms of prodromal VaD from the prodromal stages of AD (Hayden et al. 2005). The study design applied to examine the preclinical predictors of VaD and AD is illustrated in Fig. 2. Normal participants from the population who had complete diagnostic evaluations at the baseline wave of study (Wave 1) were examined three years later to determine cognitive outcomes. Within this group of well-characterized individuals, 62 cases of incident dementia developed over the interval and were diagnosed as either VaD ($n = 14$) or AD ($n = 42$) at the Wave 2 clinical assessment.

Multivariate analysis of the previous neuropsychological test results for these individuals, *three years prior to the diagnosis of dementia*, revealed distinct differences between the two groups after adjusting for age, sex, and education (see Hayden et al. 2005). The group developing incident AD had significant impairments at baseline on measures of delayed verbal recall, specifically on Logical Memory II from the Wechsler Memory Scale-Revise (WMS-R, Wechsler 1987) and on verbal recognition memory, assessed by the Word List Memory Test from the Consortium to Establish a Registry for Alzheimer's Disease (CERAD, Welsh et al. 1994). By contrast, the individuals who were later diagnosed with VaD tended to have better preserved episodic recall at baseline when contrasted to the AD group; although poor performance compared to controls. The group with later VaD diagnoses had a lower discrimination index on recognition memory when compared to normal controls. However, unlike those who developed AD, the VaD cases had a conservative response bias (low false alarm rate, low hit rate) whereas AD cases had both poor delayed recall and a high-false alarm rate on recognition discrimination. The incident VaD group also had relatively more impairment on tests of speeded performance and executive control (e.g. Trail Making test and verbal fluency) at baseline compared to both controls and their AD counterparts, although these latter findings were not significant in all analyses. The results from this nested case–control study within a large community sample are consistent with other reports deriving from clinical samples that prodromal deficits in cognition are detectable in persons who later develop VaD. Other studies have underscored specific early deficits in executive function and the CCS results are consistent with this general finding. However, the population results also suggest that tasks tapping verbal recognition performance are additionally informative in drawing distinctions between the effects of prodromal AD and those associated with vascular disease. The findings also suggest importantly that cognitive deficits are detectable in the community, a setting where deficits are likely to be relatively subtle, given that individuals are not self-identifying as having a problem and seeking medical assistance.

2.4 Refining the Clinical Diagnosis of Prodromal AD

The information from longitudinal studies in large- and small-cohort studies now amply illustrate that detectable neuropsychological deficits are observable many years before the onset of diagnosable dementia (Backman et al. 2005) and that the presence of multiple deficits, typically in episodic memory and at least one other domain of cognition is predictive of imminent progression to AD within a three-year window (Bozoki et al. 2001). Similarly, from a clinical (Goldman et al. 2001) and neuropathological perspective (Braak and Braak 1996; Price and Morris 1999; Price et al. 2001) it has long been recognized that AD is a chronic neurodegenerative disease, developing silently over a protracted period of time before clinical symptoms are manifest (Katzman 1976). A prodromal phase to AD dementia is evident in many cases and there have been concerted investigative efforts, particularly in the last 20 years, to characterize this AD prodrome as well as any associated disease biomarkers to facilitate accurate early diagnosis and the development of effective drug therapies, preventing further disease expression. A number of diagnostic criteria have been advanced in recent years to characterize intermediate stages of cognitive impairment at high risk of progressing to dementia. Among these, is the concept of "MCI" a disorder often used synonymously to mean prodromal AD (Petersen et al. 1999). As originally conceived this disorder was characterized by relatively isolated episodic memory impairments, which represented a change from a previous level of function, did not interfere with daily function, and had no identifiable medical etiology. Modifications of this MCI concept emphasized the importance of subtle change in function (i.e. in higher order, instrumental activities of daily living; Morris et al. 2001). Other clinical nomenclature has been advanced, basically describing the same clinical phenomenon albeit defining the problem somewhat differently (see Table 2).

Table 2 Classification for cognitive impairment

Disorder	Criteria
Mild Cognitive Impairment (MCI-Petersen)—amnesic MCI (Petersen et al. 1999)	1. Subjective memory complaint 2. Objective evidence of memory impairment by cognitive testing 3. Normal cognitive function otherwise 4. Intact activities of daily living 5. Absence of dementia 6. Exclusion of medical causation (e.g. stroke, depression)
Mild Cognitive Impairment CDR 0.5—functional MCI (fMCI) (Morris et al. 2001)	1. Clinical dementia rating scale of 0.5 (no required evidence quantitative memory deficit) 2. Exclusion of medical explanations for cognitive change
Mild Ambiguous/Prodromal AD (mild ambiguous) (Breitner et al. 1995)	1. History of functional impairment consistent with a CDR = 0.5 and/or 2. Objective assessment profile of mild impairment in memory and cognitive function suggesting early-stage AD 3. Exclusion of medical explanations for symptoms
Cognitive Impairment Not Dementia (CIND) (Graham et al. 1997; Tuokko et al. 2001)	1. Memory impairment (short or long term) 2. Impairment in at least one other area of cognitive functioning such as abstraction, judgment, higher cortical functions, or personality. Impairment is established by neuropsychological testing or clinical examination/interview 3. Can occur in context of medical explanations (e.g. stroke, depression)
Vascular Cognitive Impairment (vascular CIND)	1. Meet criteria for CIND above 2. Known vascular causation established either by history of CVA or by neuroimaging evidence of completed stroke
Age-Associated Cognitive Impairment Not Dementia (AACD) (Levy 1994)	1. Decline of more than one SD in any area of cognitive function in comparison with age-matched controls 2. Exclusion of medical explanations
Age-Associated Memory Impairment (AAMI) (Crook et al. 1986; Blackford and LA Rua 1989)	1. Subjective complaint of memory loss 2. Decrement in memory at least 1 SD below mean for young adults 3. Adequate intellectual function 4. Absence of global cognitive impairment 5. Exclusion of medical explanations

As work has progressed it has become clear, that these mild cognitive disorders, however defined, appear in many circumstances other than the development of AD. To learn more about the clinical pathogenesis of AD, and because it is useful to have criteria that can identify groups enriched for individuals who will shortly "convert" to AD, it is important to have efficient methods that differentiate the "true" AD prodrome from both normal cognitive aging and other cognitive disorders. Most prior work on this problem has concentrated on the identification of the AD prodrome in clinical samples. Because it is important not only to understand the phenomenology of the AD prodrome but also the factors that modify its onset or "conversion" to dementia, there is a need for improved methods for its detection and analysis *in populations*.

Recent work in the CCS has evaluated the predictive value of these different categorical terms in identifying true prodromal AD in the population. Operationalized diagnostic criteria using standardized test measures of cognition and functional decline were applied to capture cases of cognitive disorders in the population that were not sufficiently impaired for a diagnosis of AD meeting (Mayeux et al. 2011). Contrasted in the analysis was the predictive utility of MCI (single domain amnesic MCI, aMCI) (Petersen et al. 1999), questionable dementia/MCI defined as functional impairment measured as Clinical Dementia Rating Scale score of 0.5 (CDR = 0.5, Morris et al. 2001), Age-Associated Cognitive Decline (Levy 1994), Cognitive Impairment not Dementia (Rockwood et al. 1997), and Age-Associated Memory Impairment (Crook et al. 1986). Also examined was a set of criteria that formalized a clinical construct identified in 1994 in epidemiological studies of dementia as "mild/ambiguous" prodromal AD (Breitner et al. 1994b). This group of subjects had a clinical picture that strongly suggested early AD (affirmative observations beyond the exclusion of other causes of dementia). The "mild/ambiguous" prodromal AD (prAD) diagnosis was given to participants observed in the population that had *either a memory disorder along with subtle changes in language, praxis, or gnostic function, or had notable problems in instrumental functional abilities, regardless of whether there was demonstrable memory disorder.*

Three-year outcomes for the diagnostic subgroupings were contrasted to determine which was most successful at identifying people who would shortly develop AD (positive predictive value, PPV). Also assessed was accuracy in identifying those not likely to progress (negative predictive value, NPV) and relative risk (RR) for subsequent dementia compared to unimpaired individuals.

The study demonstrated that the various diagnostic criteria differed substantially in their frequency (Fig. 3), with fMCI (Morris et al. 2001) and mild/ambiguous prAD groups being much more numerous than aMCI. Most diagnostic groups also had poor stability over three years: their members reverted to cognitive normality at least as often as they remained stable or progressed to dementia. By contrast, the constructs that included indicators of functional impairment (fMCI, prAD) were more stable, with only 8–16% of their numbers reverting to normality.

Among the diagnostic constructs evaluated, mild/ambiguous prAD had the highest PPV (48 vs. 19–35% for other groups), NPV (96 vs. 85–93%), and RR (RR = 10.27 vs. 1.07–8.0). The three-year rate of reversion to unimpaired status (24%) for the prAD group was also among the lowest of the all "impaired" groups,

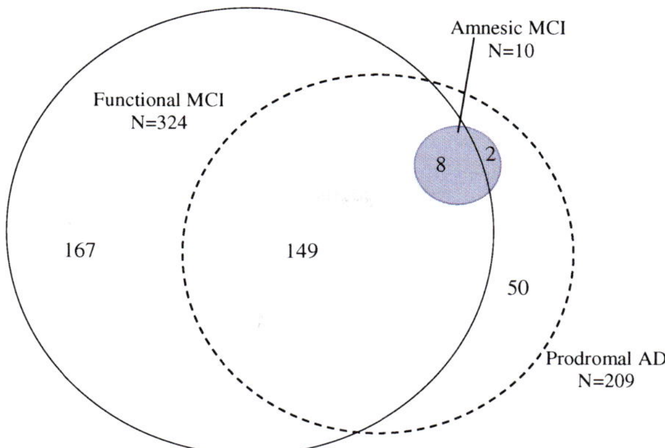

Fig. 3 Overlap between diagnostic categories of MCI and prAD (in Mayeux et al. 2011)

which ranged up to 67%. From this study, it appeared that criteria of mild/ambiguous AD which included elements of both cognitive decline and functional decline had superior performance as a predictor of imminent dementia.

In a separate study, three-year progression to dementia was assessed in individuals who had been previously clinically diagnosed as mild/ambiguous prAD (Tschanz et al. 2006). Although the original CCMS screening protocols were not designed to "capture" people with mild cognitive disorder, 120 individuals were identified clinically with such a disorder at Wave I. Among these, 51 were diagnosed as mild/ambiguous AD. Analysis of Wave II data revealed that nearly 47% of the members of this group had since "converted" to AD. We also reported a 22- to 25-fold increase risk of dementia in all cognitively impaired groups, particularly in those diagnosed as "mild/ambiguous AD, and those with an *APOE* ε4 allele.

Together these studies suggest that it is possible to identify a prodrome to AD and other forms of late-life dementia. Importantly, the work with diagnostic algorithms suggests that studies need not rely on clinician judgment for diagnostic definition. Use of standardized procedures mapped across studies as was done here (Mayeux et al. 2011) can allow some harmonization of diagnostic definitions across studies. If the apparent advantages of the prAD diagnostic algorithmic approach can be confirmed in other, more diverse samples, application of this method has implication for studies of AD prevention.

It should be noted that similar findings regarding important characteristics of AD diagnostic criteria have been reported by European groups (Dubois et al. 2007). There is ongoing work nationally and internationally to revise AD diagnostic criteria which take advantage of the population observations to improve diagnosis at the early stages of AD. Incorporating biomarkers to further enhance diagnostic efficacy, it is hoped that the criteria prove valid and have clinical utility in the research setting. If so, they could conceivably be integrated into criteria for use in general practice (DeKosky et al. 2011). For research, having clearly defined criteria for prodromal AD offers a tractable outcome that permits the examination of factors influencing early trajectories of disease and transitions from normal cognition to early symptomatic disease to fully expressed dementia. The next section deals with factors, identified in epidemiological study, that influence cognitive decline and transitions to fully expressed AD dementia.

3 Part II: Factors Modifying Cognitive Decline and AD Progress

A goal of epidemiological studies in general is to identify common factors that affect disease trends, such as environmental exposures acting to accelerate or delay disease onset. In the last decade observational studies have suggested a number of common medical factors and lifestyle practices which appear to modify the risk for AD dementia and influence cognitive decline. Perhaps the most notable of these is the genetic susceptibility gene, apolipoprotein E (*APOE*). Allelic variations in this gene are related to AD risk and symptom onset in a dose-dependent fashion, such that individuals homozygous for the ε4 allele have the most enhanced risk and onsets on average 10 years earlier than those who do not have an ε4. Those who are heterozygous for ε4 have an intermediate level of risk of AD (Corder et al. 1993). Although knowledge of *APOE* status confers information regarding lifetime risk of developing AD, genetic modification of AD risk is not a realistic option at the present time. Epidemiological studies of dementia have focused on three principal targets as potential environmental exposures that lend themselves of modification. These include an examination of:

1. *medical conditions* that when treated would confer disease protection,
2. *common exposures*, such as medications, that when taken (or avoided) reduce cognitive decline to dementia, and
3. *lifestyle habits*, such as dietary practices, activity level, or cognitive engagement, which may act to influence cognitive function over the lifespan in a positive way.

Some of the key observations from the CCS are summarized as these observations have resulted in testable hypotheses for understanding AD pathogenesis and suggest potential therapeutic directions for randomized clinical trials to prevent AD.

3.1 Medical Conditions Influencing Cognitive Decline

Among the most promising leads with respect to factors contributing to AD risk and progression are studies suggesting a relationship between vascular factors and AD pathogenesis. Vascular risk conditions have been shown to affect some of the fundamental biological processes associated with AD. For example, elevated cholesterol, a risk for CVD and stroke, has been shown to increase amyloid beta deposition in transgenic mouse models of AD (Refolo et al. 2000). Additionally, many population-based studies, including the CCS, have shown that a range of vascular conditions including hypertension (Skoog et al. 1996), diabetes (Ott et al. 1999), stroke (Honig et al. 2003), atherosclerosis (Hofman et al. 1997), and atrial fibrillation (Ott et al. 1997) increase AD risk and predict greater rates of progression (Hayden et al. 2006; Mielke et al. 2007).

In terms of underscoring mechanisms of disease, understanding whether the relationship of CVD to AD is independent of its known relationship to VaD is critical. Few studies have been able to look at this question from a population perspective. The CCS with its highly detailed diagnostic information on population ascertained dementia cases permits this type of risk analysis (Hayden et al. 2006). In the 3,264 subjects assessed as part of the second incidence wave, a number of vascular risk conditions were found to be associated with incident dementia. A history of previous stroke, diabetes, hypertension, and obesity were associated with dementia risk; however, some of the risk relationships were complex, varying by sex and whether the diagnostic outcome was VaD or AD. Hypertension and diabetes were both associated with VaD risk but not with the risk of AD. By contrast, obesity was related to AD dementia. The effects of both diabetes and obesity were most robust in women, becoming non-significant in men after adjustment for covariates of age, education, and *APOE* genotype. These results suggest that the vascular risk relationship noted in AD may be primarily due to a superimposed cerebrovascular disease component, which is having a synergistic or additive effect on dementia expression and is mediating the risk effect.

A clinically important question is whether this risk relationship observed is modifiable and would reduce the incidence of AD. A number of studies have examined this question in relationship to hypertension. These studies generally find that the strongest associations between hypertension and later dementia incidence are in instances where the condition was untreated (Launer et al. 2000; Kivipelto et al. 2001; Khachaturian et al. 2006). This observation suggests that whether due to cerebrovascular disease or AD pathology, effective identification and treatment of vascular risk factors may help buffer against cognitive decline in the aging nervous system.

To determine whether treatment of these conditions has any impact on rate of decline in individuals already affected by the disease, the CCS examined rate of change in functional abilities and overall cognition in the subset of population participants who had diagnosed dementia (Mielke et al. 2007). In AD cases, the presence of many cardiovascular risk conditions was associated with more aggressive rates of change. Atrial fibrillation, angina, myocardial infarction, and systolic hypertension were all associated with a greater rate of decline on both measures of function and cognition, measured by the Clinical Dementia Rating Scale and the Mini Mental State examination, respectively (Mielke et al. 2007). Curiously, both diabetes and coronary artery bypass graft surgery, risk factors for incident dementia, appeared to attenuate cognitive decline in individuals already diagnosed with AD, likely due to either a treatment effect or the influence of selective survival (mean age = 85 years). Timing of the medical conditions also bore a relationship to risk of decline. Although hypertension is a particularly potent risk factor for dementia when the condition is reported in midlife (Launer et al. 2000), the risk relationship of hypertension with decline was associated with age, suggesting that hypertension in later life contributes to greater rates of cognitive and functional decline in patients already experiencing dementia.

On balance, the results from epidemiological studies provide a strong rationale for tackling cardiovascular risk conditions in mid-life, both for mitigating cardiovascular events and for reducing the risk of central nervous system diseases in later life. Additionally, the results in Cache suggest that cardiovascular risk conditions may contribute to excess cognitive morbidity in patients already diagnosed with AD. Consequently, implementing treatments at early stages of the AD process may be beneficial, not only in preventing disease onset but also in potentially slowing the rate of cognitive decline in expressed disease, permitting longer periods of quality life and functional autonomy.

3.2 Medications Delaying Symptom Expression and Cognitive Decline

Identifying methods for delaying AD expression and facilitating successful cognitive aging is a priority as discussed. Attention to pharmacological compounds has resulted in some particularly promising leads, many of which have been explored in clinical trials. Compounds found associated with reduced AD incidence include: (1) non-steroidal anti-inflammatory drugs (NSAIDs) (Breitner et al. 1994a, 1995; Andersen et al. 1995; McGeer et al. 1996; Stewart et al. 1997; in 't Veld et al. 1998), (2) anti-oxidant vitamin supplements (Morris et al. 1998; Paleologos et al. 1998; Zandi et al. 2004), and (3) certain classes of antihypertensive agents (Khachaturian et al. 2006). Post-menopausal hormone replacement therapy (HRT) in women, particularly if taken early after menopause, has also been suggested to be protective, although recent safety concerns have led to a more cautionary stance (Zandi et al. 2002b). Here, we highlight some of the key results obtained and their importance in generating hypotheses regarding their timing in relation to AD onset and their effects on underlying brain mechanisms.

3.2.1 NSAIDS

Results from early work in CCS suggest that NSAID use may delay AD onset only if participants were exposed before the neurodegenerative process has reached a critical stage near the point where participants had developed prodromal AD (Zandi et al. 2002a). Separating prior users of NSAIDs from current users at baseline, reduced risk was seen only in the former and the strongest effects were in prior users with three or more years of use (hazard ratio, or HR, 0.29; CI 0.02–1.35). Randomized trials of NSAIDS to treat AD have not been successful (Aisen et al. 2003), and a recent primary prevention study in AD, ADAPT (Alzheimer's Disease Anti-Inflammatory Prevention Trial) was halted early due to safety concerns in a sister prevention trial, the Adenoma Prevention with Celecoxib Trial (APC) because of vascular morbidity (see Meinert and Breitner 2008). In results published from continued observations in ADAPT (Martin et al. 2008), it appears that the two compounds examined (celecoxib and naproxen) may accelerate decline to dementia if used in older individuals. The results appear to be inconsistent with the epidemiological studies, however the ADAPT study included some prevalent, albeit very mild cases of AD dementia who were undetected at baseline screening and were randomized to treatment. If these individuals were removed from the sample, the effect was no longer seen, suggesting that NSAIDS may accelerate dementia onset in individuals already expressing symptomatic disease. This remains to be verified, and further masked observation continues in the cohort to determine whether the active treatment is related to a reduced risk of AD in later life.

Continuing observations in CCS appear to provide some clarification, reconciling observational studies and clinical trial findings. In one study that examined use of NSAIDS on cognitive trajectories over eight years of observation, associations of NSAID use depended on when the compounds were started and on *APOE* ε4 genotype (Hayden et al. 2007). Those who started NSAID use prior to age 65 and had one or more *APOE* ε4 alleles showed greatest protection from decline compared to nonusers. Whereas those who started compound after age 65 and had no *APOE* ε4 allele showed the greater decline. These data suggest that midlife use of NSAIDs may help to prevent cognitive decline, particularly in individuals at high risk of disease by virtue of their *APOE* genotype. Starting NSAID use after age 65 may actually have the opposite effect and act to accelerate AD symptom onset.

The mechanism for these disparate effects of NSAIDS on AD expression is unclear but may be related to brain inflammatory mechanisms widely thought to influence AD pathogenesis (Weggen et al. 2001). In the extended presymptomatic period of AD neurodegeneration, NSAID use may act to suppress brain inflammatory response and delay AB deposition. However, once AB deposition is already in process, the brain inflammatory response may be crucial to AB clearance (Martin et al. 2008). These hypotheses generated from observational studies can be tested in animal models to clarify timing of compound use on AD pathogenesis. Differential effectiveness of various compounds can also be explored.

3.2.2 Anti-Oxidant Vitamins

Several observational studies have produced some evidence that antioxidants might be useful in the prevention of AD (Price and Morris 1999; Goldman et al. 2001;

Luchsinger et al. 2003). However, the early encouraging treatment trial results (Sano et al. 1997) were not reinforced by telephone cognitive screening results in the Vitamin E intervention group in the UK Heart Protection Study (1999). A trial regimen of high-dose vitamin E alone (no vitamin C allowed) was also ineffective in delaying the "conversion" of aMCI to AD (Petersen and Morris 2005).

Data from the CCS suggested a possible explanation because apparent benefits were observed only among participants who reported use of *both* water-soluble vitamin (C) and lipid-soluble vitamin E, but not among users of either vitamin alone (Zandi et al. 2004). More recent analyses of cognitive decline in CCMS cognitive screening scores from Waves I–IV again found a suggestive reduction in "normal" age-related cognitive decline in users of both vitamins, but neither alone (Wengreen et al. 2007). These findings were echoed by results from other studies relating CSF F_2-isoprostanes, a possible biomarker of Alzheimer pathogenesis, and vitamin use in mild AD patients. Only users of E + C (but not of E alone) showed an attenuation of the usual time-dependent increase in isoprostane concentrations (Quinn et al. 2004). These last findings not only support the idea of synergy between E and C but also suggest that the beneficial "effect" of E + C may extend through the AD prodrome into the mild dementia stage of AD.

3.2.3 Pharmcogenetic Effects of Medication Use and the Future of AD Prevention

Later work in CCS examined the "protective" effects of NSAIDS and vitamin use in relationship to *APOE* ε4. In one such study, sequential 3MS scores were examined (Lyketsos et al. 2006) from Waves I–III to explore the individual and conjoint association of NSAIDs and vitamins E or C and cognitive trajectory. (Fotuhi et al. 2006) A modest "effect" was demonstrated on the trajectory of cognitive decline in those who had combined use of vitamins E and C together with NSAIDs. However, this association was quite specifically related to *APOE* genotype (Fig. 4), and was apparent only in ε4 carriers. (Fotuhi et al. 2006) This dichotomy possibly reflects a preponderance of fulminate pre-clinical AD, not yet expressed, as the cause of cognitive change in those with ε4, where other causes may predominate in those without this allele. Similar arguments have been offered in the interpretation of HRT and cognitive decline in women, where an apparent "protective" effect of HRT was evident only in very old women. Once again, leading to the argument that this group's cognitive decline without HRT may have represented pre-clinical AD (Carlson et al. 2001).

It is important to emphasize that because of the safety concerns with both HRT and NSAIDS, it is *highly unlikely* at this point that additional AD prevention trials would ever be considered with these compounds to resolve issues of dosing, timing, and compound formulation. Consequently, perhaps the *only way* to resolve these important questions is through continued exploration in well designed epidemiological studies. Augmented by increasingly available electronic medical record databases, it may be possible to track medication use across extended

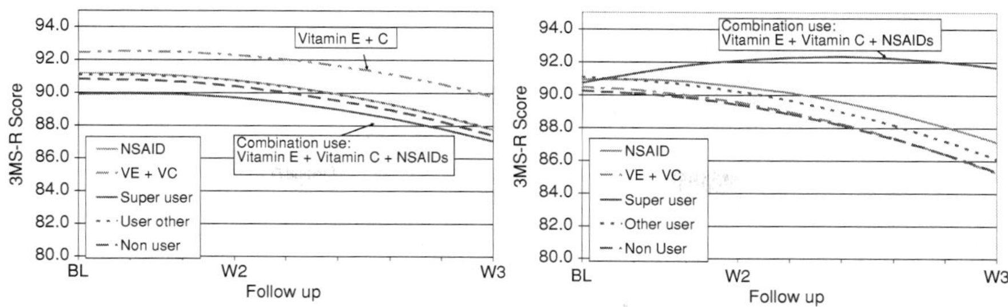

Fig. 4 *Left panel* from individuals without an APOE ε4 allele. In these individuals, consumers of Vitamins E and C show higher initial 3MS scores, which remain high. No differences seen in rate of decline between any of the groups (E and C, NSAIDS, or combined) over 8 years of follow-up. *Right panel* shows results in the APOE ε4 group, illustrating protective effect in those with combined use of NSAIDS, Vitamin E, and Vitamin C

periods of time and to then examine early medication patterns in relation to later dementia outcomes, an approach that has already met with some success in understanding the importance of HRT in relation to cognition (Whitmer et al. 2011).

3.3 Lifestyle Factors and Cognitive Decline

Due in some measure to safety concerns, the public turns to other non-pharmacological approaches to stave-off memory decline. This rationale is supported by a number of observational studies that suggest lifestyle behaviors, such as diet, exercise, and cognitive engagement have important effects on cognition and risk of dementia (see Carlson et al. 2009; Smith et al. 2010).

3.3.1 Diet and Exercise

Recent work in Cache County and other studies have shown that variations in dietary practices and nutrient intake are predictive of cognitive decline (Wengreen et al. 2005) and AD (Scarmeas et al. 2006). A number of specific nutrient components of these diets have been examined in relation to cognitive performance and shown to be protective. These components include fish oils, fruits, and vegetables. Additionally, adherence to the Mediterranean diet is associated with reduced risk of neurocognitive impairment and dementia and degree of adherence appears to protect against subsequent cognitive dysfunction in a dose-dependent fashion (Wengreen et al. in press). Moreover, when a healthy diet is combined with higher levels of physical activity, there is some evidence that these practices are additive and lead to a greater lowering of the risk of developing AD (Scarmeas et al. 2009).

Complimentary studies have focused on physical activity related to cognition. In general, the results from observational studies, again suggest that those who are active out-perform their less active counterparts on formal neuropsychological tests. Further in the limited number of interventional studies conducted to date, there is an indication that aerobic exercise is associated with better cognitive outcomes and this is particularly evident in those individuals who are experiencing cognitive disorders (Smith et al. 2010). Two recent exercise interventions in patients with mild cognitive disorders support the observation that exercise is beneficial for cognition. One study was done with older adults who had memory complaints (Lautenschlager et al. 2008). The other was conducted in patients with verified MCI (Baker et al. 2010). Both studies demonstrated improved cognitive functions following extended periods (6 months) of high intensity aerobic exercise.

Although recent evidence-based reviews exploring the overall efficacy of prevention strategies in AD have been very disappointing (Plassman et al. 2010), only very few direct interventional studies have been conducted to date, to test the utility of lifestyle interventions on cognitive and dementia outcomes. The ultimate test of whether dietary and exercise approaches are beneficial in reducing AD risk, rests in the gold standard, randomized controlled clinical trials, many of which are either underway currently or expected to launch in the next year. Future studies will benefit not only from testing whether these approaches work but also by suggesting potential biological mechanisms. Regardless of whether the relation is due to shared risk factors or to direct and indirect influences of CVD on brain pathology, exercise, and dietary interventions designed to reduce cardiovascular risk factors are plausible strategies to prevent or retard cognitive decline in vulnerable patients.

3.3.2 Cognitive Engagement

Another non-pharmacological approach explored in relation to AD onset is the notion of cognitive stimulation or "cognitive engagement." The theory examined in this context is that by continued intellectual and social activity into later life (such as reading, participating in clubs, volunteer activities, and traveling), fundamental cognitive and memory processes in areas affected by AD are activated and help to maintain brain reserve, buffering against cognitive decline (Barnes et al. 2004; Carlson et al. 2009). The CCS explored the association between engagement in a number of stimulating activities on later cognitive and functional decline in a population-based sample of incident AD (Treiber et al. 2011). Participants were asked about their daily activities using a Lifestyle Activity Questionnaire which categorizes activities by their level of processing demands. These activities were also rated by two expert judges (neuropsychologists) as either active leisure pursuits (e.g. crossword puzzles and taking courses), passive activities (watching television, listening to radio or music), or intermediate (visiting with friends, driving/using public transportation). Frequency of activities were evaluated and then examined in relationship to global cognitive decline (MMSE) and change in functional ability (CDR) over 5 time points and 2.5 years of observation. At initial assessment, 87% of the participants with mild forms of dementia were engaged in one or more active or intermediate activity. Because of moderate correlation between the number of intermediate and more demanding activities, the two categories were summed to create an aggregate variable. The results showed that individuals who were actively engaged in these stimulating activities had slower cognitive decline and better functional ability.

The results from the CCS and other observational studies suggest that engagement in cognitively stimulating activities may slow rates of dementia progression in AD. However, it is equally plausible that declines in social involvement are not leading to functional decline but rather the opposite, increasing functional impairments lead to gradual truncation in activities. Testing the directionality of the effects can be determined with clinical trials. Several such studies are underway and are encouraging both in normal vital elderly and in MCI patients (Unverzagt et al. 2009), with domain-specific improvements seen following targeted cognitive or memory interventions. Although the effects are not seen for memory training in MCI, these patients benefit to same degree as vital older adults with cognitive and speed of processing training. Additionally, involvement in these interventions contributed to enhanced feelings of mastery and self efficacy (Wolinsky et al. 2010). It remains to be seen how persistent these training effects are and whether such interventions lead to overall slower rates of dementia progression.

4 Conclusions

4.1 Preventing AD and Successful Cognitive Aging

Begun as a study of examining the epidemiology of AD and other dementias into late-old age, the CCS has provided considerable information regarding normative aging, the characteristics of early prodromal AD, and factors that may modify risk of dementia and cognitive decline. Defining prodromal AD reliably has important implications for facilitating comparisons across population studies to determine changes in incidence trends and factors influencing transitions rates in disease. By identifying members most at risk for decline to dementia within the population, early interventional efforts are possible. The information gleaned from the CCS with regard to modifiable factors has suggested possible targets for intervention and strategies. Although some of these compounds, most notably NSAIDS and HRT, have not yielded positive results when tested in clinical trials, the observations from the CCS pointing to the importance of exposure timing have provided testable hypotheses for disease pathogenesis and for critical time windows when treating the disease. The data suggest that there may be discrete windows of time when specific targeted exposures to NSAIDs or HRT may be beneficial. Started outside this window, the same treatments may have no effect or worse yet, prove deleterious.

Additional work continues in the population setting to better understand the timing issue, the relative efficacy of different classes of compounds, formulations, and doses, and the interaction of genes with these environmental factors over time. With many very old survivors in the population, the CCS also offers the opportunity to examine predictors of robust aging in the absence of dementia. Preliminary work has suggested distinctive features in these robust survivors including familial longevity, moderate physical activity, and few major health problems (Tschanz et al. 2005). Continuing work focuses on the genetic and biological factors underlying robust aging. Defining these factors in this elite group of survivors may highlight biological pathways and treatment strategies that when applied across the broader population may act to delay expression of disease in those who are predisposed to AD. By delaying the onset of AD symptoms by five years or more, one has essentially prevented the disease and in so doing has facilitated successful cognitive aging for a major sector of the vulnerable population.

References

Aisen PS, Schafer KA et al (2003) Effects of rofecoxib or naproxen vs placebo on Alzheimer disease progression: a randomized controlled trial. J Am Med Assoc 289(21):2819–2826

American Psychiatric Association (1987) Diagnostic and statistical manual of mental disorders, 3rd edn revised: DSM-III-R. American Psychiatric Association, Washington

American Psychiatric Association (1994) Diagnostic and statistical manual of mental disorders, 4th edn: DSM-IV. American Psychiatric Association, Washington

Andersen K, Launer LJ et al (1995) Do nonsteroidal anti-inflammatory drugs decrease the risk for Alzheimer's disease? The Rotterdam Study. Neurology 45(8):1441–1445

Backman L, Jones S et al (2005) Cognitive impairment in preclinical Alzheimer's disease: a meta-analysis. Neuropsychology 19(4):520–531

Baker LD, Frank LL et al (2010) Effects of aerobic exercise on mild cognitive impairment: a controlled trial. Arch Neurol 67(1):71–79

Barnes LL, Mendes de Leon CF et al (2004) Social resources and cognitive decline in a population of older African Americans and Whites. Neurology 63(12):2322–2326

Beck AT, Steer RA et al (1996) Manual for the beck depression inventory-II. Psychological Corporation, San Antonio

Benton A (1992) Benton Visual Retention Test, 5th edn. Psychological Corporation, New York

Benton AL, Hamsher K (eds) (1983) Multilingual aphasia examination. AJA Associates, Iowa City

Blackford RC, LA Rua A (1989) Criterion for diagnosing age associated memory impairment: proposed improvements from the field. Dev Neuropsychol 5:298–300

Bozoki A, Giordani B et al (2001) Mild cognitive impairments predict dementia in nondemented elderly patients with memory loss. Arch Neurol 58(3):411–416

Braak H, Braak E (1996) Evolution of the neuropathology of Alzheimer's disease. Acta Neurol Scand Suppl 165:3–12

Brayne C, Stephan BC et al (2011) A European perspective on population studies of dementia. Alzheimers Dement 7(1):3–9

Breitner JC, Gau BA et al (1994a) Inverse association of anti-inflammatory treatments and Alzheimer's disease: initial results of a co-twin control study. Neurology 44(2):227–232

Breitner JC, Welsh KA et al (1994b) Alzheimer's disease in the NAS-NRC registry of aging twin veterans. II. Longitudinal findings in a pilot series. National Academy of Sciences. National Research Council Registry. Dementia 5(2):99–105

Breitner JC, Welsh KA et al (1995) Alzheimer's disease in the National Academy of Sciences–National Research Council Registry of Aging Twin Veterans. III. Detection of cases, longitudinal results, and observations on twin concordance. Arch Neurol 52(8):763–771

Breitner JC, Wyse BW et al (1999) APOE-epsilon4 count predicts age when prevalence of AD increases, then declines: the Cache County Study. Neurology 53(2):321–331

Breteler MM, Claus JJ et al (1992) Epidemiology of Alzheimer's disease. Epidemiol Rev 14: 59–82

Brookmeyer R, Johnson E et al (2007) Forecasting the global burden of Alzheimer's disease. Alzheimers Dement 3(3):186–191

Brookmeyer R, Evans DA et al (2011) National estimates of the prevalence of Alzheimer's disease in the United States. Alzheimers Dement 7(1):61–73

Carlson MC, Zandi PP et al (2001) Hormone replacement therapy and reduced cognitive decline in older women: the Cache County Study. Neurology 57(12):2210–2216

Carlson MC, Erickson KI et al (2009) Evidence for neurocognitive plasticity in at-risk older adults: the experience corps program. J Gerontol A Biol Sci Med Sci 64(12):1275–1282

Corder EH, Saunders AM et al (1993) Gene dose of apolipoprotein E type 4 allele and the risk of Alzheimer's disease in late onset families. Science 261(5123):921–923

Crook T, Bartus RT et al (1986) Age-associated memory impairment: proposed diagnostic criteria and measures of clinical change—report of a National Institute of Mental Health Work Group. Dev Neuropsychol 2(4):261–276

Cummings JL, Mega M et al (1994) The Neuropsychiatric Inventory: comprehensive assessment of psychopathology in dementia. Neurology 44(12):2308–2314

DeKosky ST, Carrillo MC et al (2011) Revision of the criteria for Alzheimer's disease: a symposium. Alzheimers Dement 7(1):e1–e12

Dubois B, Feldman HH et al (2007) Research criteria for the diagnosis of Alzheimer's disease: revising the NINCDS-ADRDA criteria. Lancet Neurol 6(8):734–746

Folstein MF, Folstein SE et al (1975) "Mini-mental state". A practical method for grading the cognitive state of patients for the clinician. J Psychiatr Res 12(3):189–198

Fotuhi M, Zandi PP et al (2006) APOE ε4 positive elderly taking anti-oxidant vitamins E and C in combination with NSAIDs develop less cognitive decline with aging: the Cache County Study. In: 10th international conference on Alzheimer's disease and related disorders, Madrid, Spain

Goldman W, Price J et al (2001) Absence of cognitive impairment or decline in preclinical Alzheimer's disease. Neurology 56(3):361–367

Graham JE, Rockwood K et al (1997) Prevalence and severity of cognitive impairment with and without dementia in an elderly population. Lancet 349(9068):1793–1796

Hayden KM, Warren LH et al (2005) Identification of VaD and AD prodromes: the Cache County Study. Alzheimers Dement 1(1):19–29

Hayden KM, Zandi PP et al (2006) Vascular risk factors for incident Alzheimer disease and vascular dementia: the Cache County Study. Alzheimer Dis Assoc Disord 20(2):93–100

Hayden KM, Zandi PP et al (2007) Does NSAID use modify cognitive trajectories in the elderly? The Cache County Study. Neurology 69(3):275–282

Heart Protection Study Collaborative Group (1999) MRC/BHF Heart Protection Study of cholesterol-lowering therapy and of antioxidant vitamin supplementation in a wide range of patients at increased risk of coronary heart disease death: early safety and efficacy experience. Eur Heart J 20(10):725–741

Hebert LE, Scherr PA et al (2003) Alzheimer disease in the US population: prevalence estimates using the 2000 census. Arch Neurol 60(8):1119–1122

Hofman A, Ott A et al (1997) Atherosclerosis, apolipoprotein E, and prevalence of dementia and Alzheimer's disease in the Rotterdam Study. Lancet 349(9046):151–154
Honig LS, Tang MX et al (2003) Stroke and the risk of Alzheimer disease. Arch Neurol 60(12):1707–1712
in 't Veld BA, Launer LJ et al (1998) NSAIDs and incident Alzheimer's disease. The Rotterdam Study. Neurobiol Aging 19(6):607–611
Jorm AF, Jacomb PA (1989) The Informant Questionnaire on Cognitive Decline in the Elderly (IQCODE): socio-demographic correlates, reliability, validity and some norms. Psychol Med 19(4):1015–1022
Jorm AF, Jolley D (1998) The incidence of dementia: a meta-analysis. Neurology 51(3):728–733
Jorm AF, Korten AE et al (1987) The prevalence of dementia: a quantitative integration of the literature. Acta Psychiatr Scand 76(5):465–479
Kaplan E, Goodglass H et al (1978) The Boston naming test. Veterans Administration, Boston
Katzman R (1976) Editorial: the prevalence and malignancy of Alzheimer disease. A major killer. Arch Neurol 33(4):217–218
Kawas C, Segal J et al (1994) A validation study of the Dementia Questionnaire. Arch Neurol 51(9):901–906
Khachaturian AS, Zandi PP et al (2006) Antihypertensive medication use and incident Alzheimer disease: the Cache County Study. Arch Neurol 63(5):686–692
Kivipelto M, Helkala EL et al (2001) Midlife vascular risk factors and late-life mild cognitive impairment: a population-based study. Neurology 56(12):1683–1689
Kryscio RJ, Mendiondo MS et al (2004) Designing a large prevention trial: statistical issues. Stat Med 23(2):285–296
Langa KM, Plassman BL et al (2005) The aging, demographics, and memory study: study design and methods. Neuroepidemiology 25(4):181–191
Launer LJ (2011) Counting dementia: there is no one "best" way. Alzheimers Dement 7(1):10–14
Launer LJ, Brock DB (2004) Population-based studies of AD: message and methods: an epidemiologic view. Stat Med 23(2):191–197
Launer LJ, Ross GW et al (2000) Midlife blood pressure and dementia: the Honolulu-Asia aging study. Neurobiol Aging 21(1):49–55
Lautenschlager NT, Cox KL et al (2008) Effect of physical activity on cognitive function in older adults at risk for Alzheimer disease: a randomized trial. J Am Med Assoc 300(9):1027–1037
Levy R (1994) Aging-associated cognitive decline. Working Party of the International Psychogeriatric Association in collaboration with the World Health Organization. Int Psychogeriatr 6(1):63–68
Lindsay J, Sykes E et al (2004) More than the epidemiology of Alzheimer's disease: contributions of the Canadian Study of Health and Aging. Can J Psychiatry 49(2):83–91
Luchsinger JA, Tang MX et al (2003) Antioxidant vitamin intake and risk of Alzheimer disease. Arch Neurol 60(2):203–208
Lyketsos CG, Toone L et al (2006) A population-based study of the association between coronary artery bypass graft surgery (CABG) and cognitive decline: the Cache County Study. Int J Geriatr Psychiatry 21(6):509–518
Manly JJ, Bell-McGinty S et al (2005) Implementing diagnostic criteria and estimating frequency of mild cognitive impairment in an urban community. Arch Neurol 62(11):1739–1746
Manton KG, Tolley HD (1991) Rectangularization of the survival curve: implications of an ill-posed question. J Aging Health 3(2):172–193
Martin BK, Szekely C et al (2008) Cognitive function over time in the Alzheimer's Disease Anti-inflammatory Prevention Trial (ADAPT): results of a randomized, controlled trial of naproxen and celecoxib. Arch Neurol 65(7):896–905
Mayeux R, Reitz C et al (2011) Operationalizing diagnostic criteria for Alzheimer's disease and other age-related cognitive impairment—part 1. Alzheimers Dement 7(1):15–34
McGeer PL, Schulzer M et al (1996) Arthritis and anti-inflammatory agents as possible protective factors for Alzheimer's disease: a review of 17 epidemiologic studies. Neurology 47(2): 425–432

McGinnis JM, Foege WH (1993) Actual causes of death in the United States. J Am Med Assoc 270(18):2207–2212

McKhann G, Drachman D et al (1984) Clinical diagnosis of Alzheimer's disease: report of the NINCDS-ADRDA Work Group under the auspices of Department of Health and Human Services Task Force on Alzheimer's Disease. Neurology 34(7):939–944

Meinert CL, Breitner JC (2008) Chronic disease long-term drug prevention trials: lessons from the Alzheimer's Disease Anti-inflammatory Prevention Trial (ADAPT). Alzheimers Dement 4(Suppl 1):S7–S14

Miech RA, Breitner JC et al (2002) Incidence of AD may decline in the early 90s for men, later for women: the Cache County Study. Neurology 58(2):209–218

Mielke MM, Rosenberg PB et al (2007) Vascular factors predict rate of progression in Alzheimer disease. Neurology 69(19):1850–1858

Morris MC, Beckett LA et al (1998) Vitamin E and vitamin C supplement use and risk of incident Alzheimer disease. Alzheimer Dis Assoc Disord 12(3):121–126

Morris JC, Storandt M et al (2001) Mild cognitive impairment represents early-stage Alzheimer disease. Arch Neurol 58(3):397–405

Murray CJL, Michaud CM et al (1988) U.S. patterns of mortality by county and race: 1965–1994. C. f. P. a. D. Studies. Harvard, Cambridge

Norton MC, Breitner JC et al (1994) Characteristics of nonresponders in a community survey of the elderly. J Am Geriatr Soc 42(12):1252–1256

Ostbye T, Krause KM et al (2006) Ten dimensions of health and their relationships with overall self-reported health and survival in a predominately religiously active elderly population: the Cache County Memory Study. J Am Geriatr Soc 54(2):199–209

Ott A, Breteler MM et al (1997) Atrial fibrillation and dementia in a population-based study. The Rotterdam Study. Stroke 28(2):316–321

Ott A, Stolk RP et al (1999) Diabetes mellitus and the risk of dementia: the Rotterdam Study. Neurology 53(9):1937–1942

Paleologos M, Cumming RG et al (1998) Cohort study of vitamin C intake and cognitive impairment. Am J Epidemiol 148(1):45–50

Park DC, Reuter-Lorenz P (2009) The adaptive brain: aging and neurocognitive scaffolding. Annu Rev Psychol 60:173–196

Petersen RC, Morris JC (2005) Mild cognitive impairment as a clinical entity and treatment target. Arch Neurol 62(7):1160–1163, discussion 1167

Petersen RC, Smith GE et al (1999) Mild cognitive impairment: clinical characterization and outcome. Arch Neurol 56(3):303–308

Plassman BL, Langa KM et al (2007) Prevalence of dementia in the United States: the aging, demographics, and memory study. Neuroepidemiology 29(1–2):125–132

Plassman BL, Langa KM et al (2008) Prevalence of cognitive impairment without dementia in the United States. Ann Intern Med 148(6):427–434

Plassman BL, Williams JW Jr et al (2010) Systematic review: factors associated with risk for and possible prevention of cognitive decline in later life. Ann Intern Med 153(3):182–193

Plassman BL, Langa KM et al (2011) Incidence of dementia and cognitive impairment, not dementia in the United States. Ann Neurol [Epub ahead of print]

Price JL, Morris JC (1999) Tangles and plaques in nondemented aging and "preclinical" Alzheimer's disease. Ann Neurol 45(3):358–368

Price JL, Ko AI et al (2001) Neuron number in the entorhinal cortex and CA1 in preclinical Alzheimer disease. Arch Neurol 58(9):1395–1402

Quinn JF, Montine KS et al (2004) Suppression of longitudinal increase in CSF F2-isoprostanes in Alzheimer's disease. J Alzheimers Dis 6(1):93–97

Refolo LM, Malester B et al (2000) Hypercholesterolemia accelerates the Alzheimer's amyloid pathology in a transgenic mouse model. Neurobiol Dis 7(4):321–331

Reitan R (1958) Validity of the Trail Making Test as an indicator of organic brain damage. Percept Mot Skills 8:271–276

Rockwood K, Ebly E et al (1997) Presence and treatment of vascular risk factors in patients with vascular cognitive impairment. Arch Neurol 54(1):33–39

Roman GC, Tatemichi TK et al (1993) Vascular dementia: diagnostic criteria for research studies. Report of the NINDS-AIREN International Workshop. Neurology 43(2):250–260

Sano M, Ernesto C et al (1997) A controlled trial of selegiline, alpha-tocopherol, or both as treatment for Alzheimer's disease. The Alzheimer's Disease Cooperative Study. N Engl J Med 336(17):1216–1222

Scarmeas N, Stern Y et al (2006) Mediterranean diet and risk for Alzheimer's disease. Ann Neurol 59(6):912–921

Scarmeas N, Luchsinger JA et al (2009) Physical activity, diet, and risk of Alzheimer disease. J Am Med Assoc 302(6):627–637

Shipley W (1967) Shipley Institute of Living Scale. Western Psychological Services, Los Angeles

Skoog I, Lernfelt B et al (1996) 15-year longitudinal study of blood pressure and dementia. Lancet 347(9009):1141–1145

Sliwinski M, Buschke H et al (1997) The effect of dementia risk factors on comparative and diagnostic selective reminding norms. J Int Neuropsychol Soc 3(4):317–326

Smith A (1991) Symbol Digit Modalities Test. Western Psychological Services, Los Angeles

Smith PJ, Blumenthal JA et al (2010) Aerobic exercise and neurocognitive performance: a meta-analytic review of randomized controlled trials. Psychosom Med 72(3):239–252

Stewart WF, Kawas C et al (1997) Risk of Alzheimer's disease and duration of NSAID use. Neurology 48(3):626–632

Strauss E, Sherman EMS, Spreen O (eds) (2006) A Compendium of Neuropsychological Tests, 3rd edn. Oxford University Press, New York

Treiber KA, Carlson MC et al (2011) Cognitive stimulation and cognitive and functional decline in Alzheimer's disease: the Cache County Dementia Progression Study. J Gerontol B Psychol Sci Soc Sci

Tschanz JT, Welsh-Bohmer KA et al (2000) Dementia diagnoses from clinical and neuropsychological data compared: the Cache County Study. Neurology 54(6):1290–1296

Tschanz JT, Welsh-Bohmer KA et al (2002) An adaptation of the modified mini-mental state examination: analysis of demographic influences and normative data: the Cache County Study. Neuropsychiatry Neuropsychol Behav Neurol 15(1):28–38

Tschanz JT, Treiber K et al (2005) A population study of Alzheimer's disease: findings from the Cache County Study on memory, health, and aging. Care Manag J 6(2):107–114

Tschanz JT, Welsh-Bohmer KA et al (2006) Conversion to dementia from mild cognitive disorder: the Cache County Study. Neurology 67(2):229–234

Tuokko HA, Frerichs RJ et al (2001) Cognitive impairment, no dementia: concepts and issues. Int Psychogeriatr 13(Supp 1):183–202

Unverzagt FW, Smith DM et al (2009) The Indiana Alzheimer Disease Center's symposium on mild cognitive impairment. Cognitive training in older adults: lessons from the ACTIVE Study. Curr Alzheimer Res 6(4):375–383

Wechsler D (1987) WMS-R: Wechsler Memory Scale-Revised. Harcourt Brace Jovanovich, New York

Wechsler D (1997) Wechsler Adult Intelligence Scales, 3rd edn. The Psychological Corporation, San Antonio

Weggen S, Eriksen JL et al (2001) A subset of NSAIDs lower amyloidogenic Abeta42 independently of cyclooxygenase activity. Nature 414(6860):212–216

Welsh KA, Butters N et al (1994) The Consortium to Establish a Registry for Alzheimer's Disease (CERAD). Part V. A normative study of the neuropsychological battery. Neurology 44(4):609–614

Welsh-Bohmer KA, Mohs RC (1997) Neuropsychological assessment of Alzheimer's disease. Neurology 49(Suppl 3):S11–13

Welsh-Bohmer KA, Ostbye T et al (2009) Neuropsychological performance in advanced age: influences of demographic factors and Apolipoprotein E: findings from the Cache County Memory Study. Clin Neuropsychol 23(1):77–99

Wengreen HJ, Munger R et al (2005). Fruit and vegetable intake and cognitive function in the elderly: the Cache County Study on memory, health and aging. Alzheimer's Dementia J Alzheimer's Assoc 1(1): S100

Wengreen HJ, Munger RG et al (2007) Antioxidant intake and cognitive function of elderly men and women: the Cache County Study. J Nutr Health Aging 11(3):230–237

Wengreen H, Munger R et al Prospective study of DASH-and Mediterranean-style dietary patterns and age-related cognitive change. J Nutr Health Aging (in press)

Whitmer RA, Quesenberry CP et al (2011) Timing of hormone therapy and dementia: the critical window theory revisited. Ann Neurol 69(1):163–169

Wolinsky FD, Vander Weg MW et al (2010) Does cognitive training improve internal locus of control among older adults? J Gerontol B Psychol Sci Soc Sci 65(5):591–598

Zandi PP, Anthony JC et al (2002a) Reduced incidence of AD with NSAID but not H2 receptor antagonists: the Cache County Study. Neurology 59(6):880–886

Zandi PP, Carlson MC et al (2002b) Hormone replacement therapy and incidence of Alzheimer disease in older women: the Cache County Study. J Am Med Assoc 288(17):2123–2129

Zandi PP, Anthony JC et al (2004) Reduced risk of Alzheimer disease in users of antioxidant vitamin supplements: the Cache County Study. Arch Neurol 61(1):82–88

Part II
Cognitive and Emotional Perspectives in Aging

Successful Cognitive Aging

Colin A. Depp, Alexandria Harmell and Ipsit V. Vahia

Abstract Given the rapid rate of population aging, basic science and public health efforts have increasingly focused on the determinants of successful cognitive aging. In this chapter, we review the definition and biological, psychological, and environmental determinants of cognitive health in later life. Successful cognitive aging is a multi-dimensional construct that lacks a consensus operationalized definition, and has been variously conceptualized in an ipsative, normative, or criterion-referenced manner. Nevertheless, there are a number of biomarkers, at the genetic and cellular level, that provide indicators of cognitive health in aging. Functional and structural neuroimaging suggest multiple pathways to successful cognitive aging, by way of brain reserve and cognitive reserve. A number of behavioral and environmental interventions, including dietary restriction, physical activity, and cognitive stimulation, are promising avenues for extending the cognitive healthspan associated with normal aging. Thus, there is a variety of recent findings providing optimism that successful cognitive aging, howsoever defined, will be attainable by more older adults in the future.

Keywords Aging · Older adults · Cognitive · Neuropsychology · Brain · Health behavior · Lifestyle

---

C. A. Depp (✉) · I. V. Vahia
Sam and Rose Stein Institute for Research on Aging, San Diego, USA
e-mail: cdepp@ucsd.edu

C. A. Depp · I. V. Vahia
Department of Psychiatry, University of California, San Diego, USA

A. Harmell
Joint Doctoral Program in Clinical Psychology,
San Diego State University/University of California, San Diego, CA, US

Curr Topics Behav Neurosci (2012) 10: 35–50
DOI: 10.1007/7854_2011_158

Published Online: 25 January 2012

Contents

1 Introduction 36
2 Rationale for a Focus on Cognitive Health 37
3 Defining Successful Cognitive Aging 38
4 Determinants of Successful Cognitive Aging 39
4.1 Genetic Influences 39
4.2 Stress and Resilience 40
4.3 Brain Reserve and Cognitive Reserve 41
4.4 Wisdom 42
4.5 Lifestyle Behaviors 43
5 Successful Cognitive Aging in Selected Clinical Populations 46
6 Conclusions 47
References 48

1 Introduction

While the traditional focus of research on the aging brain has centered on delineating pathological from typical age-associated changes, a much smaller body of work has focused on successful cognitive aging. This literature includes research on the contributors to maintaining high cognitive performance into later life, as well as interventions that might enhance cognitive abilities beyond that typically associated with normal aging. As yet, successful cognitive aging lacks a consensus definition, and there are a number of long-standing challenges and controversies as to how to operationally define positive states of health in older age. There is no controversy, however, about the considerable public health need for better understanding how to lengthen the healthspan into older age. This imperative comes from the fact that there are now more people who are older than the age of 65 than at any time in recorded history. In fact, two-thirds of the people who have *ever* reached the age of 65 are alive right now (National Institute of Aging 2007).

Much of the work on successful aging has focused on the prevention of physical illnesses and disabilities that stem from age-associated conditions, and less has focused on defining and understanding the determinants of cognitive health, specifically. In a 2006 review of 28 quantitative studies that reported an operational definition of successful aging, only 13 of the studies included cognitive functioning as a component in their definitions (Depp and Jeste 2006). Nevertheless, there are a number of reasons for increased focus on successful *cognitive* aging as well as a number of exciting recent findings that suggest emerging avenues to maintaining brain health in older age. In this chapter, we describe research examining the constituents of successful aging, from traditional neuropsychological constructs to more esoteric ones like wisdom. We then review the evidence for the determinants of cognitive and brain reserve, including interventions aimed at altering the trajectory of normal cognitive aging. Finally, we describe the utility of applying models of successful cognitive aging to selected clinical populations.

2 Rationale for a Focus on Cognitive Health

Tracing the demographic transition to an aging society, there are several reasons for the increasing relevance of cognitive health to overall health in older age. One hundred years ago, the leading causes of morbidity and mortality were infectious diseases—as many as 40% of people did not survive beyond childhood and the mean age at death was 40 years (Fogel 2005). Improvements in access to clean drinking water, along with other practices targeting the prevention and treatments of infectious diseases, dramatically reduced the leading causes of illness. With infectious diseases less likely to cause mortality, there was a subsequent shift to primary causes of mortality to those that were substantially age-associated (e.g., cancer, heart disease). Even so, more recent treatments have begun to delay the onset and reduce the incidence of heart disease, cancer, and stroke—in Robert Fogel's classic comparison of the Civil War cohort to the baby boomers—the age at onset of heart disease was approximately 10 years later among baby boomers (Fogel 2005). Therefore, a variety of factors have lengthened both the human life span and healthspan in rapid fashion over the twentieth century.

In the past several decades, brain illnesses, particularly Alzheimer's disease, have begun to increase as leading causes of mortality (Steenland et al. 2009). At the present time, there is a much larger armamentarium of treatments and prevention strategies for cardiovascular health than for preventing cognitive decline, and so it is reasonable to expect that cognitive health may become a more potent rate-limiting factor in avoiding age-associated morbidity as the present cohort of younger and middle-aged adults ages. Thus, research identifying strategies for maintenance of cognitive health in older age will be increasingly important.

A second, more optimistic reason to focus on cognitive health is the shift in conceptualization of the aging brain, from static to malleable. Early in the twentieth century, Sigmund Freud captured the prevalent view of intractability of the aging brain in his quote that "…near or above the age of fifty the elasticity of the mental processes on which treatment depends is as a rule lacking—old people are no longer educable" (Freud 1924). Nevertheless, several decades of animal studies indicate that enriched environments are associated with evidence of greater neuroplasticity, even when experiments were conducted on animals that were already aged. Later research studies have evidenced the adaptability of the aging brain in humans. Thus, excitement has increased about the potential for altering aging trajectories, in contrast to previously entrenched beliefs.

A third reason for the importance of cognitive health comes from the broader context of successful aging, and many of the best studied interventions to prevent physical decline, reduce disability, or improve emotional and social functioning involve volitional behaviors. For example, physical activity appears to reduce the risk for cognitive decline, yet cognitive impairment reduces the likelihood of engagement in physical activity (Geda et al. 2010). Thus, various cognitive abilities are involved in the daily decisions that individuals make in regard to engaging in physical activity or to adapt to chronic illness. Therefore, successful cognitive aging will be increasingly important, and we review attempts and challenges in defining this construct in the following section.

3 Defining Successful Cognitive Aging

There are several challenges in arriving at a consensus definition of successful cognitive aging. Perhaps the closest to consensus is that proposed by the 2006 National Institutes of Health's Cognitive and Emotional Health Project (Hendrie et al. 2006). This workgroup described cognitive health with the following definition as "Not just the absence of cognitive impairment, but the development and preservation of the multi-dimensional cognitive structure that allows the older adult to maintain social connectedness, and ongoing sense of purpose, and the abilities to function independently, to permit functional recovery from illness and injury, and to cope with residual cognitive deficits." Key elements of this definition are that successful cognitive aging combines multiple cognitive domains, extending beyond traditional neuropsychological abilities, such as memory and executive functions, to more esoteric constructs such as wisdom and resilience. In addition, this definition considers central the link between cognitive health with functional independence and engagement with life.

While the CEHP's definition provides a useful focal point, there are several issues in operationalizing this, or any other, successful aging construct. In particular, successful cognitive aging can be defined in terms of thresholds, normative comparisons, or in comparison with past performance. Examples of threshold criteria used in prior studies of successful aging include having a Mini-Mental Status Examination score greater than 24. The advantage of this approach is the ease of generalization and comparability with other samples, yet the performance of cutoffs varies by a number of factors, not the least of which is age. Moreover, not all cognitive abilities decline at the same rate, and some abilities, such as vocabulary and social cognition, may actually improve with age. Therefore, at the cognitive domain level, the cutoff for some domains may change little with age, whereas for processing speed the cutoff may change a great deal.

An example of a normative criterion that has been used in the definition of successful cognitive aging includes being above the median score on neurocognitive tests. This approach may be more robust to the increasing variance in cognitive abilities across individuals with age and may be more likely to maximize power to detect differences between successful and 'unsuccessful' groups. Nevertheless, using normative criteria is disadvantageous in regard to generalizability relative to the criterion approach, as sample characteristics have a strong impact on relative performance. An additional approach to operationalization would be to define success in reference to preservation of past performance—i.e., maintaining levels of cognitive performance attained at mid-life. This approach dovetails with the portion of the CEHP definition that corresponds to maintenance of independence, yet would require longitudinal data on cognitive abilities from earlier ages. For example, Yaffe showed that individuals who maintained cognitive abilities into their 80s and 90s were less at risk for disability and death (Yaffe et al. 2010). Thus, overall, even if agreement is reached on the broad constituents of the successful cognitive aging construct, its operationalization presents many unresolved challenges.

4 Determinants of Successful Cognitive Aging

Given the caveats about attaining a precise and useful definition of successful cognitive aging, there are a number of putative biological, behavioral, and social mechanisms by which cognitive abilities can be maintained and perhaps even improved into older age. We review these determinants in the following section, beginning at the genetic and molecular level, then to the aging brain, and finally broadening into various lifestyle factors.

4.1 Genetic Influences

Although most studies on the genetic influence on aging have focused on the phenotype of longevity, a number of studies have assessed the degree to which genes influence aging-related trajectories in cognitive health. Among twin studies, there does appear to be evidence that cognitive performance in later life is heritable. In the Swedish twin study (Finkel et al. 1998), 54% of the variance in a general cognitive factor was attributable to heritability. However, it was notable that the proportion of variance attributed to genetic factors in middle-aged adults was much higher ($\sim$80%), such that the relative influence of environmental factors on cognitive health is likely to increase with age (Finkel et al. 1998). In regard to functional performance, Gurland found that 20–25% of the variance in disability was accounted for by genetic factors in a sample of 1,384 monozygotic and 1,337 dizygotic septenegerians (Gurland et al. 2004). Therefore, cognitive health in older age likely has a moderate degree of heritability, providing confidence in the potential utility of the search for specific genes that influence this phenotype. Nevertheless, the influence of genes may lessen with age and certainly environmental factors play a large, if not larger, role than genes in late-life cognitive health.

Several studies have examined candidate single nucleotide polymorphisms (SNPs) in case-control studies comparing a successful aging group to older adults who do not meet the criteria for successful aging. Glatt et al. (2007) reviewed 29 studies that examined genetic influences, and while these studies were not restricted to cognitive health as a phenotype, it is notable that the genes clustered into three groups (Glatt et al. 2007): (a) genes involved in cardiovascular health and cholesterol, lipid, and lipoprotein transport or metabolization (e.g., PON1, APOE), (b) genes involved in inflammatory processes (e.g., IL6, IL10), and (c) genes involved in cell cycling, growth, and signaling (e.g., SIRT3). Although this body of literature is quite small, these fundamental physiological processes are linked, in the related literature, with indicators of cognitive decline. For example, a large body of literature indicates the associated links with the APOE SNP (Haan et al. 1999) and inflammation (McGeer and McGeer 2003) with Alzheimer's Disease.

More recent work has employed the rapidly growing armenentarium of genomic and molecular biology to address the genetic determinants of cognitive health in later life. Zubenko et al. (2007) used a genome-wide association approach in comparing 100 cognitively intact adults 90 years or older to 100 young adults. A total of 16 markers were identified that were associated with membership in the cognitively intact older adult group, yet these were not consistent across men and women, suggesting a need for future study to examine gene by sex interactions. Additional work has investigated mitochondrial DNA and epigenetic factors, such as DNA methylation.

The advantage of an epigenetic approach is that it enables capturing the impact of environmental stress on cellular functioning, thereby potentially permitting future interventions to act at the intersection between environment (e.g., nutrition, lifestyle) and genetic markers, so as to compensate for potential deleterious effects of genes. A particularly exciting area is the epigenetic regulation of the telomere, a portion of the chromosome that appears to provide an indication of cellular aging. Shorter telomeres appear to be associated with accelerated aging, and greater environmental stress appears to be associated with telomere shortening (Aviv et al. 2003). One recent study in Ashkenazi centenarians showed that older adults who had longer telomeres had significantly better scores on the Mini-Mental Status Examination, which is remarkable considering that the population of 100-year olds has already been subjected to substantial attrition due to mortality (Atzmon et al. 2010). Therefore, the telomere represents an objectively measurable phenotype intermediate between basic physiological processes and environmental factors, and is related to cognitive health even among the oldest old.

4.2 Stress and Resilience

In general, older adults have different stressors than younger adults, as the issues that older adults face tend to be more likely to be health-related, chronic, and uncontrollable stress (e.g., bereavement, caregiving) rather than acute stressors that involve decision making (e.g., losing a job, divorce—Karel 1997). There is remarkable diversity among older adults in their response to stressors, even when the nature and type of stressor is similar across people. The impact of psychosocial stress on the aging brain has been examined for decades, and a more recent body of work has examined the characteristics of resilience to stress. It is clear that chronic unremitting stress in older adults influences a network of physiological processes that often results in neuronal degradation. Specifically, the stress-associated stimulation of the Hypothalamic-Pituitary-Adrenal axis results in the secretion of glucocorticoids such as cortisol, which is associated with damage to various brain structures, and particularly the hippocampus. In addition, stress may induce inflammatory cytokines and decrease immune response, which also result in deleterious impacts on the brain (McEwen 2000). O'Hara and Hallmayer (2007) found greater reactivity to stress, as defined by increased levels of cortisol, among carriers of the 5HTT short allele (O'Hara and Hallmayer 2007). These factors related to reduced hippocampal volume, and thus genetic variation may produce greater vulnerability to stress and associated cognitive deficits—such links may explain the well-documented links between depression and anxiety with cognitive impairment in later life.

Clearly, however, there are some individuals who do not experience the deleterious effects of stress. For some, this may be because of genetic advantages, such as the research on 5HTT SNP indicates. However, for others, behavioral coping strategies may reduce the impact of stress on the brain. Resilience has been the focus of a small body of literature, yet Lamond et al. found in a sample of 1,395 community dwelling older women that the predictors of resilience (lower levels of depression, greater optimism, social engagement) tend to be relatively stable across the life span (Lamond et al. 2008). However, the structure of resilience was slightly different from that found in younger adults, with greater positive associations between emotion-focused coping (e.g., tolerating negative affect) and freedom from depression than problem-focused coping. In regard to cognitive health, resilience represents a trait that may buffer the effects of stress on the brain, with future research necessary to understand its biological mechanisms.

4.3 Brain Reserve and Cognitive Reserve

Moving from molecular biology to the aging brain, two organizing concepts that are relevant to successful cognitive aging are brain reserve and cognitive reserve. Brain reserve is defined as the ability to withstand damage and yet continue to function, metaphorically the "hardware" of the brain. Normal aging is associated with global volumetric shrinkage in brain structures such as the caudate, cerebellum, hippocampus, and prefrontal areas (Raz et al. 2005) as well as decreased organization and integrity of white matter tracts leading to potential consequences on cognitive functioning. Additional studies have demonstrated a relationship between increasing age and reduced glucose metabolism and blood flow at rest, specifically in the frontal regions and anterior cingulate. Brain reserve theory is considered passive in that it presupposes that there is some threshold of damage required that will result in cognitive deficits when met.

Brain reserve has been most commonly used to provide an explanation describing a subset of individuals (approximately 25%) who, at autopsy, have the hallmarks of Alzheimer's disease (AD) including amyloid plaques and neurofibrillary tangles and yet who during their life do not show clinical manifestations of the disease (Snowdon 2003). This discrepancy between having extensive brain neuropathology without cognitive impairment led to the theory that perhaps having a larger brain volume, larger neurons, and more synaptic connections can act as a buffer, or protective factor in preventing or slowing down cognitive decline. Several studies provide compelling evidence in support of the brain reserve theory including a recent study by Perneczky et al. (2010) showing that larger head circumference attenuated the relationship between cerebral atrophy and cognitive functioning in 270 patients diagnosed with AD patients.

Cognitive reserve, on the other hand, is more akin to the "software" of the brain and involves active compensation rather than the passive model of brain reserve.

One potential compensatory strategy that has been noted in functional imaging studies comparing younger adults to older adults is that prefrontal activity during cognitive performance tends to be less lateralized (localized specifically to the right or left side of the brain) in older adults compared to younger adults. One theory of this qualitative difference in brain response between the two groups (referred to as Hemispheric Asymmetry Reduction in older adults, HAROLD—Cabeza et al. 2002) is that the aging brain, compared to the younger brain, uses more of its resources in an attempt to compensate for structural and functional decline. This theory has been further supported by the Scaffolding Theory of Aging and Cognition (STAC) which states that the brain utilizes complementary, alternative neural circuits with increasing age (scaffolding) in an effort to maintain or strengthen particular cognitive objectives (Park and Reuter-Lorenz 2009). It is important to note that brain reserve and cognitive reserve are not competing or mutually exclusive models, rather these are two parallel processes that help to explain that there is more than one pathway to maintaining cognitive health in older age.

4.4 Wisdom

In considering the phenotypes related to successful cognitive aging, memory and processing speed are prototypical cognitive domains that are associated with age-related declines. However, some cognitive abilities may increase with age, and such increases may contribute just as much, if not more, to the maintenance of independence. Wisdom is one cognitive ability that is commonly associated with older adults. Wisdom remains a fledgling area for neurobiological research, and it suffers from the same definitional issues as successful cognitive aging. Nevertheless, newer measures such as the three- dimensional wisdom scale (3D-WS) have resulted in both a clearer understanding of wisdom (Ardelt 1997), as well as the start of biological research in this area. A recent study employing the Delphi method—a widely used and accepted method for seeking consensus among experts within a certain topic area—to defining wisdom surveyed experts on wisdom research from around the world and concluded that wisdom is uniquely human; a form of advanced cognitive and emotional development that is experience driven; and a personal quality, albeit a rare one, that can be learned, increases with age, can be measured, and is not likely to be enhanced by taking medication (Jeste et al. 2010).

Based on this definition and a review of multiple other related sources in the literature (Jeste and Vahia 2008), Meeks and Jeste (2009) have proposed a putative neurobiological basis for wisdom. According to their proposed model, wisdom comprises six distinct domains: prosocial attitudes and behavior; social decision making/pragmatic knowledge of life; emotional homeostasis; reflection/self-understanding; value relativism/tolerance and acknowledgment of and effective dealing with uncertainty and ambiguity. Based on a comprehensive review of the literature related to these domains, the authors suggest that multiple neurotransmitters have a role in acquiring and maintaining wisdom, including dopamine (in regulating impulsivity and selflessness), serotonin (in maintaining social cooperation), norepinephrine (regulation and dampening of stress-related performance and decision making), vasopressin (for affiliative behavior in animal models) and oxytocin (for social cognition and social decision making). The authors also identify several brain regions that may be part of a circuitry involved in the process of being 'wise'. These regions were identified through multiple neuroimaging studies. The neurobiological model proposed by the authors suggests that lateral prefrontal cortex (PFC) in concert with dorsal Anterior Cingulate Cortex (ACC), Orbito-frontal Cortex (OFC) and Medial Prefrontal Cortex (MPFC), appears to have an important inhibitory effect on several brain areas associated with emotionality and immediate reward dependence (e.g, amygdala, ventral striatum). They also note that there is a complementary emotion-based subcomponent, including prosocial attitudes and behaviors that involve MPFC, PCC, OFC, superior temporal sulcus, and reward neurocircuitry. Finally, Meeks and Jeste suggest that the interplay and balance between phylogenetically older brain regions (e.g, limbic cortex) and the more recently evolved PFC is the key to maintaining wisdom. Research on wisdom shows how more esoteric concepts associated with successful aging can be deconstructed, studied using laboratory experiments, and related to brain structure and function.

4.5 *Lifestyle Behaviors*

4.5.1 Physical Activity

Physical activity is associated with a variety of health benefits, including reduced risk of mortality, physical disability, cardiovascular disease, and osteoporosis.

Although fewer in number, the results of studies examining the impact of physical activity on cognitive health are equally impressive. Experimental animal studies employing a variety of protocols and species have indicated that physical activity is associated with reduced neurodegeneration (Cotman and Berchtold 2002). In humans, observational studies indicated that greater exercise participation is associated with a reduced risk for dementia (Larson and Wang 2004). A meta-analysis of 18 studies examining physical activity interventions that enrolled samples of older adults found that physical activity was associated with increases in performance on several cognitive domains, in particular executive functioning (Kramer et al. 2006). One recent study that randomized older adults who were sedentary to aerobic exercise or a control condition found evidence for increases in brain volume in gray and white-matter regions after one year of participation (Colcombe et al. 2006). This latter study was particularly notable in that the sample consisted of older adults who were not engaged in exercise at the time of enrollment, and thus initiating exercise participation in older age may still provide benefits to cognitive health.

The mechanisms of physical activity on brain health remain unclear, as there are a number of potential pathways that are indirect (e.g., improvement in cardiovascular health, increased social engagement). However, there do appear to be more direct influences of physical activity on the brain, in particular by reducing oxidative stress and inflammation (Kramer et al. 2006). Anterior brain regions appear to be more altered by cardiovascular fitness than posterior regions (Prakash et al. 2011). Finally, it is worth noting that only a minority of older adults in the United States meet at the Center for Disease Control recommendations for daily exercise participation, and older women are the segment of the population with the single lowest rate of engagement in physical activity. Therefore, there is much potential impact of increasing physical activity at the population level.

4.5.2 Nutrition/Dietary Restriction

Among the potential interventions to extend longevity, none have generated more enthusiasm than dietary restriction. In rodents, restriction of ad libitum diets in mice by approximately 1/3 is associated with 30–40% increase in life span (Fontana et al. 2010). Smaller but still significant effects have been seen among primates, and human trials have been completed. Importantly, other age-related phenotypes, including cognitive ability, also appear to be improved by dietary restriction. A recent randomized controlled trial in humans enrolled 50 overweight adults in a three-month trial of dietary restriction. Compared to the control group and a second group that received an increase of unsaturated fatty acids, memory performance was 30% better in the dietary restriction group at post-study (Witte et al. 2009). Further sensitivity analyses within the dietary restriction group found that improvement in insulin sensitivity and inflammatory markers correlated with improvement in memory functioning. The exact mechanisms by which dietary restriction could extend the life span or improve neurocognitive ability are unknown and actively debated. However, it does appear that dietary restriction slows the metabolic rate and, as such, may reduce the oxidative stress associated with metabolic processes.

Dietary restriction and physical activity (as well as cognitive stimulation, described next) may share similar mechanisms of action in reducing neuronal vulnerability. According to Mattson and Magnus (Mattson and Magnus 2006), these three activities introduce a mild stressor, which causes the brain to release neurotrophic factors (e.g., BDNF), which, in turn, promote synaptogenesis. This process is called "hormesis" and is analogous to a vaccine, in which degraded pathogens are introduced to stimulate the immune response to develop antibodies. Thus, quite different behaviors (e.g., reduction in calories, engaging in cognitively stimulating activity) could produce similar positive effects on the brain.

There are a host of nutritional products marketed toward healthy brain aging. The current state of the evidence ranges from negative (e.g., Ginko Biloba) to inconclusive (e.g., Fish oil, reservatrol—Daffner 2010). However, there is evidence that vitamin deficiencies and dehydration are remediable risk factors for cognitive impairment. The list of conditions that result from vitamin deficiencies is expansive, but a growing body of research has recently emerged showing possible links between deficiencies in vitamins such as vitamin D, K, and B_{12} in older adults and adverse cognitive outcomes. Also strongly encouraged by nutritionists is the intake of foods rich in antioxidants.

4.5.3 Cognitive Stimulation

There have been a number of observational studies that have linked participation in cognitively stimulating activities, such as recreational activities (e.g., engagement in puzzles, games) and cognitively demanding vocations, with reduced risk of dementia. Conversely, a number of studies have found that greater engagement in less cognitively stimulating activities, particularly television watching, is associated with greater risk of dementia. This collection of studies has fed into the so-called "use it or lose it" hypothesis in regard to the influence of activity participation on the prevention of cognitive decline. There is some controversy as to whether such claims are overstated, as Salthouse (2006) has argued that no patterns of cognitive activities to date have proven to change the *rate* of cognitive decline (Salthouse 2006). It is also difficult to quantify the amount of cognitive stimulation associated with activities, which may differ across people within the same activity class. Moreover, it is difficult to parcel out the effect of selection, as people who gravitate toward and subsequently participate in cognitively stimulating tasks may have greater cognitive resources to begin with. Nevertheless, even if the true effect of cognitive stimulation is marginal, there may be additional benefits of engaging in cognitively stimulating activities (e.g., associated physical or social activity) that may occur.

Although cognitive training to alter the course of normal age-related decline has been a field of study for many years, there have been several recent developments.

For one, the ACTIVE trial, completed in 2006, was the largest trial of cognitive training to date, enrolling 2,802 older adults (Willis et al. 2006). Participants were older adults without evidence of cognitive impairments, and they were randomized to one of four types of cognitive training, each targeting a different cognitive ability: reasoning and problem-solving, memory, attention, and processing speed. The primary outcomes of the trial were performance on cognitive tests, and secondary or distal outcomes were functional measures. The results suggested that cognitive training was associated with statistically significant improvements in cognitive ability, although restricted to the domain that was trained. Improvements in functional measures were far more subtle, yet in the case of reasoning and problem solving, persisted for up to 5 years after randomization (Willis et al. 2006).

A second recent study shows the emergence of computerized approaches in the delivery of cognitive training (Mahncke et al. 2006). Finally, a number of non-traditional approaches to enhancing cognitive abilities have been evaluated—these do not engage individuals in specific cognitive training exercises, rather they provide standardized cognitively stimulating activities. Examples include narrative writing, acting, volunteering, and group problem solving (Park et al. 2007). Increases in cognitive stimulation may also be attainable from reducing cognitively sedentary behavior. For example, television use is higher among older adults than younger and middle-aged adults despite the observation that older adults enjoy television less than younger people (Depp et al. 2010).

5 Successful Cognitive Aging in Selected Clinical Populations

Depression: A number of studies have linked depression with risk for cognitive impairment in older adults. While the prevalence rates for major depression in older adults are lower than in younger adults, subsyndromal depressive (SSD) symptoms may in fact be three times more prevalent (Meeks et al. 2011). In one study, SSD was found to significantly increase the prospective likelihood of cognitive impairment and dementia at one-year follow-up (Han et al. 2006). There are direct and indirect pathways by which depression may impair cognitive abilities. Direct effects of depression on the aging brain may be through increasing cellular stress as described above. Indirect effects may be in interfering with activities that foster successful cognitive aging, as depressed older adults are less likely to engage in physical activity, and more likely to engage in sedentary solitary behavior. Thus, treatment for depression, if successful, may influence the likelihood of attaining successful cognitive aging through multiple pathways.

HIV Infection: Due to the advent of more advanced anti-retroviral treatments, the life expectancy of people living with HIV infection has increased dramatically over the past decade. The first surviving generation to be exposed to the HIV virus is currently entering older adulthood. Malaspina et al. (2011) examined the predictors of successful cognitive aging, using an operational definition which included freedom from impairment on neurocognitive testing and lack of endorsement of subjective confusion (Malaspina et al. 2011). Patients who met these criteria were less likely to exhibit depressive symptoms, and they were also more likely to endorse greater engagement with treatment and adherence to antiretroviral medications. Indicators of viral load did not differ between successful and unsuccessful groups. This study provides an exemplar for examining the characteristics of "survivors" of the HIV epidemic and the future work may delineate how these individuals have managed to avoid accumulating cognitive impairment over the course of their illness.

Schizophrenia: One of the core features of the illness is pervasive cognitive impairment, with average cognitive deficits at approximately one standard deviation below the mean of healthy comparison subjects (Dickinson et al. 2007). Although this general trend in differential cognitive functioning is found between patients and healthy individuals, there is enormous heterogeneity in both the level and pattern of cognitive deficits within patient populations. This substantial intragroup variability in cognitive functioning is further evidenced by the interesting discovery that an estimated 20–25% of schizophrenia patients have "neuropsychologically normal" profiles (Palmer et al. 2009). It does not appear as though patients without cognitive impairment have less severe psychopathologic symptoms, and thus preservation of brain function is not simply a function of lower disease severity. Rather, the evidence of greater educational attainment among non-impaired patients suggests that a number of hypotheses about brain reserve described above may also apply to schizophrenia.

6 Conclusions

In this chapter, we sought to describe a diverse body of literature on the determinants of successful cognitive aging. Due to the aging of the population and the relative paucity of effective means of slowing the rate of cognitive aging, cognitive health will likely have an increasing impact on the health and independence of older adults. There is no consensus definition of successful cognitive aging, let alone successful aging. Nevertheless, it is generally agreed upon that successful cognitive aging is multi-dimensional and extends beyond performance in traditional cognitive domains to more socioemotional constructs such as wisdom. It is unclear whether the operationalization of this construct should be based upon normative, ipsative, or threshold criteria—or some combination. The NIH CEHP project and the NIH Toolbox project represent attempts to create a common language in defining cognitive health and a core set of instruments to measure this construct, respectively.

Despite the lack of consensus, there is marked convergence among basic physiological markers such as genes, and environmental factors such as stress, that implicate fundamental aging processes in the effect of potentially modifiable aspects of aging. Moreover, the concepts of brain and cognitive reserve show how early and late-life experiences can influence cognitive aging trajectories and indicate that multiple pathways are available to successful cognitive aging. Several lifestyle factors, including physical activity, cognitive stimulation, and dietary restriction appear to relate to improvement in cognitive functioning, even among older adults who initiate these activities in later life. Although involving unique behaviors, these lifestyle factors may share common pathways to stimulating neurotrophic factors to reduce neuronal vulnerability. Finally, there is some potential utility in examining the successful cognitive aging construct in illnesses that typically accompany cognitive impairments. It is clear that much needs to be learned in regard to defining and enhancing cognitive health, yet it is equally clear that the next several decades will generate a host of new hypotheses and potential strategies for maintaining cognitive abilities further into later life.

References

Ardelt M (1997) Wisdom and life satisfaction in old age. J Gerontol B Psychol Sci Soc Sci 52(1):P15–P27

Atzmon G, Cho M et al (2010) Genetic variation in human telomerase is associated with telomere length in Ashkenazi centenarians. Proc Nat Acad Sci U S A 107:1710–1717

Aviv A, Levy D et al (2003) Growth, telomere dynamics and successful and unsuccessful human aging. Mech Ageing Dev 124(7):829–837

Cabeza R, Anderson ND et al (2002) Aging gracefully: compensatory brain activity in high-performing older adults. Neuroimage 17(3):1394–1402

Colcombe SJ, Erickson KI et al (2006) Aerobic exercise training increases brain volume in aging humans. J Gerontol A Biol Sci Med Sci 61(11):1166–1170

Cotman CW, Berchtold NC (2002) Exercise: a behavioral intervention to enhance brain health and plasticity. Trends Neurosci 25(6):295–301

Daffner KR (2010) Promoting successful cognitive aging: a comprehensive review. J Alzheimer's Dis 19(4):1101–1122

Depp CA, Schkade DA et al (2010) Age, affective experience, and television use. Am J Prev Med 39(2):173–178

Depp CA, Jeste DV (2006) Definitions and predictors of successful aging: a comprehensive review of larger quantitative studies. Am J Geriatr Psychiatry 14(1):6–20

Dickinson D, Ramsey ME et al (2007) Overlooking the obvious: a meta-analytic comparison of digit symbol coding tasks and other cognitive measures in Schizophrenia. Arch Gen Psychiatry 64(5):532–542

Finkel D, Pedersen NL et al (1998) Longitudinal and cross-sectional twin data on cognitive abilities in adulthood: the Swedish adoption/twin study of aging. Dev Psychol 34(6): 1400–1413

Fogel RW (2005) Changes in the physiology of aging during the twentieth century. NBER work paper series w 11233, National Bureau of Economic Research, Cambridge

Fontana L, Partridge L et al (2010) Extending healthy life span—from yeast to humans. Science 328(5976):321–326

Freud S (1924) On psychotherapy. Hogarth Press, London

Geda YE, Roberts RO et al (2010) Physical exercise, aging, and mild cognitive impairment: a population-based study. Arch Neurol 67(1):80–86

Glatt SJ, Chayavichitsilp P et al (2007) Successful aging: from phenotype to genotype. Biol Psychiatry 62(4):282–293

Gurland BJ, Page WF et al (2004) A twin study of the genetic contribution to age-related functional impairment. J Gerontol A Biol Sci Med Sci 59(8):859–863
Haan MN, Shemanski L et al (1999) The role of APOE Ïμ4 in modulating effects of other risk factors for cognitive decline in elderly persons. JAMA J Am Med Assoc 282(1):40–46
Han L, McCusker J et al (2006) The temporal relationship between depression symptoms and cognitive functioning in older medical patients—prospective or concurrent? J Gerontol A Biol Sci Med Sci 61(12):1319–1323
Hendrie H, Albert M et al (2006) The NIH cognitive and emotional health project: report of the critical evaluation study committee. Alzheimer's Dementia J Alzheimer's Assoc 2(1):12–32
Jeste DV, Ardelt M et al (2010) Expert consensus on characteristics of wisdom: a Delphi method study. Gerontologist 50(5):668–680
Jeste DV, Vahia IV (2008) Comparison of the conceptualization of wisdom in ancient Indian literature with modern views: focus on the Bhagavad Gita. Psychiatry 71(3):197–209
Karel MJ (1997) Aging and depression: vulnerability and stress across adulthood. Clin Psychol Rev 17(8):847
Kramer AF, Erickson KI et al (2006) Exercise, cognition, and the aging brain. J Appl Physiol 101(4):1237–1242
Lamond AJ, Depp CA et al (2008) Measurement and predictors of resilience among community-dwelling older women. J Psychiatr Res 43(2):148–154
Larson EB, Wang L (2004) Exercise, aging, and Alzheimer disease. Alzheimer Dis Assoc Disord 18(2):54–56
Mahncke HW, Connor BB et al (2006) Memory enhancement in healthy older adults using a brain plasticity-based training program: a randomized, controlled study. Proc Natl Acad Sci U S A 103(33):12523–12528
Malaspina L, Woods SP et al (2011) Successful cognitive aging in persons living with HIV infection. J Neurovirol 17(1):110–119
Mattson MP, Magnus T (2006) Ageing and neuronal vulnerability. Nat Rev Neurosci 7(4): 278–294
McEwen BS (2000) Allostasis, allostatic load, and the aging nervous system: role of excitatory amino acids and excitotoxicity. Neurochem Res 25(9–10):1219–1231
McGeer EG, McGeer PL (2003) Inflammatory processes in Alzheimer's disease. Prog Neuropsychopharmacol Biol Psychiatry 27(5):741–749
Meeks TW, Vahia IV et al (2011) A tune in "a minor" can "b major": a review of epidemiology, illness course, and public health implications of subthreshold depression in older adults. J Affect Disord 129(1–3):126–142
Meeks TW, Jeste DV (2009) Neurobiology of wisdom: a literature overview. Arch Gen Psychiatry 66(4):355–365
National Institute on Aging (2007) Why population aging matters: a global perspective. National Institute on Aging, Bethesda
O'Hara R, Hallmayer JF (2007) Serotonin transporter polymorphism and stress: a view across the lifespan. Curr Psychiatry Rep 9(3):173–175
Palmer B, Dawes S et al (2009) What do we know about neuropsychological aspects of Schizophrenia? Neuropsychol Rev 19(3):365–384
Park DC, Reuter-Lorenz P (2009) The adaptive brain: aging and neurocognitive scaffolding. Annu Rev Psychol 60:173–196
Park DC, Gutchess AH et al (2007) Improving cognitive function in older adults: nontraditional approaches. J Gerontol B Psychol Sci Soc Sci 62(1):152
Perneczky R, Wagenpfeil S et al (2010) Head circumference, atrophy, and cognition: implications for brain reserve in Alzheimer disease. Neurology 75(2):137–142
Prakash RS, Voss MW et al (2011) Cardiorespiratory fitness and attentional control in the aging brain. Front Hum Neurosci 4:229
Raz N, Lindenberger U et al (2005) Regional brain changes in aging healthy adults: general trends, individual differences and modifiers. Cereb Cortex 15(11):1676–1689
Salthouse TA (2006) Mental exercise and mental aging. Perspect Psychol Sci 1(1):68–87

Snowdon DA (2003) Healthy aging and dementia: findings from the nun study. Ann Intern Med 139(2):450–454

Steenland K, MacNeil J et al (2009) Recent trends in Alzheimer disease mortality in the United States, 1999–2004. Alzheimer Dis Assoc Disord 23(2):165–170

Willis SL, Tennstedt SL et al (2006) Long-term effects of cognitive training on everyday functional outcomes in older adults. JAMA 296(23):2805–2814

Witte AV, Fobker M et al (2009) Caloric restriction improves memory in elderly humans. Proc Nat Acad Sci 106(4):1255–1260

Yaffe K, Lindquist K et al (2010) The effect of maintaining cognition on risk of disability and death. J Am Geriatr Soc 58(5):889–894

Zubenko GS, Hughes HB III et al (2007) Genome survey for loci that influence successful aging: results at 10-cM resolution. Am J Geriatr Psychiatry 15(3):184–193

Behavioral Neuroscience of Emotion in Aging

Alfred W. Kaszniak and Marisa Menchola

Abstract Recent research on emotion and aging has revealed a stability of emotional experience from adulthood to older age, despite aging-related decrements in the perception and categorization of emotionally relevant stimuli. Research also shows that emotional expression remains intact with aging. In contrast, other studies provide evidence for an age-related decrease in autonomic nervous system physiological arousal, particularly in response to emotionally negative stimuli, and for shifts in central nervous system physiologic response to emotional stimuli, with increased prefrontal cortex activation and decreased amygdala activation in aging. Research on attention and memory for emotional information supports a decreased processing of negative emotional stimuli (i.e., a decrease in the negativity effect seen in younger adults), and a relative increase in the processing of emotionally positive stimuli (positivity effect). These physiological response and attentional/memory preference differences across increasingly older groups have been interpreted, within socioemotional selectivity theory, as reflecting greater motivation for emotion regulation with aging. According to this theory, as persons age, their perceived future time horizon shrinks, and a greater value is placed upon cultivating close, familiar, and meaningful relationships and other situations that give rise to positive emotional experience, and avoiding, or shifting attention from, those people and situations that are likely to elicit negative emotion. Even though there are central nervous system structural changes in emotion-relevant brain regions with aging, this shift in socioemotional

A. W. Kaszniak (✉) · M. Menchola
Department of Psychology, University of Arizona,
1503 E. University, Tucson, AZ 85721, USA
e-mail: kaszniak@email.arizona.edu

M. Menchola
Department of Family and Community Medicine,
University of Arizona, 1450 N Cherry, Tucson, AZ 85719, USA

Curr Topics Behav Neurosci (2012) 10: 51–66
DOI: 10.1007/7854_2011_163

Published Online: 11 September 2011

selectivity, and perhaps the decreased autonomic nervous system physiological arousal of emotion with aging, facilitate enhanced emotion regulation with aging.

Keywords Emotion · Aging · Neuroscience · Socioemotional selectivity

Contents

1 Conceptualizing and Measuring Emotion 53
2 Emotion in Aging 55
2.1 Emotional Experience and Appraisal 55
2.2 Emotional Expression 58
2.3 ANS and CNS Physiological Change 59
2.4 Emotion Regulation 61
3 Summary and Conclusions 61
References 62

During the mid-twentieth century, theorists (e.g., Banham 1951) posited that emotional well-being declines in aging, similar to many aspects of biological and psychological functioning that were being studied at the time. However, subsequent empirical studies of adult age-group differences in emotion have not supported such theories. Aging does not appear to be associated with increased emotional distress: Surveys employing standard diagnostic criteria have found lower rates of depression and anxiety among older adults than in younger or mid-life adults (for review, see O'Donnell and Kaszniak 2011). Self-report of negative affect has also been found to be lower among older adults than middle-aged and younger adults (Lawton et al. 1992). Compared to younger adults, older adults report a lower frequency of negative emotion, and a similar intensity of both positive and negative emotion in their daily lives (Carstensen et al. 2000). On average, resilience, sense of coherence, emotion-regulation, and overall hedonic well-being appear to increase in older age, whereas experienced stress appears to decrease (Carstensen, et al. 2000; Charles and Carstensen 2010; Nilsson et al. 2010; Stone et al. 2010). Given the many biological changes of aging, and the evidence for age-related reduction in such cognitive domains as attention, working memory, and other aspects of executive functioning (e.g., Gazzaley et al. 2005; Salthouse 1990; Salthouse et al. 2003), the apparent relative age-resilience of emotional well-being may seem surprising. The present chapter explores age-related psychological and behavioral neuroscience research on emotion in an effort to better understand this apparent paradox. This review will be limited to relevant and representative studies of healthy younger, middle-aged, and older adults, and will not consider emotion-related clinical phenomena in aging, such as depression and anxiety. Emotion in age-related neurological disorders, such as Alzheimer's disease (e.g., Allender and Kaszniak 1989; Burton and Kaszniak 2006), will also not be reviewed. Some reference will be made to questionnaire and survey studies of emotion in daily life, since such studies provide rich sources of hypotheses.

However, a primary emphasis will be placed on laboratory experiments, due to their necessity for both examining the neurobiological correlates of emotion and unpacking the multiple neurobehavioral processes that characterize emotion.

1 Conceptualizing and Measuring Emotion

In order to understand recent psychological and behavioral neuroscience research on aging and emotion, it is helpful to begin with a framework for conceptualizing emotion and the ways in which it is elicited and measured in the laboratory. Many theorists and investigators view emotion as a psychological state or construct detected through observation of several associated, though distinct components, identified and operationalized through measures of self-reported experience, action dispositions, physiology, and self-regulation. Some view these components as indicators of a latent trait (i.e., emotion as *causing* measurable changes in these indicators), while others "…treat emotion not as causing these disturbances, but as emerging from them" (Clore and Ortony 2008, p. 630). The purpose of the present chapter does not necessitate a theoretical commitment in regard to whether measurable emotion components are "effect indicators" or "causal indicators" (Bollen and Lennox 1991). However, the present authors would note that viewing components as constituents rather than consequences of emotion is more consistent with some plausible neurobiological models (for review, see Clore and Ortony 2008).

The first of these components can be referred to as the *experience or feeling of emotion,* occurring in conjunction with the *appraisal* of emotion-eliciting stimuli. The term appraisal is used to describe an individual's conscious or automatic judgment of the significance of an object or event, relevant to the person's needs, motivations, and perceived coping resources (Lazarus 1966). Events appraised as challenges or affordances in this self-relevance assessment are those that elicit emotion, initiating the process of mobilizing the individual for potential action. Emotion-eliciting stimuli employed in laboratory experiments have included evocative images, sounds, films, directed facial action, music, imagery or episode recall, and dyadic interaction, among others (for comprehensive reviews, see Coan and Allen 2007). Several of these types of stimuli will be described in the following discussion of aging-related research. Although appraisal of such stimuli may be conscious, and therefore reportable, there exists a body of research demonstrating the importance of non-conscious aspects of appraisal (Bargh and Williams 2007). The appraisal of an emotional stimulus and the conscious experience of emotional feelings may be conceptually and methodologically very difficult to separate (Clore and Ortony 2000; Kaszniak 2001; Nielsen and Kaszniak 2007; Stein and Hernandez 2007), in that conscious or non-conscious appraisal appears to be a necessary condition for emotional experience or feeling to occur. Other variables (e.g., feedback from visceral changes, and information contained the context within which an appraised stimulus occurs.; for full discussion, see Nielsen and Kaszniak 2007) also theoretically contribute to emotional feeling or experience, and these variables can be difficult to measure. For the purposes of the present chapter, these multiple processes will be considered as a single aspect, and described as conscious emotional experience. This aspect of emotion is perhaps the most methodologically amenable to study (i.e., self-report), and accordingly many empirical studies of emotion and aging have assessed this component.

A second emotion component is emotional expression, which can be considered as a type of *action disposition*. Lang (1995) proposed that emotions occur when something important is happening to the organism, but responsive actions are inhibited. This pause in behavior is presumably when action dispositions begin to form. Action dispositions therefore represent a preparation for action, and not the actions themselves. Action dispositions can include skeletal muscle preparation for approach or avoidance behaviors and facial or other bodily expressions of emotion, which can be assessed by either structured observation (e.g., of facial expression; Cohn et al. 2007) or physiological measures such as facial muscle electromyography (EMG; Lang et al. 1993).

The third component of emotion is the *physiologic change* that occurs in response to an emotionally salient stimulus. Arousal related to emotion is thought to be detectible in several aspects of autonomic nervous system (ANS) activity, and examples of psychophysiological measures reflecting ANS activity include skin-conductance response, heart rate, blood pressure, and skin temperature (Cacioppo and Tassinary 1990; Jennings and Yovetich 1991; Larsen et al. 2008). In general, measures of ANS (particular sympathetic ANS) arousal correlate with self-report of increasing arousal in emotional experience (Lang et al. 1993). The typical paradigm for assessing physiologic arousal in the laboratory is to record the relevant measure (e.g., heart rate) while inducing emotion by either exposing research participants to emotionally salient stimuli or having them engaged in emotionally evocative imagery or event recall. In addition to psychophysiological measures of ANS activity, an increasing number of studies employing neuroimaging measures of localized central nervous system (CNS) response to emotional stimuli have been reported over the past two decades. This body of research has shown peak positron emotion tomography (PET) and functional magnetic resonance imaging (fMRI) activations in the brainstem, hypothalamic, and paralimbic areas relating to emotional experience, and in amygdalar complex and posterior cortex relating to emotional perception. Further, pleasant emotional experience has been found to be related to midline brainstem, hypothalamic, and ventromedial frontal regions, and negative experiences to brainstem (periacqueductal gray), insular cortex, striatum, and orbital cortical regions (for meta-analysis, see Wager et al. 2008). Such neuroimaging observations support and extend inferences concerning emotion-important brain regions gained from earlier studies of persons with focal brain damage (e.g., Bechara et al. 1999; Kaszniak et al. 1999; Tranel and Damasio 1994; Zoccolotti et al. 1982).

A fourth component is *emotion regulation,* which has been defined as the processes by which a person influences which emotions occur, when they occur, and how the emotion is experienced and expressed (Gross 1998a, 1998b). Gross (1998b) has proposed a process model of emotion regulation in which specific strategies can be differentiated according to the time at which they occur. Since emotions are multi-componential processes that unfold over time, strategies aimed at the regulation of emotion may involve changes in the appraisal/experiential, action disposition, and/or physiological components at different points in the process (Gross 2002; Gross and Levenson 1993). *Antecedent-focused strategies* refer to those implemented before the emotion response becomes fully activated. For example, a person might employ the antecedent-focused strategy of avoiding certain people or situations likely to elicit an emotion, or focusing attention on only those aspects of the situation likely to elicit desired, and not undesired emotion. Another antecedent-focused strategy, *reappraisal*, involves construing an emotion-eliciting situation in alternative, perhaps even non-emotional terms. For example, someone viewing a bloody scene of an accident victim might reappraise the image as that of posed actors with fake blood. *Response-focused strategies* are used once the emotion episode is underway, after the action dispositions have been generated. For example, the response-focused regulation strategy of expressive suppression involves the inhibition of ongoing emotion-expressive (e.g., facial) behavior. Reviews of the growing research literature on emotion regulation can be found in Gross (2007).

2 Emotion in Aging

2.1 Emotional Experience and Appraisal

As noted by Charles and Carstensen (2007), one plausible explanation for why older adults in survey studies do not generally report greater levels or frequency of emotional distress than younger adults is that their experience of emotion might be diminished or less intense with aging. However, laboratory experiments focusing on emotional experience have typically not found evidence that the conscious experience of emotion significantly diminishes across the adult age-range.

In examining the relevant research, it is important to distinguish between studies that measure younger and older adults' emotional experience/appraisal, and those that measure perceptual discrimination or categorization of emotionally-relevant stimuli (e.g., naming emotions expressed in faces and matching images and sounds according to whether both are sad, happy, or angry.). Some investigators who have assessed perception or categorization of emotional stimuli have found adult age-group differences. Early studies described consistent decrements from younger to older groups on several tasks of emotional stimulus perception in both visual and auditory modalities (Malatesta et al. 1987; Oscar-Berman et al. 1990). Older adults may also have greater difficulty distinguishing between certain types of negative facial expressions of emotions (Sullivana and Ruffman 2003), and identifying facial expressions, emotional voice quality (Dupuis and Pichora-Fuller 2010; Mitchell 2007; Ryan et al. 2010), and some emotional bodily expressions (for meta-analysis of emotion expression in aging across several modalities, see Rufman et al. 2008). Such emotion recognition deficits among older individuals may be independent of performance on non-emotional tasks in the same modality, such as facial gender recognition (Sullivan and Ruffman 2004). Older adults may also process emotional faces differently than younger adults, relying on more widespread neocortical networks (Tessitore et al. 2005). However, these age-group differences in perception or categorizing of emotional stimuli do not necessarily imply differences in emotional experience. It should also be noted that there is some evidence for relative preservation with aging of the ability to discriminate positive facial expressions (e.g., happiness, surprise), despite older adults' disadvantage in the perception of negative expressions, such as sadness, anger, and fear (Mathersul et al. 2009; Ortega and Phillips 2008; Slessor et al. 2010; Sullivan and Ruffman 2004). This difference in the pattern of age relationships in the perception/discrimination of positive versus negative expressions may be related to a *positivity effect* among older adults in attention to emotional stimuli (see section on Emotion Regulation below). Across several attention and memory studies, although younger adults show preference for negative, versus positive, emotional stimuli, older adults either do not show, or have smaller such negativity preferences (for meta-analysis, see Murphy and Isaacowitz 2008).

In order to illustrate a commonly used experimental design in research on emotional experience and other emotion components, it is useful to describe in some detail a study conducted in the present authors' laboratory (Reminger et al. 2000). Reminger et al. compared vision-, education-, and gender-matched, healthy, non-depressed, and cognitively intact younger (mean age = 26.4 years; range = 18–48) and older adults (mean age = 68.4; range = 57–81) in their response to emotional images selected from the International Affective Picture System (IAPS; Lang et al. 2005). The IAPS allows selection of images according to young adult normative ratings of emotional valence (the dimension of pleasant or positive, to unpleasant or negative emotional experience) and arousal (from low to high intensity) elicited by the images. Ratings of emotional experience valence and arousal in response to each IAPS image in the Reminger et al. study were made using the Self-Assessment Manikin (SAM; Lang 1980). The SAM employs cartoon-like figures on a visual analog scale designed to minimize the effects that language could have in reporting emotional experience in response to the IAPS images. The valence and arousal dimensions of the SAM are scaled with five figures for each of these dimensions, and the option is given to make ratings between two figures, thus providing a scale ranging from 1 to 9 for each dimension. The measurement of emotional experience with the SAM is motivated by factor-analytic studies of evaluative responses to a large variety of verbal, visual, and sound emotional stimuli, wherein valence and arousal dimensions have been found to account for most of the variance (see Russell 1980). Reminger et al. (2000) found no differences between the older and younger groups on averaged SAM ratings of experienced emotional valence or arousal in response to each IAPS image (Fig. 1).

In the Reminger et al. study, the older and younger groups showed an identical magnitude and pattern of SAM-reported emotional experience (i.e., more positive experienced valence in response to the normatively rated pleasant IAPS images,

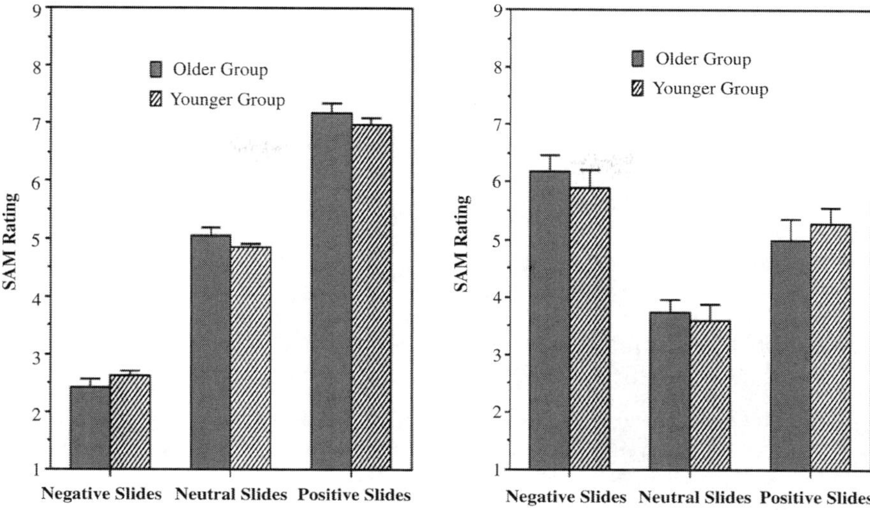

Fig. 1 Emotional valence (*left*) and arousal (*right*) ratings of older and younger adults in response to emotionally salient images. SAM = Self-Assessment Manikin ratings. Graphs are based upon data reported by Reminger et al. (2000)

and more negative experience in response to the unpleasant images, compared to neutral images; higher experienced arousal to both the pleasant and unpleasant images, compared to neutral images). These results are similar to those reported by investigators utilizing other kinds of laboratory emotional stimuli, including film clips, remembered emotional events, and spouse interaction about emotionally charged conflicts (Levenson et al. 1991; Magai et al. 2006; Malatesta and Kalnok 1984; Tsai et al. 2000). For the smaller number of studies that have reported adult age-group differences in response to laboratory emotion-induction procedures, older adults tend to report stronger, not weaker, emotional experience (e.g., Kliegel et al. 2007; Kunzmann and Grühn 2005; and Kunzmann and Richter 2009). When older adults do report stronger experience in response to laboratory emotion induction, they may also be more effective than younger adults in repairing negative mood after it is induced (Kliegel et al. 2007). There is also some evidence from laboratory emotion-induction studies to suggest that older adults may experience more complex emotion (i.e., several simultaneous emotions), when compared to younger adults (Charles 2005). Overall, available research does not support the hypothesis that the intensity of emotional experience, elicited in the laboratory, declines with aging.

In the Reminger et al. (2000) study, participants also completed the Profile of Mood States Questionnaire (POMS; adjective rating scales used to ascertain a person's current mood state; McNair et al. 1992). Despite the equivalence of younger and older adults in their SAM-reported emotional experience in response to IAPS images in this study, analyses of POMS scores showed significant differences across age groups. For the POMS total mood disturbance score, collapsed across the two consecutive assessment sessions, the older group endorsed significantly fewer items indicating negative mood. Thus, within the same sample of participants studied by Reminger and colleagues, older adults reported less current negative mood, although reporting emotional experience identical to that of younger participants in response to laboratory emotional stimuli. It would thus appear that the lower levels of emotional distress in older, compared to younger adults, cannot be explained by any age-group differences in the type (valence) or intensity (arousal) of emotional experience stimulated within controlled laboratory conditions.

2.2 Emotional Expression

Overall, studies of facial emotion expression have also revealed more similarities than differences between younger and older adults. For example, older and younger participants have not been found to differ in ability to pose facial emotion expressions (Moreno et al. 1990), in the frequency of spontaneous facial expressions during relived-emotion tasks (Levenson et al. 1991) or in facial response to emotional films (Tsai et al. 2000). Reminger et al. (2000), in the study described above, also reported evidence for adult age-group similarity in stimulus-elicited emotional expression. In addition to emotional experience measures, Reminger et al. examined facial expression in their older and younger adult groups. Bilateral facial skin-surface EMG recordings were obtained from the zygomatic (mid-cheek) and corrugator (region just above the eyebrows) muscle areas during presentation of the IAPS images. Despite having sufficient statistical power, no significant differences were found on any of the facial EMG measures (Fig. 2) to the different IAPS image types (positive, neutral,and negative). The authors concluded that emotional expression, similar to emotional experience, remains invariant from younger to older adulthood.

Bailey and Henry (2009) report a similar lack of difference between older and younger adults' facial EMG responses to angry and happy facial expression images (i.e., facial expression mimicry), presented either supraliminally or subliminally (through backward masking of the faces). The expression of emotion through affective prosody (the melodic contour of speech that communicates emotion), when reading or repeating emotional sentences, also has been found to not differ between younger and older adults (Dupuis and Pichora-Fuller 2010).

Reminger et al. (2000) also found a greater left versus right zygomatic EMG response to the IAPS images for both younger and older participants. Another study similarly failed to detect any age-related differences in the greater left-sided facial expression of emotion, or in the hemispace bias of stimulus preference for perception of emotional stimuli (Moreno et al. 1990). These results thus also suggest age invariance in the greater contribution of the right hemisphere to the expression of emotion, consistent with neuropsychological research that has failed to support a hypothesis of differential hemispheric change in aging (Kaszniak and Newman 2000).

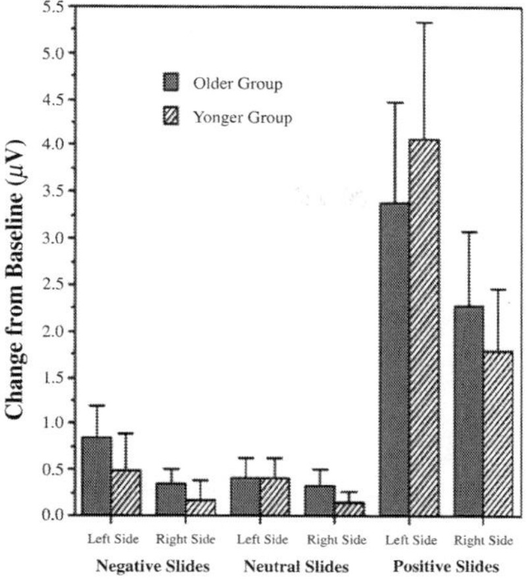

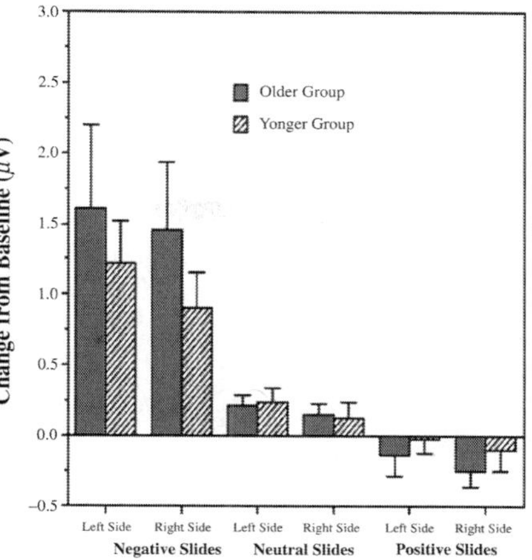

Fig. 2 Facial electromyographic (EMG) responses from the zygomatic (*left*) and corrugator (*right*) face regions, in response to viewing emotionally salient images. Left side = recording from left side of face; Right side = recording from right side of face. Graphs are based upon data reported by Reminger et al. (2000)

2.3 ANS and CNS Physiological Change

The magnitude of ANS reactivity in emotion appears to be one emotion component where older and younger adults more reliably differ, at least for some ANS indices. Levenson et al. (1991) examined emotion-specific changes in ANS functioning in younger and older adult groups. In this study, emotion was elicited through posed facial expressions and relived past emotional experiences, measuring heart rate and finger temperature change as indices of ANS activity. It was found that overall magnitudes of ANS changes were smaller in the older group, although the patterns of activity were similar for the two age groups, across the two emotion-induction techniques. The magnitude of cardiovascular activity change is reduced in older, compared to younger adults, when emotion is elicited by happy films (Tsai et al. 2000), when viewing negative pictures (Smith et al. 2005), when recalling both emotionally negative and positive autobiographical memories (Labouvie-Vief et al. 2003), and even during conflict-related discussions with one's spouse (Levenson et al. 1994). A meta-analysis of emotion and cardiovascular reactivity studies (Uchino et al. 2010) concluded that increasing adult age is associated with lower heart rate reactivity but higher systolic blood pressure reactivity during emotionally evocative tasks (e.g., Uchino et al. 2006). Age-related reduction in heart rate reactivity has been hypothesized to contribute to older adults better ability to regulate emotion than younger adults (Cacioppo et al. 1997; Levenson, 2000), as described in the following section.

Studies of emotion-related CNS physiology have also documented several adult age-group differences. Comparing functional MRI measures of regional brain activation during the viewing of emotional faces, Gunning-Dixon et al. (2003) found older, compared to younger adults to have less activity in the amygdala region and more in the prefrontal cortex. This apparent relative shift toward frontal region activation, and away from amygdala region activation with increasing adult age seemed puzzling to Gunning-Dixon, et al., in light of research demonstrating structural brain changes with aging to be greatest in the prefrontal cortex (for review, see Raz 2000), while the amygdala appears relatively better preserved with age (Grieve et al. 2005). However, Mather et al. (2004) interpreted such age-related differences in regional brain activation as reflecting an increased motivation to regulate emotion, with frontal areas inhibiting the amygdala, despite the relative age-related structural changes in these brain areas.

Wood and Kisley (2006) examined event-related potentials (ERP; stimulus-locked electrical brain responses derived from signal-averaged electroencephalograms) to positive, neutral, and negative IAPS emotional images. The late positive potential of the ERP was smaller in older than in younger adults in response to positive and negative images. Further, the older adults did not show the negativity effect (i.e., great ERP response to negative images) found for the younger adults. This age-related reduction in the negativity effect can be interpreted as consistent with an emotion-regulatory shift in older age away from negative and toward more positive emotion. Studies utilizing fMRI measures of regional brain response and connectivity have found similar evidence. For example, Jacques et al. (2010) found older and younger adults to make similar valence evaluations in response to emotional pictures, although the older adults experienced negatively valenced pictures as being less negative. The right amygdala (a limbic brain structure involved in threat vigilance and other aspects of brain emotional response) was similarly activated in fMRI. However, measures of the functional connectivity of this area with the rest of the brain differed between the age groups. Compared to the younger adults, the older adults showed greater functional connectivity between the right amygdala and the ventral anterior cingulate cortex, interpreted by the authors as possibly reflecting increased emotional regulation with aging, inhibiting negative emotion. Conversely, the older adults demonstrated decreased functional connectivity, compared to the younger adults, between the right amygdala and the posterior brain regions, possibly reflecting decreased perceptual processing (see above section on experience and appraisal) of the emotional stimuli.

There is also some evidence supporting a difference between older and younger adults in regional brain response to the elicitation of differently valenced emotion. Leclerc and Kensinger (2008) compared older and younger adults in an fMRI study examining regional brain response to emotionally positive, neutral, and negative images. The negative, in contrast to the positive images activated the ventromedial prefrontal cortex (VMPFC) more in the younger adult group, while the positive, in contrast to the negative images activated the VMPFC more strongly for the older adults. The authors interpreted these age-group differences as contributing to the relatively greater negativity effect (increased attention and memory for negative emotional stimuli) reported across several studies in younger adults, and the relatively greater positivity effect in older adults.

2.4 Emotion Regulation

As noted above, age-related changes in *emotion regulation* have been offered as explanations for adult age-group differences in ANS and CNS physiologic changes of emotion, as well as for an advantageous decrease in the frequency of daily life negative emotion experience with increasing adult age (Gross et al. 1997; also see Charles and Carstensen 2007, for review). What might explain such aging-related changes in emotion regulation?

Carstensen and her colleagues have studied emotion and aging in the context of socioemotional selectivity theory, which postulates that reduction in social contacts in old age are likely a result of healthy "pruning" of less emotionally important or rewarding relationships (Charles and Carstensen 2010), and an increased focus on situations and familiar people that increase the probability of positive emotion experience. In a series of studies, Carstensen and her colleagues have shown that the perception of available time is the critical component in the move toward focusing on rewarding relationships to the exclusion of others, regardless of age (for review, see Charles and Carstensen 2007). Thus, all adults maximize their use of the perceived time available to them, and this "pruning" appears to be an adaptive strategy for older adults who typically perceive that they are closer to the end of the life cycle than younger adults. Implicit in this theory is the assumption that the experience and appraisal of emotionally salient stimuli (in this case, other persons) continue to play an important role and function well in older age, as supported by the relevant research reviewed above. Carstensen et al. (1997) originally proposed that age-related improvements in emotional well-being could be accounted for by increased use of the antecedent emotion regulation strategy of situation selection, with older adults more frequently avoiding distressing social environments than do younger adults. Younger adults, on average, tend to prefer social and other situations that provide variety and novelty, building relationships and skills that are likely to have instrumental value into the distant perceived future (Charles and Carstensen 2007).

3 Summary and Conclusions

In summary, the research literature on emotion and aging provides evidence for age invariance in adulthood for both the nature (valence) and intensity (arousal) of emotional experience, despite aging-related decrements in the perception and categorization of emotionally relevant stimuli. Similarly, research supports the conclusion that emotional expression remains intact with aging. In contrast, other studies provide evidence for an age-related decrease in ANS physiological arousal, particularly in response to emotionally negative stimuli, and for shifts in CNS physiologic response to emotional stimuli toward increased prefrontal cortex activation and decreased amygdala activation with increasing adult age. Research on attention and memory for emotional information supports a decreased processing of negative emotional stimuli (i.e., a decrease in the negativity effect seen in younger adults), and a relative increase in the processing of emotionally positive stimuli (positivity effect).

These physiological response and attentional/memory preference differences across increasingly older groups have been theoretically interpreted as reflecting greater motivation for emotion regulation with aging. Some investigators have also suggested that the decreased physiological arousal in response to negative emotional stimuli with aging might itself facilitate enhanced emotion regulation. Socioemotional selectivity theory has provided a theoretical context to understand this increased emotion regulation motivation. For younger persons, the future appears to stretch out nearly endlessly, and a premium is consequently placed on attending to and processing novelty and diversity in social and other situations. Since the challenges posed by such preferences often give rise to negative emotion, negativity effects are observed in attention and memory studies, and younger adults are relatively more likely to experience a higher frequency of negative emotion in daily life. As persons age, their perceived future time horizon shrinks, and a greater value is placed upon cultivating close, familiar, and meaningful relationships and other situations that give rise to positive emotional experience, and avoiding, or shifting attention from, those less intimate people and less familiar situations that are likely to elicit negative emotion. Even though there are CNS structural changes in emotion-relevant brain regions with aging, this shift in socioemotional selectivity, and perhaps the decreased ANS physiological arousal of emotion with aging, facilitate enhanced emotion regulation with aging. This enhanced emotion regulation in turn results in fewer experiences of negative emotion in daily life, despite the accumulation of events (e.g., illness and loss of loved ones) that might be expected to increase negative emotion with aging.

References

Allender JA, Kaszniak AW (1989) Processing of emotional cues in patients with dementia of the Alzheimer type. Int J Neurosci 46:147–155

Bailey PE, Henry JD (2009) Subconscious facial expression mimicry is preserved in older adulthood. Psychol Aging 24:995–1000

Banham KM (1951) Senescence and the emotions: a genetic theory. J Genet Psychol 78:175–183

Bargh JA, Williams LE (2007) The nonconscious regulation of emotion. In: Gross JJ (ed) Handbook of emotion regulation. Guilford Press, New York, pp 429–445

Bechara A, Damasio H, Damasio AR, Lee GP (1999) Differential contributions of the human amygdala and ventromedial prefrontal cortex to decision making. J Neurosci 19:5473–5481

Bollen KA, Lennox R (1991) Conventional wisdom on measurement: a structural equation perspective. Psychol Bull 110:305–314

Burton KW, Kaszniak AW (2006) Emotional experience and facial expression in Alzheimer's disease. Aging Neuropsychol Cognit 13:636–651
Cacioppo JT, Tassinary LG (eds) (1990) Principles of psychophysiology: physical, social and inferential elements. Cambridge University Press, New York
Cacioppo JT, Berntson GG, Klein DJ, Poehlmann KM (1997) Psychophysiology of emotion across the lifespan. In: Schaie KW, Lawton MP (eds) Annual review of gerontology and geriatrics (Focus on emotion and adult development), vol 17. Springer, New York, pp 27–74
Carstensen LL, Gross J, Fung HH (1997) The social context of emotional experience. In: Schaie KW, Lawton MP (eds) Annual review of gerontology and geriatrics (Focus on emotion and adult development), vol 17. Springer, New York, pp 325–352
Carstensen LL, Pasupathi M, Mayr U, Nesselroade J (2000) Emotional experience in everyday life across the adult life span. J Pers Soc Psychol 79:644–655
Charles ST (2005) Viewing injustice: age differences in emotional experience. Psychol Aging 20:159–164
Charles ST, Carstensen LL (2007) Emotion regulation and aging. In: Gross JJ (ed) Handbook of Emotion Regulation. Guilford Press, New York, pp 307–327
Charles ST, Carstensen LL (2010) Social and emotional aging. Annu Rev Psychol 61:383–409
Clore GL, Ortony A (2000) Cognition in emotion: always, sometimes, or never? In: Lane RD, Nadel L (eds) Cognitive neuroscience of emotion. Oxford University Press, New York, pp 24–61
Clore GL, Ortony A (2008) Appraisal theories: how cognition shapes affect into emotion. In: Lewis M, Haviland-Jones JM, Feldman Barrett L (eds) Handbook of emotions, 3rd edn. New York, Guilford, pp 618–627
Coan JA, Allen JJB (eds) (2007) Handbook of emotion elicitation and assessment. Oxford University Press, New York
Cohn JF, Ambadar Z, Ekman P (2007) Observer-based measurement of facial expression with the facial action coding system. In: Coan JA, Allen JJB (eds) Handbook of emotion elicitation and assessment. Oxford University Press, New York, pp 203–221
Dupuis K, Pichora-Fuller MK (2010) Use of affective prosody by young and older adults. Psychol Aging 25:16–29
Gazzaley A, Cooney JW, Rissman J, D'Esposito M (2005) Top-down suppression deficit underlies working memory impairment in normal aging. Nat Neurosci 8:1298–1300
Grieve SM, Clark CR, Williams LM, Peduto AJ, Gordon E (2005) Preservation of limbic and paralimbic structures in aging. Hum Brain Mapp 25:391–401
Gross JJ (1998a) The emergent field of emotion regulation: an integrative review. Rev Gen Psychol 2:271–299
Gross JJ (1998b) Antecedent- and response-focused emotion regulation: divergent consequences for experience, expression, and physiology. J Pers Soc Psychol 24:224–237
Gross JJ (2002) Emotion regulation: affective, cognitive, and social consequences. Psychophysiology 39:281–291
Gross JJ (ed) (2007) Handbook of emotion regulation. Guilford, New York
Gross JJ, Levenson RW (1993) Emotional suppression: physiology, self-report, and expressive behavior. J Pers Soc Psychol 64:970–986
Gross JJ, Carstensen LL, Pasupathi M, Tsai J, Skorpen CG, Hsu AYC (1997) Emotion and aging: experience, expression, and control. Psychol Aging 12:590–599
Gunning-Dixon FM, Gur RC, Perkins AC, Schroeder L, Turner T, Turetsky BI (2003) Age-related differences in brain activation during emotional face processing. Neurobiol Aging 24:285–295
Jennings JR, Yovetich N (1991) The aging information processing system: autonomic and nervous system indices. In: Jennings JR, Coles MGH (eds) Handbook of the psychophysiology of information processing: an integration of central and autonomic nervous system approaches. Wiley, Chichester, pp 706–727
Kaszniak AW (2001) Emotion and consciousness: current research and controversies. In: Kaszniak AW (ed) Emotions, qualia, and consciousness. World Scientific, London, pp 3–21

Kaszniak AW, Newman MC (2000) Toward a neuropsychology of cognitive aging. In: Qualls SH, Abeles N (eds) Psychology and the aging revolution: how we adapt to longer life. American Psychological Association, Washington, DC, pp 43–67

Kaszniak AW, Reminger SL, Rapcsak SZ, Glisky EL (1999) Conscious experience and autonomic response to emotional stimuli following frontal lobe damage. In: Hameroff S, Kaszniak AW, Chalmers D (eds) Toward a science of consciousness III: the third Tucson discussions and debates. MIT Press, Cambridge, pp 201–213

Kliegel M, Jäger T, Phillips LH (2007) Emotional development across adulthood: differential age-related emotional reactivity and emotion regulation in a negative mood induction procedure. Int J Aging Hum Dev 64:217–244

Kunzmann U, Grühn D (2005) Age differences in emotional reactivity: the sample case of sadness. Psychol Aging 20:47–59

Kunzmann U, Richter D (2009) Emotional reactivity across the adult life span: the cognitive pragmatics make a difference. Psychol Aging 24:879–889

Labouvie-Vief G, Lumley MA, Jain E, Heinze H (2003) Age and gender differences in cardiac reactivity and subjective emotion responses to emotional autobiographical memories. Emotion 3:115–126

Lang PJ (1980) Behavioral treatment and bio-behavioral assessment: computer applications. In: Sidowski JB, Johnson JH, Williams TA (eds) Technology in mental health care delivery systems. Ablex, Norwood, pp 119–137

Lang PJ (1995) The emotion probe: studies of motivation and attention. Am Psychol 50:372–385

Lang PJ, Greenwald MK, Bradley MM, Hamm AO (1993) Looking at pictures: affective, facial, visceral, and behavioral reactions. Psychophysiology 30:261–273

Lang PJ, Bradley, MM, Cuthbert, BN (2005) International affective picture system (IAPS): affective ratings of pictures and instruction manual. Technical report A-6, University of Florida, Gainesville

Larsen JT, Berntson GG, Poehlmann KM, Ito TA, Cacioppo JT (2008) The psychophysiology of emotion. In: Lewis M, Haviland-Jones JM, Feldman Barrett L (eds) Handbook of emotions. Guilford, New York, pp 180–195

Lawton MP, Kleban MH, Rajagopal D, Dean J (1992) Dimensions of affective experience in three age groups. Psychol Aging 7:171–184

Lazarus RS (1966) Psychological stress and the coping process. McGraw-Hill, New York

Leclerc CM, Kensinger EA (2008) Age-related differences in medial prefrontal activation in response to emotional images. Cognitive Affective Behav Neurosci 8:153–164

Levenson RW (2000) Expressive, physiological, and subjective changes in emotion across adulthood. In: Qualls SH, Ables N (eds) Psychology and the aging revolution: how we adapt to longer life. American Psychological Association, Washington, DC, pp 123–140

Levenson RW, Carstensen LL, Friesen WV, Ekman P (1991) Emotion, physiology, and expression in old age. Psychol Aging 6:28–35

Levenson RW, Carstensen LL, Gottman J (1994) The influence of age and gender on affect, physiology and their interrelations: a study of long-term marriages. J Pers Soc Psychol 67:56–68

Magai C, Consedine NS, Krivoshekova YS, Kudadjie-Gyarnfi E, McPherson R (2006) Emotional experience and expression across the adult life span: insights from a multimodal assessment study. Psychol Aging 21:303–317

Malatesta CZ, Kalnok M (1984) Emotional experience in younger and older adults. J Gerontol 39:301–308

Malatesta CZ, Izard CE, Culver C, Nicolich M (1987) Emotion communication skills in young, middle-aged, and older women. Psychol Aging 2:193–203

Mather M, Canli T, English T, Whitfield S, Wais P, Ochsner K et al (2004) Amygdala responses to emotionally valenced stimuli in older and younger adults. Psychol Sci 15:259–263

Mathersul D, Palmer DM, Gur RC, Gur RE, Cooper N, Gordon E, Williams LM (2009) Explicit identification and implicit recognition of facial emotions: IICore domains and relationships with general cognition. J Clin Exp Neuropsychol 31:278–291

McNair DM, Lorr M, Droppleman L (1992) EdITS manual for the profile of mood states. San Diego (revised)

Mitchell RLC (2007) Age-related decline in the ability to decode emotional prosody: primary or secondary phenomenon? Cognit Emot 21:1435–1454
Moreno CR, Borod JC, Welkowitz J, Alpert M (1990) Lateralization for the expression and perception of facial emotion as a function of age. Neuropsychologia 28:199–209
Murphy NA, Isaacowitz DM (2008) Preferences for emotional information in older and younger adults: a meta-analysis of memory and attention tasks. Psychol Aging 23:263–286
Nielsen L, Kaszniak AW (2007) Conceptual, theoretical, and methodological issues in inferring subjective emotional experience: recommendations for researchers. In: Allen JJB, Coan J (eds) The Handbook of emotion elicitation and assessment. Oxford University Press, New York, pp 361–375
Nilsson KW, Leppert J, Simonsson B, Starrin B (2010) Sense of coherence and psychological well-being: improvement with age. J Epidemiol Community Health 64:347–352
O'Donnell RM, Kaszniak AW (2011) Affective disorders in late life. Generations 35(3)
Ortega V, Phillips LH (2008) Effects of age and emotional intensity on the recognition of facial emotion. Exp Aging Res 34:63–79
Oscar-Berman M, Hancock M, Mildworf B, Hunter N, Weber DA (1990) Emotional perception and memory in alcoholism and aging. Alcohol Clin Exp Res 14:383–393
Raz N (2000) Aging of the brain and its impact on cognitive performance: Integration of structural and functional findings. In: Craik FIM, Salthouse TA (eds) The handbook of aging and cognition, 2nd edn. Erlbaum, Mahwah, pp 1–90
Reminger SL, Kaszniak AW, Dalby P (2000) Age-invariance in the asymmetry of stimulus-evoked emotional facial muscle activity. Aging Neuropsychol Cognit 7:156–168
Rufman T, Henry JD, Livingstone V, Phillips LH (2008) A meta-analytic review of emotion recognition and aging: implications for neuropsychological models of aging. Neurosci Biobehav Rev 32:863–881
Russell JA (1980) A circumplex model of affect. J Pers Soc Psychol 39:1161–1178
Ryan M, Murray J, Ruffman T (2010) Aging and the perception of emotion: processing vocal expressions alone and with faces. Exp Aging Res 36:1–22
Salthouse TA (1990) Working memory as a processing resource in cognitive aging. Dev Rev 10:101–124
Salthouse TA, Atkinson TM, Berish DE (2003) Executive functioning as a potential mediator of age-related cognitive decline in normal adults. J Exp Psychol Gen 132:566–594
Slessor G, Miles LK, Bull R, Phillips LH (2010) Age-related changes in detecting happiness: discriminating between enjoyment and nonenjoyment smiles. Psychol Aging 25:246–250
Smith DP, Hillman CH, Duley AR (2005) Influences of age on emotional reactivity during picture processing. J Gerontol B Psychol Sci Soc Sci 60:P49–P56
Stein NL, Hernandez MW (2007) Assessing understanding and appraisals during emotional experience. In: Coan JA, Allen JJB (eds) Handbook of emotion elicitation and assessment. Oxford University Press, New York, pp 298–317
St Jacques P, Dolcos F, Cabeza R (2010) Effects of aging on functional connectivity of the amygdala during negative evaluation: a network analysis of fMRI data. Neurobiol Aging 31:315–327
Stone AA, Schwartz JE, Broderick JE, Deaton A (2010) A snapshot of the age distribution of psychological well-being in the United States. Proc Natl Acad Sci 107:16489–16493
Sullivan S, Ruffman T (2003) Emotion recognition deficits in the elderly. Int J Neurosci 114:403–432
Sullivan S, Ruffman T (2004) Emotion recognition deficits in the elderly. Int J Neurosci 114: 403–432
Tessitore A, Hariri AR, Fera F, Smith WG, Das S, Weinberger DR, Mattay VS (2005) Functional changes in the activity of brain regions underlying emotion processing in the elderly. Psychiatry Res Neuroimaging 139:9–18
Tranel D, Damasio H (1994) Neuroanatomical correlates of electrodermal skin conductance responses. Psychophysiology 31:427–438
Tsai JL, Levenson RW, Carstensen LL (2000) Autonomic, expressive, and subjective responses to emotional films in older and younger Chinese American and European American adults. Psychol Aging 15:684–693

Uchino BN, Berg CA, Smith TW, Pearce G, Skinner M (2006) Age-related differences in ambulatory blood pressure during daily stress: evidence for greater blood pressure reactivity with age. Psychol Aging 21:231–239

Uchino BN, Birmingham W, Berg C (2010) Are older adults less or more physiologically reactive? A meta-analysis of age-related differences in cardiovascular reactivity to laboratory tasks. J Gerontol Psychol Sci 65B:154–162

Wager TD, Feldman Barrett L, Bliss-Moreau E, Lindquist KA, Duncan S, Kover H, Joseph J, Davidson M, Mize J (2008) The neuroimaging of emotion. In: Lewis M, Haviland-Jones JM, Feldman Barrett L (eds) Handbook of emotions. Guilford, New York, pp 249–271

Wood S, Kisley MA (2006) The negativity bias is eliminated in older adults: age-related reduction in event-related brain potentials associated with evaluative categorization. Psychol Aging 21:815–820

Zoccolotti P, Scabini D, Violani C (1982) Electrodermal responses in patients with unilateral brain damage. J Clin Neuropsychol 4:143–150

Studying the Impact of Aging on Memory Systems: Contribution of Two Behavioral Models in the Mouse

Aline Marighetto, Laurent Brayda-Bruno and Nicole Etchamendy

Abstract In the present chapter, we describe our own attempts to improve our understanding of the pathophysiology of memory in aging. First, we tried to improve animal models of memory degradations occurring in aging, and develop common behavioral tools between mice and humans. Second, we began to use these behavioral tools to identify the molecular/intracellular changes occurring within the integrate network of memory systems in order to bridge the gap between the molecular and system level of analysis. The chapter is divided into three parts (i) modeling aging-related degradation in declarative memory (DM) in mice, (ii) assessing the main components of working memory (WM) with a common radial-maze task in mice and humans and (iii) studying the role of the retinoid cellular signaling path in aging-related changes in memory systems.

Keywords Declarative memory · Working memory · Mouse · Human · Fos imaging

Contents

1 Modeling Aging-Related Declarative Memory Decline in Mice 69
1.1 Aging-Related Relational/Declarative Memory Deficit: Involvement of an Hippocampal Impairment in Encoding 70
1.2 Translating the Relational/Declarative Memory Model to Humans 73
1.3 Conclusion 74

A. Marighetto (✉) · L. Brayda-Bruno
Neurocentre Magendie-Inserm U862, 146 Rue Leo Saignat, 33077 Bordeaux-Cedex, France
e-mail: a.marighetto@cnic.u-bordeaux1.fr

N. Etchamendy
NutriNeuro Bordeaux University, avenue des facultés, 33406 Talence, France

Curr Topics Behav Neurosci (2012) 10: 67–89
DOI: 10.1007/7854_2011_151

Published Online: 31 July 2011

2 A Common Radial-Maze Task to Study Working Memory in Mice and in Humans 74
2.1 Description of the Working Memory Radial-Maze Task 75
2.2 Working Memory Organization Deteriorates More than Working Memory Maintenance in Aged Mice and Humans 77
2.3 Role of the Hippocampus and Prefrontal Cortex in Mice 79
2.4 Concluding Remarks on the Aging-Related Deterioration in Relational/Declarative Memory and Working Memory in Our Models 81
3 Implication of the Retinoid Signaling Pathway in Memory Deficits in Aging 82
3.1 Retinoid Signaling has a Role in Adult Brain Plasticity and Memory 82
3.2 Hypofunction in Retinoid Signaling Contributes to Aging-Related Degradation in Hippocampal Function, Relational/Declarative Memory Formation and Working Memory Organization 83
4 Conclusion 85
References 85

Memory problems are present in numerous neuropsychiatric disorders. However, it is essentially in relation with aging that pathologies of memory have become a major health problem in developed societies. Indeed, memory dysfunction is at the core of Alzheimer's disease whose incidence is rapidly increasing. Consequently, there is an urgent need for therapeutics and preventive strategies to address the human and financial cost of memory-related diseases.

However, to identify the neurobiological bases of aging-related memory alterations, and thereby provide predictive targets for the development of therapeutics or preventive strategies, research is facing two major obstacles.

The first obstacle relates to the forms of memory which are concerned (Gabrieli 1996; Grady 2008; Friedman et al. 2007), and difficulty of studying them in animals. First, in normal aging, there is a progressive decline in short-term/working memory (WM), the capability to temporarily store information for the execution of an act in the near-term. Because WM deterioration has consequences on almost all kinds of cognitive abilities, WM is intensively explored and conceptualized in humans (Baddeley 1996; Baddeley et al. 1999; Bunge et al. 2000; Repovs and Baddeley 2006). Unfortunately, this conceptualization has no correspondence in animals, and this insufficiency in concepts and methods relating research in humans and animals represents a barrier to the advancement of knowledge on the neurobiological bases of WM decline in aging. Second, in normal aging and early Alzheimer's disease, long-term memory deteriorates preferentially in its declarative component. Declarative memory (DM) is the conscious memory of events and facts adequately related to their spatio-temporal context, and that can be expressed verbally or in any controlled and explicit manner (Squire and Zola 1996; Cohen et al. 1997). Because DM is typically human and requires consciousness and verbalization, the development of animal models of DM encounters difficulties. These difficulties perpetuate a gap between research in humans and animals.

The second difficulty concerns the transition between levels of analysis. Thus, DM and WM disturbances in aging are characterized by functional alterations within and between brain structures (Gabrieli 1996; Grady 2008; Friedman et al. 2007; Narayanan et al. 2005) as well as by the alteration of specific intracellular mechanisms (Burke and Barnes 2006, 2010; Lund et al. 2004). Unfortunately, the systemic and cellular/molecular levels of analysis remain largely unrelated.

In conclusion, in order to improve our understanding of the pathophysiology of WM and DM in aging, we should try first to improve animal models of these memory degradations, and develop common behavioral tools between animals and humans. Then we should use these behavioral tools to identify the molecular/intracellular changes occurring within the integrate network of memory systems in order to bridge the gap between the molecular and system level of analysis.

In this chapter, we report on our attempts to (i) model the preferential degradation of DM in mice, (ii) assess the main components of WM with a common tool between mice and humans, and (iii) study the role of the retinoid cellular signaling path in aging-related changes in memory systems by using these behavioral tools.

1 Modeling Aging-Related Declarative Memory Decline in Mice

To study DM in animals, it is necessary to define what DM is without consciousness and verbalization that are central to the definition and study of DM in humans. Two complementary strategies are possible. The most frequent one consists in considering that DM is the memory lost after hippocampal lesion since the hippocampus is necessary for DM (Squire 1992). The other strategy consists in defining elementary psychological properties of DM, outside consciousness and verbalization. Thus, the characteristic flexibility of DM expression is proposed as a necessary and sufficient condition to define this type of memory in the relational theory developed by Cohen et al. (1997). DM flexibility can be exemplified in the use and comparison of separately acquired information to make a decision in a novel or changed situation. Conversely, procedural memory is characterized by its rigidity; it is learned and expressed by response repetition in an unchanged situation. According to the relational theory, flexible expression of DM relies on relational representation of past events; it requires a systematic processing of incoming information in memory based on organization of associations and relations among items (Eichenbaum et al. 1988, 1989, 1992).

Combining the above-mentioned two strategies, we designed a behavioral procedure assessing flexible and hippocampus-dependent memory in the mouse using the radial maze, and developed a model of aging-related preferential degradation in relational/declarative memory (Marighetto et al. 1999; Etchamendy et al. 2001, 2003a; Marighetto et al. 2000, 2008a–c) (R/DM) (cf. Fig. 1). In the first stage of this test (stage 1, learning) aged mice as well as young hippocampectomized mice can acquire separate information about each arm rewarding valence (food/no food) by repetition, showing an intact procedural memory. However, during the second stage of the test (stage 2, flexibility probe) when the learning situation is changed the same mice are not able to use this acquired information in a new context, showing an impairment in flexible memory expression. According to the relational theory, the selective deficit seen in old mice would result from a defect in the relational processing of separate information, and represent a model for specific degradation in R/DM.

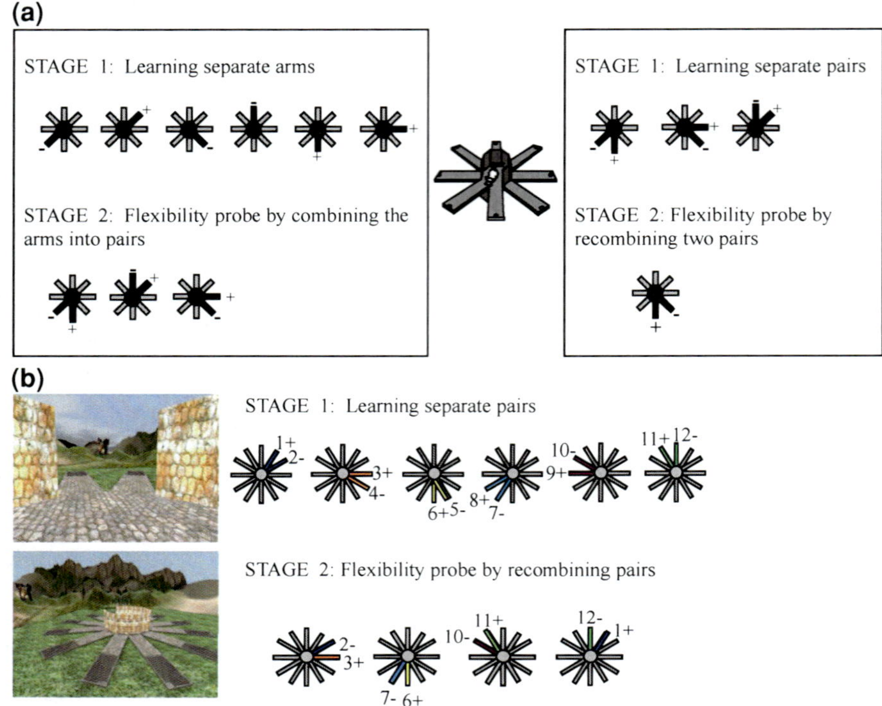

Fig. 1 **a** Schematic representation of the two versions of the radial-maze procedure testing characteristic flexibility of R/DM in the mouse [see in the text and in Marighetto et al. (1999) for detailed description]. **b** Schematic representation of the virtual radial-maze procedure testing R/DM flexibility in humans (Etchamendy et al. 2011)

As described, we then attempted to validate our mouse model by showing its similarity with the aging-related DM decline in humans. First, one aspect of DM decline in old humans is a deficiency in memory encoding. This is why we have tried to verify that the impairment in R/DM observed in aged mice was due to a problem in memory encoding. Then, we tested whether we would be able to detect an alteration in R/DM in aged humans using the same parameter of flexibility as we used in mice.

1.1 Aging-Related Relational/Declarative Memory Deficit: Involvement of an Hippocampal Impairment in Encoding

Deficiency in DM encoding has been shown in aged humans (Grady et al. 1999). In psychological terms, encoding degradation is described as reduced associative capabilities (Craik 1990; Rabinowitz et al. 1982). It is a "binding" problem resulting in impoverished memory representation lacking spatio-temporal contextualization (Chalfonte and Johnson 1996). In neurobiological terms, aging-related DM defect appears to be associated with alterations in prefrontal and hippocampal activation during memory encoding (Morcom et al. 2003).

According to the relational theory, the R/DM deficit seen in our aged mice would result from a defect in hippocampal function in binding learning events during encoding. The postulate is that the hippocampus is engaged in memory encoding irrespective of the learning task's cognitive demand. It "spontaneously" binds and relates events as they income in memory, thereby encoding them into a complex/relational representation that sustains the characteristic flexibility of DM expression (Cohen et al. 1999). Hippocampal function would be altered during the stage 1 of our test even though learning performance of aged and hippocampectomized mice looks normal during this phase of the task. Hippocampal impairment during learning would result in an altered binding rending the memory rigid and the performance of the animal altered during the second stage of the testing evaluating flexibility.

To test the relational interpretation of the deficit seen in aged mice, we first studied the role of the hippocampus in R/DM encoding in young mice and then tried to identify alteration in hippocampal activity in R/DM encoding in the aged mice.

1.1.1 The Hippocampus is Spontaneously and Causally Engaged in R/DM Encoding

To validate the relational view of hippocampal function in R/DM encoding, we first demonstrated the causal implication of the hippocampus in the encoding phase of R/DM. Namely, in young mice, we tested the effects of hippocampal temporary inactivation during memory encoding only (i.e. during the stage 1) on subsequent R/DM expression assessed in the stage 2. We found that disrupting hippocampal activity during the first stage of our test did not modify the performance in this learning task. However, it disrupted the formation of flexible memory representation of the learning events, inducing a performance deficit in the second stage of our test assessing memory flexibility (Mingaud et al. 2007). Thus hippocampal function during encoding sustains the formation of R/DM representation and disrupting this function produces the same selective deficit in flexible R/DM expression as the one seen in aged animals.

Then, we showed that the hippocampus was *spontaneously* engaged in learning by analyzing encoding related-activity patterns among memory systems. Namely, we combined temporary inactivation of the hippocampus with functional neuro-imaging, using *c-fos* mRNA expression as an indicator of cellular activity following a training session in stage 1 (Mingaud et al. 2007). This experiment led to two principal observations. First, in normal conditions, learning did result in significant activation of the hippocampus (in particular the CA1 field). Since the

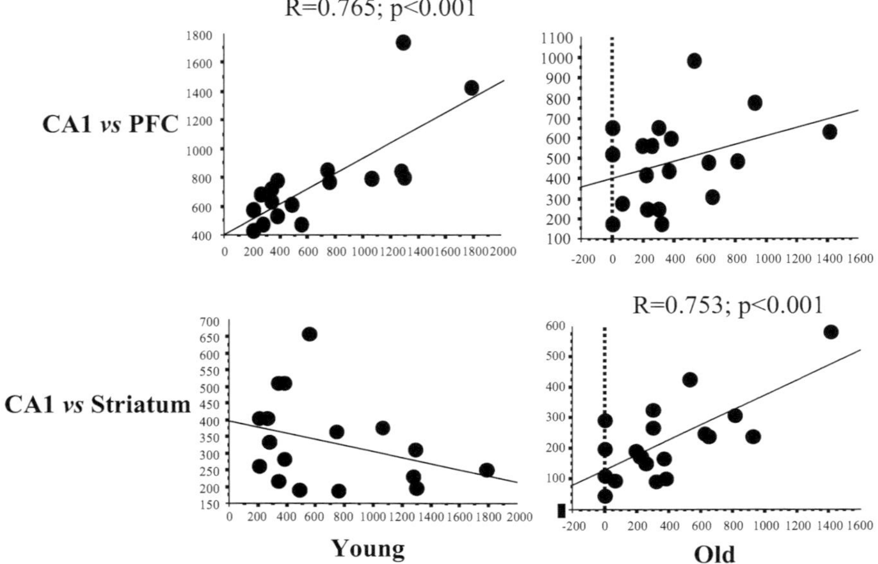

Fig. 2 Inter-structure correlations in Fos levels induced by R/DM encoding in young and aged mice. There is a positive correlation between activities in CA1 and prefrontal cortex for the young group, CA1 and striatum for the old one

learning task does not depend on hippocampal integrity, our finding supports the relational view that the hippocampus is recruited spontaneously, i.e. whether the learning task is hippocampus-dependent or not. Second, learning under hippocampal inactivation was associated with an abnormal activation of the dorso-lateral striatum, a structure believed to sustain procedural learning (Schmidtke et al. 2002). In line with previous observations on functional interaction between multiple memory systems (Sadeh et al. 2011; Poldrack and Packard 2003), our observations suggest that learning "spontaneously" relies on the hippocampus. However, the striatum-based procedural memory system can substitute in case of hippocampal dysfunction, hence resulting in inflexible memory formation.

1.1.2 Aging-Related Alteration in Hippocampal (and Prefrontal) Activities in R/DM Encoding

We identified a defect in hippocampal and prefrontal activities associated with R/DM encoding in aged mice. This identification was achieved by immunohistochemical detection of Fos protein expression among principal memory systems after a learning session (i.e. in stage 1). Aged mice exhibited a significant reduction of learning-induced Fos activation in the CA1 field of the hippocampus and prefrontal cortex compared to the young group. By contrast, learning-related activities in the DG of the hippocampus and the dorso-medial striatum were normal (Mingaud et al. 2008).

In another experiment, analyses of inter-structure correlations in Fos activity further demonstrated between age difference in memory system activities. Indeed, a positive correlation between the hippocampal CA1 and the prefrontal cortex was found in the young mice but not in the aged ones. A positive correlation was found instead between CA1 and striatal activities in the aged group (unpublished findings, see Fig. 2).

As a whole, our Fos neuroimaging findings suggest two main conclusions. First, the recruitment of hippocampo—prefrontal circuits in learning sustains the formation of flexible R/DM representation. This conclusion is in agreement with functional neuroimaging studies in humans showing that successful DM encoding correlates with coordinated activities in the prefrontal cortex and hippocampus (Addis and McAndrews 2006; Blumenfeld and Ranganath 2006; Brassen et al. 2006; Davachi and Wagner 2002; Takashima et al. 2006). Second, alteration in prefronto-hippocampal activities prevents R/DM formation in aged animals. In these mice, inflexible procedural memory only can be formed, likely relying on apparently normal activity in the striatum and its functional connexion with the hippocampus.

In summary, the aging-related R/DM deficit seen in aged mice is linked to hippocampal and prefrontal dysfunction in memory encoding, thereby resembling the deficit in DM expression seen in human senescence (Grady 2008; Grady et al. 1999; Grady and Craik 2000).

1.2 Translating the Relational/Declarative Memory Model to Humans

If our mouse model is valid, it should also be able to show R/DM deficit in aged humans as well. Using a virtual radial maze (Bohbot et al. 2007) we have adapted our R/DM paradigm to human subjects. Then, we tested young and aged volunteers and analyzed fMRI activity induced by the task as a first attempt to validate our mouse model.

The similarity between mice and humans was striking when young and aged volunteers were submitted to the same testing procedure as the one used in mice through a virtual environment version of the radial arm maze (Etchamendy et al. 2011). First, like aged mice, aged volunteers could acquire the initial task and learn separate information about the rewarding value of each arm (stage 1), but they failed in using this information when the testing situation was modified in the flexibility probe (stage 2). Second, fMRI data in young volunteers, showed that activation in the (dorsolateral) prefrontal cortex and hippocampus in memory encoding (i.e. in stage 1 learning) correlated with subsequent performance in the flexibility probe. Conversely, activation in the caudate nucleus of the striatum was seen in "inflexible" subjects in the final trials of stage 1.

Thus, translation of the mouse behavioral procedure to humans has provided initial behavioral and neuroimaging evidence which so far support the validity of the R/DM model. Further development of the translational approach is needed to test the model, and in particular demonstrate its predictive validity.

1.3 Conclusion

The work using the R/DM model illustrates the valuable interest of using radial-maze learning tasks to explore aging-related decline in memory function. Indeed, the selectivity of the deficit seen in aged mice in the flexibility test of the R/DM task represents clear evidence for the specificity of the effect of aging in that task. It thereby rules out changes in motivation, locomotion, or sensorial capacities that also accompany normal senescence as potential explaining factors. Finally, the previous success in translating the R/DM task for mice to humans through a virtual radial-maze encouraged us to continue the same approach in the exploration of WM decline in aging.

2 A Common Radial-Maze Task to Study Working Memory in Mice and in Humans

As mentioned in the introduction section, normal aging is accompanied by a progressive decline in WM. We believe that a prerequisite to improve our understanding of the neurobiological bases of this decline is the development of common behavioral WM tasks between rodents and humans. At the psychological level, WM refers to two principal capabilities: the short-term maintenance and the organization of information (reducing interference). These capabilities are reduced in aging and it is well established that their alteration is associated with changes in prefrontal function (Gabrieli 1996; Narayanan et al. 2005; Dunnett et al. 1990; Flicker et al. 1989; Olson et al. 2006; Rypma et al. 2001; Mitchell et al. 2000a, 2004; Salthouse et al. 1989). There is also evidence though that hippocampal hypo-activation can be correlated to the aging-related decline in WM organization, namely relational binding in WM (Mitchell et al. 2000b). However, the molecular bases of the aging-related deterioration in WM maintenance and organization remain to be established. Behavioral tools assessing the two principal components of WM in both mice and humans may help providing a global picture of WM degradation in aging from molecular/cellular changes to alteration in system function.

2.1 Description of the Working Memory Radial-Maze Task

Taking advantage of the preceding development of the R/DM task for mice and its successful adaptation to humans, we developed a common tool for assessment of WM maintenance and organization in mice and humans using the radial maze.

The WM radial-maze task resembles the R/DM task in its material structure but it taxes the capability to retain changing (trial-dependent) information over intervals varying from 0 second to several minutes, and to organize this information to reduce interference between successive trials. The arms of the maze are combined into pairs, but contrary to the R/DM task, the reward location within each pair varies between trials according to an alternating rule. Namely, for each novel presentation of one particular pair, the reward is located in the arm which was not chosen by the subject in the preceding trial with the same pair. Hence, the subject must remember which arm was visited in one specific trial until the next trial with the same pair, and so on. The task therefore taxes maintenance capability, but also organizational capability to reduce interference produced by the trials with concurrent pairs as well as by the repetition of trials with one specific pair.

In this WM test, as illustrated in Fig. 3, the difficulty level varies between trials according to a combination of factors which affect WM maintenance, organization or both. First, in one n trial, the difficulty of remembering the previous trial with the same pair (n-1) must increase as the amount of interposed trials between n and n-1 (with the other pairs) increases. Indeed, these intervening trials impact at the same time on the retention time, i.e. the maintenance demand, and on the amount of interference produced by concurrent pairs, i.e. the organizational demand. Second, in any n trial, the subject must choose on the basis of n-1 and not confound n-1 with the preceding trial with the same pair, n-2. Thus, there is potential (proactive) interference by n-2 which affects selectively the organizational demand. This organizational demand must depend on the distance (i.e. amount of interposed trials) between n-1 and n-2, the shorter the distance, the bigger the interference. Third, the duration of inter-trial intervals (waiting-time) affects retention time only, the longer the inter-trial intervals, the higher the maintenance demand.

In Summary, the difficulty of any one trial is expected to vary according to a combination of opposite modulation by proactive interference and maintenance demand. These characteristics make this test potentially adapted to evaluate which one of the maintenance and organizational capabilities is the more severely affected in aged subjects. Indeed, when the difficulty increases in terms of maintenance (by prolonging the inter-trial interval), proactive interference is expected to diminish, and vice versa. Hence, if the maintenance capability is the more affected in aging, difference in performance between young and aged subjects should increase as the inter-trial interval is prolonged, whereas the opposite should be observed if the organizational capability is the more deteriorated.

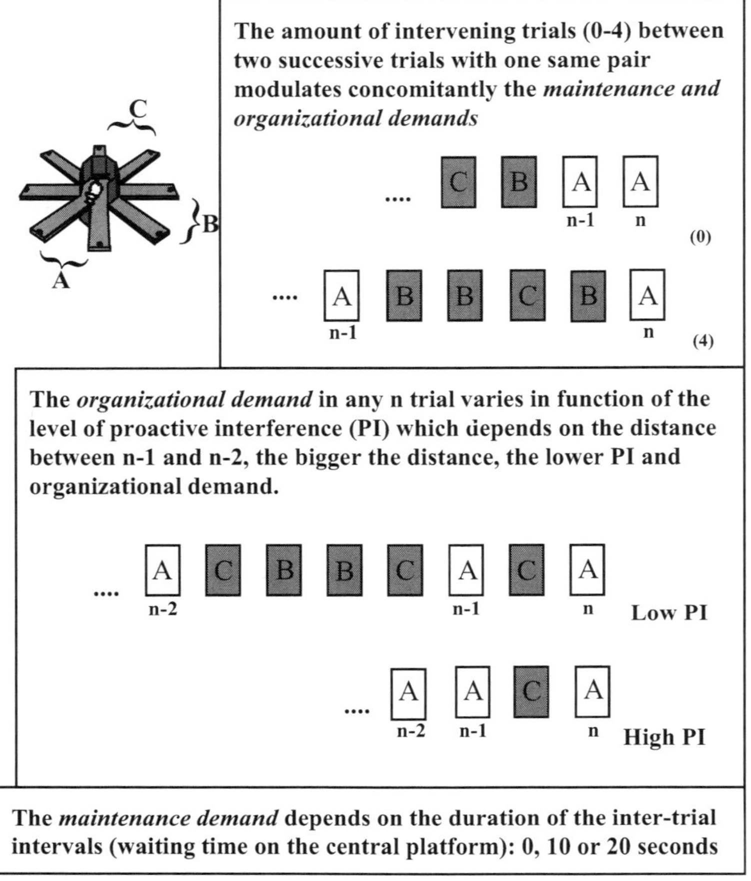

Fig. 3 The WM procedure taxes organizational and maintenance capabilities. Each session consists in alternate presentations of the three pairs of arms, A, B,and C in a controlled sequence (23 trials in total) and with a constant inter-trial interval (ITI of either 0, 10 or 20 s duration). The food location within each pair varies according to an alternation rule. Namely, on the first trial with each pair, both arms are rewarded, then in any *n* trial, the baited arm is always the one which was not chosen by the mouse in the preceding trial with the same pair (*n*-1). The difficulty varies first according to the amount of interposed trials between *n* and *n*-1 (0–4). These interposed trials prolong the retention time and potentially produce retroactive interference that increases the organizational demand. Second, the difficulty varies in function of proactive interference (PI) potentially produced by the trial preceding *n*-1 with the same pair, i.e. *n*-2: the shorter the distance between *n*-1 and *n*-2, the bigger the PI interference and therefore the organizational demand. Third, the ITI duration affects retention time, the longer the inter-trial intervals, the higher the maintenance demand

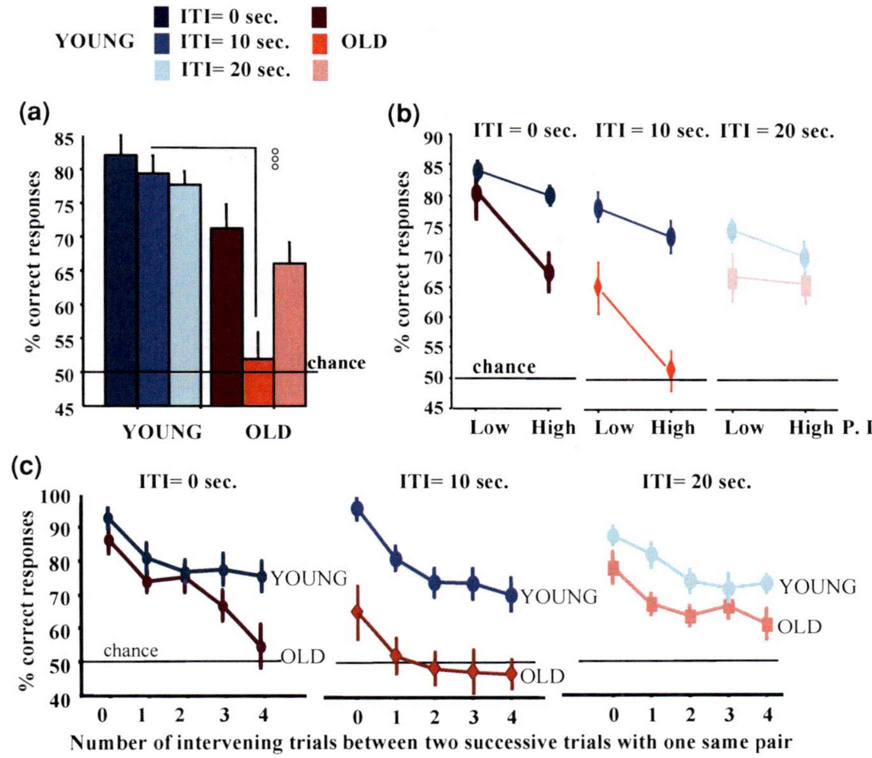

Fig. 4 Performance of young and old mice in the WM task. **a** Overall performance of each age group in the different conditions of inter-trial interval (ITI) which modulates the maintenance demand. The aged mice exhibited lower performance than the young ones under short ITI only because their performance paradoxically increased when the ITI was prolonged from 10 to 20 s. **b** Performance of each group in each inter-trial interval condition as a function of the proactive interference (PI) level which modulates the organizational demand. The aged mice exhibited lower performance than the young group under high PI and short ITI only. The modulation of performance by PI disappeared under the longer (20s.)- ITI condition in the aged mice, likely because under this delay, the *n*-2 trial was most of the time forgotten at *n* trial time. **c** Performance of each group in each ITI condition as a function the number of intervening trials between two successive trials with the same pair. The diminution of performance with increasing amount of intervening trials was comparable between the two ages

2.2 *Working Memory Organization Deteriorates More than Working Memory Maintenance in Aged Mice and Humans*

We first explored WM maintenance and organization in young and aged mice, and found that the organizational component of WM was the more deteriorated in the aged mice. Indeed, an aging-related deficit in performance was observed under

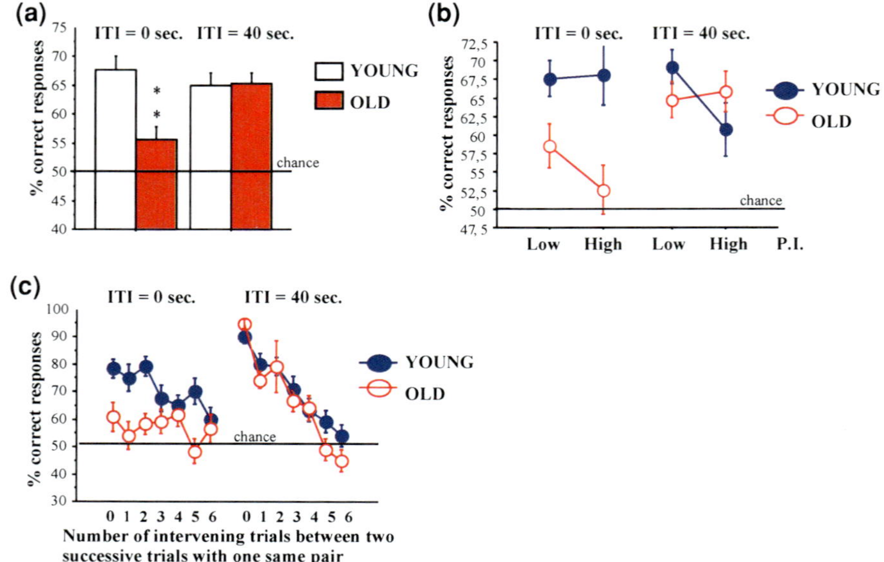

Fig. 5 Performance of young and old human volunteers in the WM task. **a** Overall performance of each age group under the different ITI conditions. As previously seen in aged mice, in the aged group the performance was lower than in the young one under the short ITI condition only because the performance of old volunteers augmented under the long-delay condition. **b** Performance of each group in each ITI condition as a function of the proactive interference (PI). As seen in aged mice, the aged group displayed lower performance than the young one under the high PI and short-delay condition. Under the long-ITI condition, there was no modulation of performance by PI level in the aged group. As suggested above for aged mice, this is probably due to the fact that, under the long-ITI condition, the potentially interfering trials (*n*-2) were often forgotten (at n trial time). **c** Performance of each group in each inter-trial interval condition as a function the number of intervening trials between two successive trials with the same pair. The performance curves were quite similar between the two ages

the conditions of short inter-trial intervals, i.e. when the demand on WM maintenance was lower, but the demand on WM organization related to proactive interference was higher. Namely, we can observe in Fig. 4a that the performance of aged mice paradoxically increased as the (inter-trial) delay was augmented to 20 s, which corresponded to the disappearance of proactive interference effect in the aged group (Fig. 4b). Finally, the diminution of performance by increasing intervening trials between two successive trials with a same pair (Fig. 4c) was roughly parallel between the two ages, whichever the delay condition. Nevertheless, aged mice displayed overall lower performance than their younger counterparts when trained under short delays only. In conclusion, it is essentially as a result of insufficient organization of proactively interfering trials in WM that aged mice were impaired in this task.

When our radial-maze WM task for mice was translated to young and aged-human subjects, the similarity of the effect of aging with the one seen in mice was striking and confirmed that WM organization deteriorates more than WM maintenance. Indeed, the same paradoxical amelioration of performance with longer retention intervals as the one seen in aged mice was observed in aged humans (Fig. 5). In addition, aged subjects displayed lower performance than the young subjects under short delay and high-proactive interference only. Finally, like in mice also, the diminution of performance with increasing number of intervening trials was similar between the two ages.

In conclusion, the paradoxical amelioration of performance with longer retention intervals demonstrated that both WM maintenance and WM organization deteriorated in aging in mice, and this deterioration was more severe for the organizational component than for the maintenance one. Interestingly, a paradoxical facilitation of WM performance was also induced by genetic suppression of hippocampal DG neurogenesis (Saxe et al. 2007), suggesting that the deficit seen in our aged subjects may be related to neurogenesis defect (Jinno 2011) and associated hippocampal dysfunction.

2.3 *Role of the Hippocampus and Prefrontal Cortex in Mice*

To examine the role of the hippocampus and prefrontal cortex in WM maintenance and organization and its deterioration with aging, we studied first the effects of a hippocampal/prefrontal lesion in young mice, second WM-related activity in critical brain regions using immunohistochemical detection of Fos protein after WM testing in young and aged mice.

As detailed successively in the following, neither the lesion of the hippocampus nor the lesion of the prefrontal cortex did produce the same effect as aging in the WM task, but Fos neuroimaging of testing revealed a functional alteration in the aged hippocampus which correlated with the memory impairment.

Clearly differing from the effect of aging, hippocampal lesion in young mice resulted in severe deterioration in WM maintenance. As it can be seen in Fig. 6a the performance of hippocampectomized mice was strictly identical to the one seen in controls in conditions of minimal maintenance demand (0 intervening trial and 0–10 s inter-trial delay). However, the performance of these mice dropped to chance level as soon as the retention delay was of at least 20 s (20 s. inter-trial interval or one intervening trial between two successive presentations of the same pair). As a logical correlate of this accelerated WM decay in the lesioned mice, there was no effect of proactive interference. In conclusion, the pattern of memory alteration induced by hippocampal lesion was different from the one seen in aged mice.

Lesion of the pre-limbic and infra-limbic cortex induced a subtle deficit in WM organization, which was therefore different from the one seen in aged mice. Indeed, performance in mice with a prefrontal cortex lesion were almost identical to the performance of controls, except when considering proactive interference

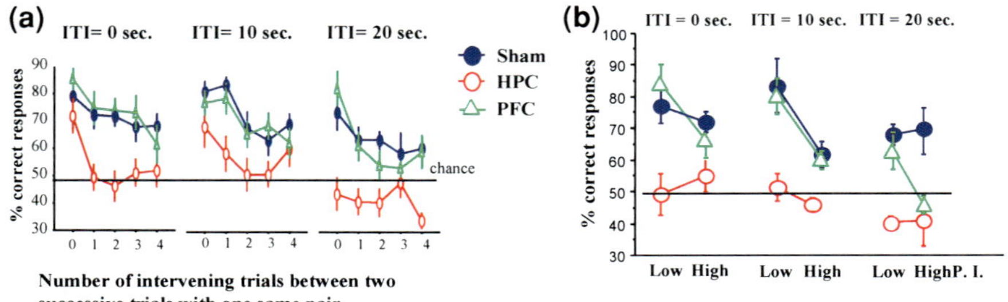

Fig. 6 Prefrontal cortex and hippocampal lesion effects on the performance in the WM task in young mice. **a** Performance of each group in each ITI condition as a function the number of intervening trials between two successive trials with the same pair. The hippocampal lesion induced accelerated forgetting: performance was normal at 0 and 10 s delays (no intervening trial) but dropped to chance levels at delays of 20 s and more (1–4 intervening trials). **b** Performance of each group in each ITI condition as a function of the proactive interference level. There was no effect of PI in mice with hippocampal lesion and a slight impairment in mice with prefrontal lesion under the high-PI and long-ITI condition only

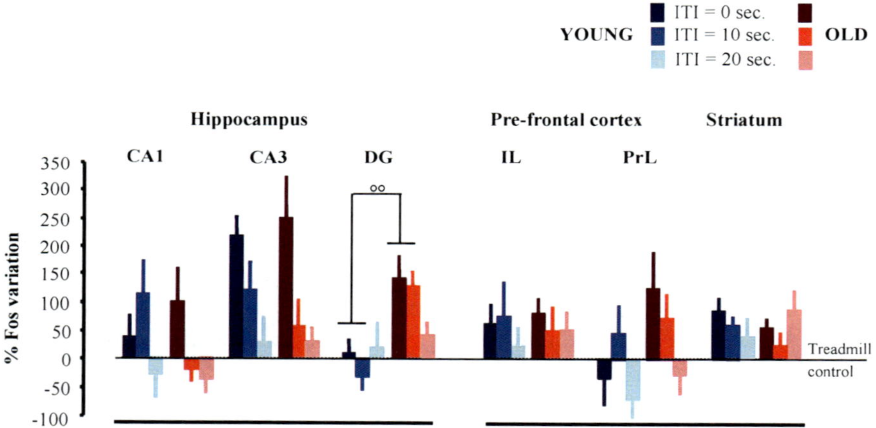

Fig. 7 Percentage of Fos variation from control levels induced by the WM task under the different ITI conditions in hippocampal (CA1, CA3, DG), prefrontal (infra-limbic and pre-limbic cortex), and striatal structures in young and aged mice. The control levels were those measured in treadmill-trained young and old mice serving as an "activity without learning "control condition

effect under the long-delay condition, which revealed a deficit in WM organization in these mice (Fig. 6b).

Fos neuroimaging revealed an alteration in WM-related activity in the aged hippocampus, specifically in the DG. Interestingly, hippocampal DG was more activated in the aged group than in the young, but only under the condition of short delays corresponding to the higher demand on WM organization (Fig. 7).

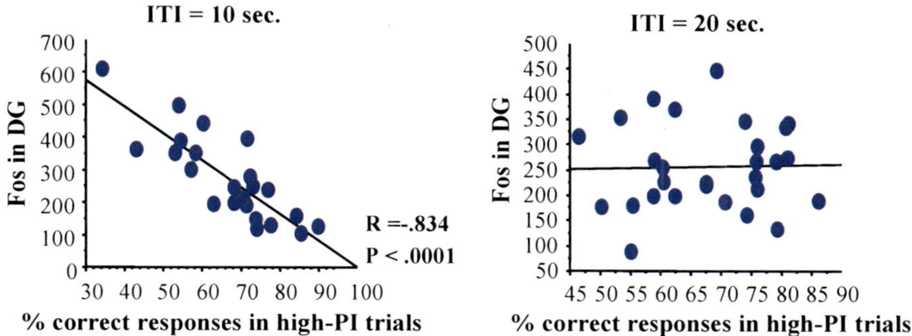

Fig. 8 Negative correlation between DG Fos level after WM testing under the short- (10 s) ITI condition (corresponding to high-organizational demand) and the performance in trials with high-proactive inference level

Furthermore, in this condition of high-organizational demand, there was a negative correlation between the overall WM performance and the DG Fos level after testing (Fig. 8). This correlation suggested that DG over-activation was involved in the deterioration in WM organization. This conclusion was further supported by the observation that aging did not significantly change WM-related activities in other studied areas.

In conclusion, the lesion experiment showed that the hippocampus is necessary for WM maintenance whereas the prefrontal cortex plays a role in WM organization under longer-retention intervals. In parallel, Fos neuroimaging showed that hippocampal activity (specifically in the DG) is also related to WM organization since the deficit seen in aged mice under the high-proactive interference condition correlates with an hyper-activation of hippocampal DG activity.

2.4 *Concluding Remarks on the Aging-Related Deterioration in Relational/Declarative Memory and Working Memory in Our Models*

First, the striking similarity of the behavioral and neurofunctional alterations induced by aging between mice and humans suggests that our radial-maze tests may represent valuable tools to bridge the gap between research on memory in rodents and in humans. Thus, our tests may help to uncover the cellular/molecular bases of aging-related decline in R/DM and WM.

Second, our Fos neuroimaging observations complement the large literature on hippocampal dysfunction in aging by showing that differential changes in hippocampal activity are associated with specific memory impairments. Namely, we found that CA1 hypoactivation correlated with the impairment in R/DM encoding, whereas DG over-activity correlated with the defect in WM organization.

These observations further highlight the need to search for intra cellular/molecular mechanisms of memory decline within the integrated network of memory systems in activity.

3 Implication of the Retinoid Signaling Pathway in Memory Deficits in Aging

We have used the above-described behavioral tools to demonstrate that the cellular signaling pathway controlled by vitamin A was involved in aging-related memory disturbances. In the following, we describe our experiments in aged mice after a brief synthesis of the literature on the implication of the retinoid signaling in adult brain plasticity and memory.

3.1 Retinoid Signaling has a Role in Adult Brain Plasticity and Memory

Most of the vitamin A functions are carried out by its metabolite retinoic acid, RA (Malik et al. 2000; Marill et al. 2003; Blomhoff and Blomhoff 2006). RA regulates gene expression in numerous cells and tissues by binding to nuclear RA receptors (RAR α, β and γ) and retinoid X receptors (RXR α, β and γ). RA receptors act as transcription factors which regulate the expression of a large spectrum of genes (Kastner et al. 1995; Mangelsdorf et al. 1995; Lefebvre et al. 2005).

Retinoid signaling path has been essentially studied for its critical involvement in brain development (Maden et al. 1998a, b). However, it has recently come to light that this signaling path continues to play a role in areas of the mature brain which can be modified by life experiences, in particular learning and memory (rev. in (Mey and McCaffery 2004; McCaffery et al. 2006; Lane and Bailey 2005). Indeed, the large spectrum of genes regulated by RA receptors includes genes coding for proteins involved in the mechanisms of adult brain plasticity, which are believed to sustain learning and memory. Furthermore, the distribution of RA receptors within the brain is compatible with a role in learning and memory, even though effective activity of RA receptors in the adult brain remains uncertain in many of concerned regions (Krezel et al. 1999).

Functional implication of retinoid signaling in cellular and synaptic hippocampal plasticity as well as learning and memory capabilities has been demonstrated by genetic, nutritional ,and pharmacological manipulation of this signaling pathway in rodents. Namely, knock-out of RARβ or RXRβ/γ receptors (Chiang et al. 1998; Wietrzych et al. 2005), vitamin A deprivation (Etchamendy et al. 2003b; Cocco et al. 2002; Misner et al. 2001; Bonnet et al. 2008), and administration of RA at high doses (Sakai et al. 2004; Crandall et al. 2004) have negative consequences on hippocampal LTP and LTD and hippocampus-dependent memory.

Dysfunction in retinoid signaling has been hypothesized to be involved in neurodevelopmental pathologies such as schizophrenia and late onset of Alzheimer's disease (Goodman 1998, 2006; Goodman and Pardee 2003; Ruano et al. 2008; Palha and Goodman 2006). The hypothesis is essentially based on the fact that genes related to the pathologies are frequently also retinoid-related genes, and on epidemiological evidence suggesting a link between pathology prevalence and vitamin A nutrional intake. Regarding Alzheimer's disease, the retinoid hypothesis has found support in the experimental studies showing that nutritional deprivation in vitamin A can induce beta-amyloid deposits in normal rats (Corcoran et al. 2004) whereas RA can prevent amyloid pathology in a genetic model (Ding et al. 2008). In fact, an increasing body of evidence on the actions of RA indicates that it may have therapeutic properties ideally served for the treatment of neurodegenerative diseases such as Alzheimer's disease (rev in Lee et al. 2009).

As developed below, we have provided the first experimental evidence that hypofunction in retinoid signaling contributes to memory decline in aging. The hypothesis was initially based on the observation that, in the mouse, aging was accompanied by a reduction in retinoid signaling expression (Enderlin et al. 1997).

3.2 Hypofunction in Retinoid Signaling Contributes to Aging-Related Degradation in Hippocampal Function, Relational/Declarative Memory Formation and Working Memory Organization

The first evidence that retinoid signaling hypoexpression was involved in the memory decline in aging was provided by a pharmacological study in the R/DM model (Etchamendy et al. 2001). Indeed, systemic administration of a moderate dose of RA to aged mice restored the hypoexpression of RA receptors and certain of their target genes to young adult levels in the hippocampus. At the same dose, RA also attenuated the deficit in hippocampal LTP seen in aged mice and selectively alleviated their specific deficit in R/DM expression. The implication of RA receptors in these beneficial effects of RA was demonstrated by the fact that co-administration of a RA receptor antagonist prevented the restoration of both retinoid signaling hypoexpression and R/DM deficit. Finally, the finding of beneficial effect of RA administration on the aging-related decline in hippocampus-dependent memory was replicated in rats by another research group (Brouillette and Quirion 2008a).

The retinoid hypothesis was further supported by showing that nutritional and genetic disruption of retinoid signaling expression in young mice could produce an aging-like pattern of changes in gene expression and memory. First, RARβ/RXRγ knock-out produced the same selective deficit in the radial-maze R/DM test as the one seen in aged mice (Mingaud et al. 2008). Second, post-weaning vitamin A deprivation was also capable of inducing an aging-like deficit in R/DM. Importantly, the emergence of this memory deficit was concomitant to the appearance of an aging-like reduction in the hippocampal expression of one of the plasticity-related target genes of RA receptors, RC3 (Etchamendy et al. 2003b). Again, converging evidence were provided by another research group (Brouillette and Quirion 2008b).

Interpretation of the above findings was that down-regulation of the retinoid-target-gene expression would primarily affect neuronal function relying on a precisely regulated intracellular machinery such as synaptic plasticity and associative properties of the hippocampus, supposedly required for R/DM formation. Hence, the normalization of retinoid-target-gene expression by RA treatment in aged mice would restore hippocampal function based on adjusted protein and molecule levels. Conversely, moderate down-regulation of retinoid signaling expression by vitamin A deprivation as well as over-activation of this cellular pathway by RA in young animals (Crandall et al. 2004) would disrupt the same hippocampal function.

The view that hippocampal function in memory relies on precise regulation of retinoid signaling and moderate down-regulation of this pathway in aging contributes to the decline in certain forms of memory was further supported by studying the effects of nutritional vitamin A supplementation at different ages (Mingaud et al. 2008). First, a life-long vitamin A enriched diet prevented the aging-related diminution of retinoid-target-gene expression in the hippocampus without changing the retinoid signaling expression in the striatum. Concomitlantly, the same diet prevented the aging-related degradation in R/DM and in WM organization, without affecting other aspects of learning and memory in the aged mice. Second, vitamin A supplementation limited to old age also maintained normal expression levels of retinoid genes and memory function, while the same nutritional manipulation tended to reduce retinoid gene expression and WM organization in young mice. Finally, beneficial effects of vitamin A supplementation against the deterioration of specific memory components were accompanied by restoration of the hypoactivation of hippocampal (and to a lesser extent prefrontal) Fos expression in R/DM encoding, while Fos activities in striatal and amygdalar areas were unchanged.

Taken as a whole, our molecular, systemic and behavioral findings have contributed to demonstrate that retinoid signaling affects functional activity of memory systems, likely through the regulation of intracellular machinery which underlies specific information processing. Indeed, although the spectrum of retinoid-target genes is potentially very large, changes in retinoid signaling expression were found to selectively affect specific functions of the hippocampus in learning and memory. Namely, moderate variations in retinoid signaling expression as either experimentally induced by vitamin A deprivation or naturally occurring in aging were associated with specific changes in R/DM encoding and WM organization.

4 Conclusion

Our work has provided new behavioral tools to explore the neurobiological bases of memory degradation in aging and opened a research avenue for the development of nutritional therapeutics and preventive strategies targeting retinoid signaling.

Regarding behavioral tools, our findings support the view that it is possible to study memory function and dysfunction with similar methods and concepts between mice and humans. Such standardization should help bridging the gap between research in animals and humans, and thereby improve our understanding of memory decline in aging. In particular, common assessment of specific memory processes between mice and humans should enable integrating molecular/intracellular events to memory system function.

As a first attempt to integrate molecular/cellular and system levels, our neuroimaging studies based on *cfos* gene/Fos protein expression have provided new elements of information on hippocampal dysfunction in aging. Namely, we found that differential alterations in learning-related activities in the hippocampus correlated with specific impairments in WM organization and R/DM encoding in aged mice. In line with an increasing literature which questions the dogma of separation between long-term DM and short-term/WM systems (Olson et al. 2006; Hannula et al. 2006; Kumaran 2008; Jonides et al. 2008; Quinette et al. 2006; Ranganath and Blumenfeld 2005; Shrager et al. 2008; Hannula and Ranganath 2008), our findings point to potential interaction between WM and DM in the hippocampus and the necessity of understanding how does aging potentially affect such interaction.

Finally, analyzing the effects of retinoid signaling pathway manipulations in young and aged mice on molecular, systemic and behavioral markers of R/DM and WM degradation has enabled us to demonstrate the critical implication the vitamin A cellular signaling in hippocampal function. The finding that nutritional vitamin A supplementation can prevent the aging-related deterioration in hippocampal function complements a currently increasing amount of literature on the potential interest of targeting retinoid signaling in future therapeutical attempts against memory disturbances in aging.

References

Addis DR, McAndrews MP (2006) Prefrontal and hippocampal contributions to the generation and binding of semantic associations during successful encoding. Neuroimage 33:1194–1206

Baddeley A (1996) The fractionation of working memory. Proc Natl Acad Sci U S A 93: 13468–13472

Baddeley A, Cocchini G, Della Sala S, Logie RH, Spinnler H (1999) Working memory and vigilance: evidence from normal aging and Alzheimer's disease. Brain Cogn 41:87–108

Blomhoff R, Blomhoff HK (2006) Overview of retinoid metabolism and function. J Neurobiol 66:606–630

Blumenfeld RS, Ranganath C (2006) Dorsolateral prefrontal cortex promotes long-term memory formation through its role in working memory organization. J Neurosci 26:916–925

Bohbot VD, Lerch J, Thorndycraft B, Iaria G, Zijdenbos AP (2007) Gray matter differences correlate with spontaneous strategies in a human virtual navigation task. J Neurosci 27:10078–10083

Bonnet E et al (2008) Retinoic acid restores adult hippocampal neurogenesis and reverses spatial memory deficit in vitamin A deprived rats. PLoS One 3: e3487

Brassen S, Weber-Fahr W, Sommer T, Lehmbeck JT, Braus DF (2006) Hippocampal-prefrontal encoding activation predicts whether words can be successfully recalled or only recognized. Behav Brain Res 171:271–278

Brouillette J, Quirion R (2008a) Transthyretin: a key gene involved in the maintenance of memory capacities during aging. Neurobiol Aging 29:1721–1732

Brouillette J, Quirion R (2008b) The common environmental pollutant dioxin-induced memory deficits by altering estrogen pathways and a major route of retinol transport involving transthyretin. Neurotoxicology 29:318–327

Burke SN, Barnes CA (2010) Senescent synapses and hippocampal circuit dynamics. Trends Neurosci 33:153–161

Bunge SA, Klingberg T, Jacobsen RB, Gabrieli JDA (2000) A resource model of the neural basis of executive working memory. Proc Natl Acad Sci U S A 97:3573–3578

Burke SN, Barnes CA (2006) Neural plasticity in the ageing brain. Nat Rev Neurosci 7:30–40

Chalfonte BL, Johnson MK (1996) Feature memory and binding in young and older adults. Mem Cognit 24:403–416

Chiang MY et al (1998) An essential role for retinoid receptors RARbeta and RXRgamma in long-term potentiation and depression. Neuron 21:1353–1361

Cocco S et al (2002) Vitamin A deficiency produces spatial learning and memory impairment in rats. Neuroscience 115:475–482

Cohen NJ, Poldrack RA, Eichenbaum H (1997) Memory for items and memory for relations in the procedural/declarative memory framework. Memory 5:131–178

Cohen NJ et al (1999) Hippocampal system and declarative (relational) memory: summarizing the data from functional neuroimaging studies. Hippocampus 9:83–98

Corcoran JP, So PL, Maden M (2004) Disruption of the retinoid signalling pathway causes a deposition of amyloid beta in the adult rat brain. Eur J Neurosci 20:896–902

Craik FI (1990) Changes in memory with normal aging: a functional view. Adv Neurol 51: 201–205

Crandall J et al (2004) 13-cis-retinoic acid suppresses hippocampal cell division and hippocampal-dependent learning in mice. Proc Natl Acad Sci U S A 101:5111–5116

Davachi L, Wagner AD (2002) Hippocampal contributions to episodic encoding: insights from relational and item-based learning. J Neurophysiol 88:982–990

Ding Y et al (2008) Retinoic acid attenuates beta-amyloid deposition and rescues memory deficits in an Alzheimer's disease transgenic mouse model. J Neurosci 28:11622–11634

Dunnett SB, Martel FL, Iversen SD (1990) Proactive interference effects on short-term memory in rats: II. Effects in young and aged rats. Behav Neurosci 104:666–670

Eichenbaum H, Fagan A, Mathews P, Cohen NJ (1988) Hippocampal system dysfunction and odor discrimination learning in rats: impairment or facilitation depending on representational demands. Behav Neurosci 102:331–339

Eichenbaum H, Mathews P, Cohen NJ (1989) Further studies of hippocampal representation during odor discrimination learning. Behav Neurosci 103:1207–1216

Eichenbaum H, Otto T, Cohen NJ (1992) The hippocampus–what does it do? Behav Neural Biol 57:2–36

Enderlin V et al (1997) Age-related decreases in mRNA for brain nuclear receptors and target genes are reversed by retinoic acid treatment. Neurosci Lett 229:125–129

Etchamendy N et al (2001) Alleviation of a selective age-related relational memory deficit in mice by pharmacologically induced normalization of brain retinoid signaling. J Neurosci 21:6423–6429

Etchamendy N, Desmedt A, Cortes-Torrea C, Marighetto A, Jaffard R (2003a) Hippocampal lesions and discrimination performance of mice in the radial maze: sparing or impairment depending on the representational demands of the task. Hippocampus 13:197–211

Etchamendy N et al (2003b) Vitamin A deficiency and relational memory deficit in adult mice: relationships with changes in brain retinoid signalling. Behav Brain Res 145:37–49
Etchamendy N, Konishi K, Pike GB, Marighetto A, Bohbot VD (2011) Evidence for a virtual human analog of a rodent relational memory task: a study of aging and fMRI in young adults. Hippocampus. Jun 8. doi:10.1002/hipo.20948
Flicker C, Ferris SH, Crook T, Bartus RT (1989) Age differences in the vulnerability of facial recognition memory to proactive interference. Exp Aging Res 15:189–194
Friedman D, Nessler D, Johnson R Jr (2007) Memory encoding and retrieval in the aging brain. Clin EEG Neurosci 38:2–7
Gabrieli JD (1996) Memory systems analyses of mnemonic disorders in aging and age-related diseases. Proc Natl Acad Sci U S A 93:13534–13540
Goodman AB (1998) Three independent lines of evidence suggest retinoids as causal to schizophrenia. Proc Natl Acad Sci U S A 95:7240–7244
Goodman AB (2006) Retinoid receptors, transporters, and metabolizers as therapeutic targets in late onset Alzheimer disease. J Cell Physiol 209:598–603
Goodman AB, Pardee AB (2003) Evidence for defective retinoid transport and function in late onset Alzheimer's disease. Proc Natl Acad Sci U S A 100:2901–2905
Grady CL (2008) Cognitive neuroscience of aging. Ann NY Acad Sci 1124:127–144
Grady CL, Craik FI (2000) Changes in memory processing with age. Curr Opin Neurobiol 10:224–231
Grady CL, McIntosh AR, Rajah MN, Beig S, Craik FI (1999) The effects of age on the neural correlates of episodic encoding. Cereb Cortex 9:805–814
Hannula DE, Ranganath C (2008) Medial temporal lobe activity predicts successful relational memory binding. J Neurosci 28:116–124
Hannula DE, Tranel D, Cohen NJ (2006) The long and the short of it: relational memory impairments in amnesia, even at short lags. J Neurosci 26:8352–8359
Jinno S (2011) Decline in adult neurogenesis during aging follows a topographic pattern in the mouse hippocampus. J Comp Neurol 519:451–466
Jonides J et al (2008) The mind and brain of short-term memory. Annu Rev Psychol 59:193–224
Kastner P, Mark M, Chambon P (1995) Nonsteroid nuclear receptors: what are genetic studies telling us about their role in real life? Cell 83:859–869
Krezel W, Kastner P, Chambon P (1999) Differential expression of retinoid receptors in the adult mouse central nervous system. Neuroscience 89:1291–1300
Kumaran D (2008) Short-term memory and the human hippocampus. J Neurosci 28:3837–3838
Lane MA, Bailey SJ (2005) Role of retinoid signalling in the adult brain. Prog Neurobiol 75:275–293
Lee HP et al (2009) All-trans retinoic acid as a novel therapeutic strategy for Alzheimer's disease. Expert Rev Neurother 9:1615–1621
Lefebvre P et al (2005) Transcriptional activities of retinoic acid receptors. Vitam Horm 70:199–264
Lund PK et al (2004) Transcriptional mechanisms of hippocampal aging. Exp Gerontol 39:1613–1622
Maden M, Sonneveld E, van der Saag PT, Gale E (1998a) The distribution of endogenous retinoic acid in the chick embryo: implications for developmental mechanisms. Development 125:4133–4144
Maden M, Gale E, Zile M (1998b) The role of vitamin A in the development of the central nervous system. J Nutr 128:471S–475S
Malik MA, Blusztajn JK, Greenwood CE (2000) Nutrients as trophic factors in neurons and the central nervous system: role of retinoic acid. J Nutr Biochem 11:2–13
Mangelsdorf DJ et al (1995) The nuclear receptor superfamily: the second decade. Cell 83:835–839
Marighetto A et al (1999) Knowing which and knowing what: a potential mouse model for age-related human declarative memory decline. Eur J Neurosci 11:3312–3322

Marighetto A et al (2000) Further evidence for a dissociation between different forms of mnemonic expressions in a mouse model of age-related cognitive decline: effects of tacrine and S 17092, a novel prolyl endopeptidase inhibitor. Learn Mem 7:159–169

Marighetto A et al (2008a) The AMPA modulator S 18986 improves declarative and working memory performances in aged mice. Behav Pharmacol 19:235–244

Marighetto A et al (2008b) Comparative effects of the dopaminergic agonists piribedil and bromocriptine in three different memory paradigms in rodents. J Psychopharmacol 22: 511–521

Marighetto A et al (2008c) Comparative effects of the alpha7 nicotinic partial agonist, S 24795, and the cholinesterase inhibitor, donepezil, against aging-related deficits in declarative and working memory in mice. Psychopharmacology (Berl) 197:499–508

Marill J, Idres N, Capron CC, Nguyen E, Chabot GG (2003) Retinoic acid metabolism and mechanism of action: a review. Curr Drug Metab 4:1–10

McCaffery P, Zhang J, Crandall JE (2006) Retinoic acid signaling and function in the adult hippocampus. J Neurobiol 66:780–791

Mey J, McCaffery P (2004) Retinoic acid signaling in the nervous system of adult vertebrates. Neuroscientist 10:409–421

Mingaud F et al (2007) The hippocampus plays a critical role at encoding discontiguous events for subsequent declarative memory expression in mice. Hippocampus 17:264–270

Mingaud F et al (2008) Retinoid hyposignaling contributes to aging-related decline in hippocampal function in short-term/working memory organization and long-term declarative memory encoding in mice. J Neurosci 28:279–291

Misner DL et al (2001) Vitamin A deprivation results in reversible loss of hippocampal long-term synaptic plasticity. Proc Natl Acad Sci U S A 98:11714–11719

Mitchell KJ, Johnson MK, Raye CL, Mather M, D'Esposito M (2000a) Aging and reflective processes of working memory: binding and test load deficits. Psychol Aging 15:527–541

Mitchell KJ, Johnson MK, Raye CL, D'Esposito M (2000b) fMRI evidence of age-related hippocampal dysfunction in feature binding in working memory. Brain Res Cogn Brain Res 10:197–206

Mitchell KJ, Johnson MK, Raye CL, Greene EJ (2004) Prefrontal cortex activity associated with source monitoring in a working memory task. J Cogn Neurosci 16:921–934

Morcom AM, Good CD, Frackowiak RS, Rugg MD (2003) Age effects on the neural correlates of successful memory encoding. Brain 126:213–229

Narayanan NS et al (2005) The role of the prefrontal cortex in the maintenance of verbal working memory: an event-related FMRI analysis. Neuropsychology 19:223–232

Olson IR, Page K, Moore KS, Chatterjee A, Verfaellie M (2006) Working memory for conjunctions relies on the medial temporal lobe. J Neurosci 26:4596–4601

Palha JA, Goodman AB (2006) Thyroid hormones and retinoids: a possible link between genes and environment in schizophrenia. Brain Res Rev 51:61–71

Poldrack RA, Packard MG (2003) Competition among multiple memory systems: converging evidence from animal and human brain studies. Neuropsychologia 41:245–251

Quinette P et al (2006) The relationship between working memory and episodic memory disorders in transient global amnesia. Neuropsychologia 44:2508–2519

Rabinowitz JC, Ackerman BP, Craik FI, Hinchley JL (1982) Aging and metamemory: the roles of relatedness and imagery. J Gerontol 37:688–695

Ranganath C, Blumenfeld RS (2005) Doubts about double dissociations between short- and long-term memory. Trends Cogn Sci 9:374–380

Repovs G, Baddeley A (2006) The multi-component model of working memory: explorations in experimental cognitive psychology. Neuroscience 139:5–21

Ruano D et al (2008) Association of the gene encoding neurogranin with schizophrenia in males. J Psychiatr Res 42:125–133

Rypma B, Prabhakaran V, Desmond JE, Gabrieli JD (2001) Age differences in prefrontal cortical activity in working memory. Psychol Aging 16:371–384

Sadeh T, Shohamy D, Levy DR, Reggev N, Maril A (2011) Cooperation between the hippocampus and the striatum during episodic encoding. J Cogn Neurosci 23(7):1597–1608 [Epub 28 Jul 2010]
Sakai Y, Crandall JE, Brodsky J, McCaffery P (2004) 13-cis Retinoic acid (accutane) suppresses hippocampal cell survival in mice. Ann N Y Acad Sci 1021:436–440
Salthouse TA, Mitchell DR, Skovronek E, Babcock RL (1989) Effects of adult age and working memory on reasoning and spatial abilities. J Exp Psychol Learn Mem Cogn 15:507–516
Saxe MD et al (2007) Paradoxical influence of hippocampal neurogenesis on working memory. Proc Natl Acad Sci U S A 104:4642–4646
Schmidtke K, Manner H, Kaufmann R, Schmolck H (2002) Cognitive procedural learning in patients with fronto-striatal lesions. Learn Mem 9:419–429
Shrager Y, Levy DA, Hopkins RO, Squire LR (2008) Working memory and the organization of brain systems. J Neurosci 28:4818–4822
Squire LR (1992) Memory and the hippocampus: a synthesis from findings with rats, monkeys, and humans. Psychol Rev 99:195–231
Squire LR, Zola SM (1996) Structure and function of declarative and nondeclarative memory systems. Proc Natl Acad Sci U S A 93:13515–13522
Takashima A et al (2006) Successful declarative memory formation is associated with ongoing activity during encoding in a distributed neocortical network related to working memory: a magnetoencephalography study. Neuroscience 139:291–297
Wietrzych M et al (2005) Working memory deficits in retinoid X receptor gamma-deficient mice. Learn Mem 12:318–326

Functional Neuroimaging Studies in Normal Aging

Leslie M. Guidotti Breting, Elizabeth R. Tuminello and S. Duke Han

Abstract With an expanding aging population, it is increasingly important to gain a better understanding of the changes in cognition and neural integrity that occur in normal aging. The advent of non-invasive functional neuroimaging techniques has spurred researchers to examine cognition and neural functioning in healthy older adults. A significant amount of research has been produced since this time and has led to influential theories of aging such as the hemispheric asymmetry reduction for older adults (HAROLD) model and the compensatory recruitment hypothesis. This chapter discusses advances in our understanding of normal aging achieved through the use of functional neuroimaging. Research examining age-related changes in domains such as attention, memory, and executive functioning, as well as imaging of the resting-state and the influences of genetic risk factors (e.g., APOE genotype), are discussed. In conclusion, limitations of the current literature and important avenues for future research are proposed.

Keywords Aging · Functional magnetic resonance imaging (fMRI) · Default-mode network · APOE

L. M. Guidotti Breting
Department of Psychiarty and Behavioral Sciences,
NorthShore University HealthSystem, Evanston, IL, USA

E. R. Tuminello
Department of Psychology, Loyola University Chicago, Chicago, IL, USA

S. Duke Han (✉)
Department of Behavioral Sciences, Rush University Medical Center,
1653 W. Congress Parkway, Chicago, IL 60612-3244, USA
e-mail: Duke_Han@rush.edu

Curr Topics Behav Neurosci (2012) 10: 91–111
DOI: 10.1007/7854_2011_139

Published Online: 13 July 2011

Contents

1 Introduction 92
2 The Use of Neuroimaging in Studies of Normal Aging 93
  2.1 Visual Perception and Attention 94
  2.2 Episodic Memory 95
  2.3 Temporal Gradient of Memory 95
  2.4 Executive Functioning 96
  2.5 Emotion Processing and Emotion Perception 97
  2.6 Resting-State fMRI, the Default-Mode Network, and Aging 97
  2.7 Aging and the Apolipoprotein E ε4 Allele 99
3 Limitations and Future Directions of fMRI Studies of Normal Aging 104
4 Conclusion 105
References 105

1 Introduction

It is estimated that, by the year 2030, 20% of the United States population (approximately 71 million people) will be over the age of 65 (Centers for Disease Control and Prevention and The Merck Company Foundation 2007). As the population continues to age at an increasing rate, it is necessary to develop a more complete understanding of the cognitive and neurological changes that accompany the aging process. In order to conduct such research, investigators must first define normal aging.

Two views on normal aging have been proposed: (1) the biological perspective and (2) the lifespan development perspective (Smith and Bondi 2008). Research conducted from the biological perspective has suggested that cognitive abilities such as memory, processing speed, and cognitive flexibility peak between 18 and 30 years of age, after which time the process of normal aging begins (Salthouse 2009). Thus, according to the biological perspective, normal aging is associated with declines in cognitive domains including memory, reasoning, and spatial abilities that begin in mid-life (Salthouse 1988). In contrast, the lifespan development perspective views aging and concomitant changes in functional status as normal processes that occur throughout the lifespan. Importantly, this perspective avoids labeling these age-related changes as abnormal (Baltes 1987). Most of the neuroimaging studies of normal aging discussed in this chapter have adhered to the lifespan development perspective and, as such, include participants with common age-associated diseases such as high blood pressure. Due in part to such differences in the definition of normal aging and differences in sample selection, the findings of research examining normal aging have been occasionally viewed as difficult to reconcile. We present a selected review of the literature focusing on current trends in the field of functional neuroimaging of normal aging.

2 The Use of Neuroimaging in Studies of Normal Aging

The advent of non-invasive functional neuroimaging spurred researchers to examine cognitive processes in the healthy brain. Less than two decades ago, the first neuroimaging study of aging was conducted by Grady et al. (1994) using positron emission tomography (PET), which found age-related changes in regional cerebral blood flow (rCBF) patterns during visual processing and matching of faces and locations. Other early studies also typically employed PET (see Cabeza and Nyberg 1997, for a review) or used block-design approaches with functional magnetic resonance imaging (fMRI). More recent neuroimaging studies have generally moved away from the use of PET and block-design fMRI in favor of imaging the blood oxygenation level dependent (BOLD) signal from the entire brain using an event-related fMRI design. The BOLD signal is examined as an indirect qualitative measure of neural activity, so that BOLD signal changes are generally interpreted as changes in neural activity (van der Zwaag et al. 2009). This chapter will largely focus on event-related fMRI studies (for a review of studies of aging utilizing PET, see Raz (2000)). A significant amount of research using fMRI to examine normal aging has been conducted in the short period of time since its advent. This chapter will discuss what we have learned about normal aging from these studies and will propose limitations and future directions for this line of research.

In light of age-related differences in neural activity observed by early studies of normal aging, several theories have been proposed. It has been noted that the asymmetrical recruitment exhibited by young adults during episodic memory tasks, termed the hemispheric encoding/retrieval asymmetry (HERA) model (Nyberg et al. 1996; Tulving et al. 1994), is reduced in older adults, leading to the proposal of the hemispheric asymmetry reduction for older adults (HAROLD) model (Cabeza 2002). Two theories that may account for the reductions in hemispheric asymmetry observed in older adults are the *compensatory recruitment* hypothesis (Cabeza 2002) and the *dedifferentiation* hypothesis (Li and Lindenberger 1999).

The compensatory recruitment hypothesis proposes that older adults exhibit enhanced and more widespread neural activation during cognitive tasks in order to compensate for age-related declines in order to maintain previous performance levels (Cabeza 2002). In order to differentiate between increases in neural activity due to compensatory strategies versus less efficient use of neural resources, evidence for the compensatory hypothesis is commonly cited in studies in which increases in neural activity are accompanied by equal or enhanced performance on cognitive tasks.

In contrast, the dedifferentiation hypothesis proposes that pathological processes produce age-related difficulties in performing a particular cognitive function. This cognitive deficit leads to the recruitment of a less specialized neural mechanism to perform that function, as opposed to the highly specialized mechanism formerly in place (Cabeza 2002). Such decreased specialization manifests an increased correlation among diverse cognitive measures and/or neural patterns of activity. The exact brain mechanism responsible for dedifferentiation is unknown, but it is considered an example of a primary degenerative event associated with aging rather than a compensatory mechanism.

Cabeza (2001a) and Reuter-Lorenz et al. (1999) have stated that compensation and dedifferentiation are not mutually exclusive processes and may co-occur. Cabeza suggests that combining cognitive functions through dedifferentiation could counteract cognitive decline associated with aging. He further argues that decreased hemispheric asymmetry may also serve a compensatory role and may account for the increase in correlations observed across cognitive measures in aging by leading to more similarity across various tasks. In contrast to the compensatory and dedifferentiation hypotheses, it is also possible that age-related increase in neural activity are simply due to the presence of greater neural noise (Beason-Held et al. 2008). To more closely examine these theories, this chapter will discuss the findings of functional neuroimaging studies of normal aging across several cognitive domains (see also Cabeza 2001b or Daselaar et al. 2006 for earlier reviews).

2.1 Visual Perception and Attention

Some of the earliest neuroimaging studies of the relationship between neural and cognitive aspects of aging examined visual processing and attention (Grady et al. 1994; Madden et al. 1996, 1997). Research suggests that, for younger adults, attentional processes are mediated by widely distributed neural networks, with critical components located in prefrontal, deep gray matter, and parietal regions (Kastner and Ungerleider 2000). Neuroimaging studies of older adults have reported an extensive and complex pattern of age-related change in brain structure and function, including changes in domains such as visual attention (Leonards et al. 2000). In fact, older adults have shown a decline in task-related activation of visual sensory cortex (Buckner et al. 2000; Madden et al. 2004). In some tasks, this decline is also accompanied by increased activation of other components of the frontoparietal attentional network, which has been interpreted as compensatory recruitment of cortical regions outside the task-relevant pathway (Cabeza et al. 2004). More recently, an fMRI study examined the neural mechanisms underlying visual-spatial working memory to find that older participants activated dorsolateral prefrontal cortex (PFC) regions bilaterally, while young subjects recruited these areas only in the left hemisphere (Piefke et al. 2010). This finding of an age-related reduction in hemispheric asymmetry in the PFC coincides with predictions made by the HAROLD model, and could also be consistent with either the compensatory recruitment or dedifferentiation hypotheses. This study also observed age-related functional reorganizations in parieto-occipital regions, and with increasing working memory demands, a reversal of typical patterns of prefrontal hemispheric asymmetry for older adults.

2.2 Episodic Memory

Memory loss is the complaint most often associated with aging (Jonker et al. 2000). For decades studies of normal aging have focused on episodic memory by examining, encoding ,and recalling of information after a delay. These studies have demonstrated that changes in memory abilities across time are a part of the normal aging process. Furthermore, healthy older adults tend to be less efficient at encoding information and also have more difficulty recalling information after a delay than younger adults (e.g., Petersen et al. 1992; Sliwinski and Buschke 1999). Research examining predictors for Alzheimer's disease (AD) report a significant decline in episodic memory prior to the development of a dementia (e.g., Bondi et al. 1999; Chen et al. 2001). Neuroimaging studies have an increasingly more important role in examining age-related neural changes in episodic memory functioning. Numerous studies have now examined the neural correlates of episodic memory functioning in normal aging, which are described below.

Many functional neuroimaging studies have reported that, relative to young participants, older individuals demonstrate greater and more widespread encoding- and retrieval-related cortical activity, consistent with the compensatory recruitment hypothesis (Cabeza et al. 1997; Grady et al. 2005; Madden, et al. 1999; Maguire and Frith 2003). In addition, as proposed by the HAROLD model, older adults have been found to exhibit reduced hemispheric asymmetry during episodic memory tasks compared to their younger counterparts. While episodic encoding is typically left-lateralized in young participants, older adults display reduced left prefrontal activation during these tasks (Logan et al. 2002; Morcom et al. 2003; Stebbins et al. 2002). Additionally, in contrast to the typically right-lateralized episodic retrieval in young adults, older adults commonly show increased left prefrontal activation during retrieval (Grady and Craik 2000). Several other fMRI studies utilizing episodic memory tasks have also corroborated the HAROLD model. For example, Grady et al. (1995) employed a face-encoding task and found that older adults displayed reduced activation in the left prefrontal cortex and temporal regions. Similarly, studies examining encoding of word pairs have reported reduced left prefrontal and occipitotemporal activation in older adults (Anderson et al. 2000; Cabeza et al. 1997; Stebbins et al. 2002).

2.3 Temporal Gradient of Memory

Recent years have been characterized by growing interest in long-term memory consolidation and the time course for medial temporal lobe (MTL) involvement (Douville et al. 2005; Haist et al. 2001; Nielson et al. 2006). Long-term memory consolidation has been investigated, for example, using fMRI paradigms involving famous faces and/or names of individuals from different decades. Temporally graded changes for older adults compared to young adults have been found for the hippocampus and the entorhinal cortex (Douville et al. 2005; Haist et al. 2001). An event-related fMRI study examining recognition of famous names from different epochs showed extensive networks of activation including posterior cingulate, right hippocampus, temporal lobe, and left prefrontal regions (Nielson et al. 2006). The study also suggested that older adults may use compensatory recruitment to support task performance, even when task performance accuracy is high. Similar studies have observed enhanced neural activity in the posterior cingulate, right middle frontal, right fusiform, and left middle temporal cortices for recent versus remote famous names in older adults (Woodard et al. 2007). Empirical support is also growing for the notion that different types of memory (e.g., autobiographical memory) are associated with different patterns of neural activity and memory retrieval for young and older adults (Maguire and Frith 2003; Denkova et al. 2006).

2.4 Executive Functioning

The corpus of tasks and procedures that fall under the rubric of executive functions is vast and include inhibition, working memory, problem solving, multi-tasking, sequencing, abstract thinking, and attentional capacity (see Han et al. 2008 for a brief review). Due to the complexity of the executive functions cognitive domain, there is currently not a single conceptual definition or a single task to measure the entire construct. Important to aging, executive functioning skills are applicable to the activities of daily living in older adults in that they are necessary to do such activities as completing complex tasks (i.e., driving or cooking), planning projects, sequencing tasks, prioritizing, and self-monitoring. Traditionally, executive functions have been associated with the frontal cortex, but it is also dependent on the integrity of frontal-subcortical and frontal-parietal systems (Cummings 1993).

Functional neuroimaging investigations of executive functioning in older adults have often focused on inhibitory control, commonly utilizing Stroop interference (Langenecker et al. 2004) and match/nonmatch tasks (Lamar et al. 2004). Among these studies, there is a general tendency for older adults to display more activation during the interference condition than younger adults (Langenecker et al. 2004; Lamar et al. 2004). While such findings may be the result of compensatory recruitment processes, studies have also found that the Stroop interference effect is often more pronounced in older adults due to a decline in inhibitory control (Milham et al. 2002). As previously mentioned, such declines in task performance complicate the interpretation of the task-related increases in activation observed in older adults. Accordingly, several studies have attempted to clarify these issues using fMRI. Langenecker et al. (2004) studied the Stroop task in healthy younger and older adults, finding that older adults displayed increased interference-related activation in several frontal areas, including the left inferior frontal gyrus. Another study utilizing the Stroop task also reported that older adults exhibited more extensive activation of ventral visual processing areas, temporal regions, and anterior inferior regions (Milham et al. 2002). More recently, a study using a go-no-go task found that, despite demonstrated integrity of older adults' hemodynamic response relative to younger adults, older adults still exhibited the typical pattern of increased prefrontal cortex activation (Nielson et al. 2004). Overall, regardless of the specific executive task employed in the fMRI paradigm, results have generally found greater recruitment of prefrontal regions, particularly the left inferior frontal gyrus, in older adults than young adults (Lamar et al. 2004; Langenecker et al. 2004; Nielson et al. 2002; Milham et al. 2002).

2.5 Emotion Processing and Emotion Perception

A growing area of interest in the normal aging and functional neuroimaging literature is the examination of emotion perception. In one of the first fMRI studies of emotion and aging investigators found less left amygdala activity for negative faces in older adults when compared to young adults (Iidaka et al. 2002), and since that study others have replicated this finding (Tessitore, et al. 2005; Erk et al. 2008). Neuroimaging researchers have consistently observed older adults showing greater recruitment of the frontal cortex, particularly in the medial frontal regions, and less activation in the amygdala for negative stimuli (St. Jacques et al. 2009). St. Jacques et al. (2009) have termed this pattern of change in emotion with aging the fronto-amygdalar age-related differences in emotion (FADE). Despite age-related declines in many cognitive domains (Dennis and Cabeza 2008), it appears that healthy aging has less of an impact on emotion perception (Kensinger 2008). In fact, some suggest that aging is characterized by "superior" emotional regulation and the ability to exert control over emotional responses (Ochsner and Gross 2005). As emotional valence of information plays a role in the ability to remember information, only recently have imaging studies begun to examine the neural basis of memory for emotional stimuli in aging. Murty et al. (2009) have found age-related effects during memory retrieval, with older adults showing a reduction in activation of the amygdala during memory retrieval of negative versus neutral stimuli and an age-related increase in right dorsolateral prefrontal cortex. St. Jacques et al. (2009) have suggested that the prefrontal cortex increase could reflect the posterior-anterior shift in aging, which is frequently observed in non-emotional domains such as self-referential processing (Dennis and Cabeza 2008).

2.6 Resting-State fMRI, the Default-Mode Network, and Aging

While much functional neuroimaging research in aging has focused on neural activity during task performance, recent work using fMRI has begun exploring functional networks active while the brain is at rest, termed resting-state fMRI. An attractive quality of resting-state imaging is the relative speed and ease with which participants can be scanned and data gathered. Resting-state fMRI research is based upon the finding that spontaneous, low-frequency BOLD fluctuations are highly correlated among certain spatially distributed brain regions at rest (Cole et al. 2010). Most interpret these low-frequency BOLD fluctuations as indicators of spontaneous neuronal activity as opposed to manifestations of physiological alterations independent of neuronal activity. In support of this contention, the widespread neural networks identified in this manner have been found to conform to functional networks identified in task-dependent paradigms (Cole et al. 2010). Also, functional networks identified using resting-state fMRI mirror structural connectivity among these regions (Damoiseaux and Greicius 2009; Greicius et al. 2009). However, some debate of the utility of resting-state fMRI exists (see Morcom and Fletcher 2007, for a critical discussion).

Despite the growing popularity of examining functionally connected networks at rest, several factors exist which complicate the interpretation of resting-state fMRI data. First, while the low-frequency BOLD oscillations examined are distinguishable from frequencies produced by physiological factors, including respiratory and cardiovascular activity (Cordes et al. 2000, 2001), physiological and scanning artifacts may still impact resting-state activation patterns (Cole et al. 2010). Researchers must account for these potential confounding factors to maximize the signal-to-noise ratio and limit the frequency of spurious findings (Cole et al. 2010). Second, the interpretation of findings from resting-state fMRI paradigms relies upon the cognitive processes evoked during scanning, as the lack of a structured task during "rest" does not preclude the occurrence of spontaneous cognitive processes (Cole et al. 2010; Morcom and Fletcher 2007). Furthermore, it has been found that the relationship between individual regions, or nodes, within and across networks may fluctuate over time, introducing variability into resting-state data (Chang and Glover 2010). Finally, the choice of statistical approach for analyzing resting-state fMRI data, including model driven seed-based correlation analyses and data driven independent components analysis (ICA), has been found to impact results (Cole et al. 2010; Koch et al. 2010).

Although such methodological issues are still being elucidated, studies utilizing resting-state fMRI have produced many relatively stable findings. Most notably, this line of research along with task-related fMRI has led to the identification of a default-mode network (DMN), a spatially-distributed network of brain regions that are active during rest and whose activity is attenuated during goal directed activity (e.g., Greicius et al. 2003; Raichle et al. 2001). Brain regions commonly associated with the DMN include the precuneus/posterior cingulate cortex, medial prefrontal cortex, lateral- parietal regions, and inferior temporal gyrus (Broyd et al. 2009; Damoiseaux et al. 2006). While developmental studies indicate that the DMN is relatively stable among young adults (Damoiseaux et al. 2006), studies with older groups have found significant changes in the DMN over time.

Research has suggested that the functional connectivity of the DMN is altered in normal aging. Several studies have found that, compared to younger adults, older adults demonstrate reduced functional connectivity within the DMN at rest and during task performance, particularly in the posterior cingulate cortex (Andrews-Hanna et al.

2007; Koch et al. 2010; Voss et al. 2010). Research with cognitively intact older adults has also found positive associations between functional connectivity of the DMN and cognitive performance, including executive functioning, episodic memory, and processing speed (Andrews-Hanna et al. 2007; Chen et al. 2009; Voss et al. 2010). Studies of aging utilizing resting-state fMRI have also revealed that older adults involve more areas in the DMN, including orbital frontal, parahippocampal, and lateral temporal regions, than younger adults (Greicius et al. 2004). Taken along with demonstrated reductions in DMN functional connectivity and their associations with cognitive performance, these findings may suggest that older adults recruit additional regions for DMN processing in order to compensate for age-related declines.

While these studies suggest that the functioning of the DMN changes across the lifespan, one limitation of such research is that it predominantly relies on cross-sectional designs, which can be influenced by age-related changes in neurovascular coupling. Other research has utilized longitudinal designs to circumvent this issue and to more specifically examine whether DMN integrity changes in advanced stages of healthy aging (Beason-Held et al. 2009). One such study using PET followed healthy older adults over an eight year period and found that, while typical DMN regions including the anterior cingulate, posterior cingulate, and hippocampus, remained stable with age, other areas beyond the DMN showed age-related changes (Beason-Held et al. 2009). The authors therefore suggest that changes in the DMN may plateau with advancing age and may only continue to become disrupted through disease processes such as those that occur in AD. Support for this contention is provided by additional studies that found disruption in DMN connectivity in cognitively intact older adults to be associated with amyloid pathology (Hedden et al. 2009; Sheline et al. 2010; Sperling et al. 2009) and reduced white matter integrity (Andrews-Hanna et al. 2007; Chen et al. 2009).

Other studies of aging have found that older adults show less task-related deactivations in the DMN than expected from studies of young adults, particularly with increasing task difficulty (Grady et al. 2006; Lustig et al. 2003; Persson et al. 2007; Sambataro et al. 2010; Stevens et al. 2008). Older adults' decrease in deactivation of the DMN during goal directed activity has been conceptualized as a failure to efficiently shift neural resources from intrinsic neural processes to task-related processing (Sambataro et al. 2010). Importantly, research has suggested that successful deactivation of the DMN during cognitive tasks is associated with superior task performance (e.g., Stevens et al. 2008). Thus, the failure of task-related DMN activation observed in older adults may be one mechanism underlying cognitive declines associated with normal aging and neurodegenerative disorders such as AD.

2.7 Aging and the Apolipoprotein E ε4 Allele

Apolipoprotein E (APOE) is a protein that plays an important role in cholesterol metabolism and synaptogenesis in the brain (Beffert et al. 1998). The gene that codes for APOE has three variants, or alleles: ε2, ε3, and ε4. The ε4 allele has been consistently associated with a higher risk of developing Alzheimer's disease (Corder et al. 1993; Saunders et al. 1993). It has also been linked to the neuropathological hallmarks of AD, including both beta-amyloid plaques and neurofibrillary tangles (Namba et al. 1991). As a result, many studies have examined the APOE ε4 allele as a biomarker for AD in healthy older adults. The first functional neuroimaging study to examine normal aging and APOE was conducted by Small et al. (1995) using PET with a group of cognitively intact older adults with a family history of AD. They found that APOE ε4 carriers had lower parietal lobe metabolism than those without the allele, suggesting that APOE ε4 may impact neural activity even in the absence of effects on cognition.

Since the study of Small et al. (1995), fMRI has been applied to the investigation of memory-related activation in individuals with genetic or familial risk for AD. Bookheimer et al. (2000) were the first to use fMRI to demonstrate an APOE-related difference in activation patterns during a memory task. They found that ε4 carriers showed greater activation in left-hippocampal, parietal, and prefrontal regions during the memory encoding phase and greater hippocampal activation during the recall phase, which were interpreted as support for compensatory mechanisms among ε4 carriers. Several other fMRI studies have also demonstrated a pattern of increased neural activity in cognitively intact older adult ε4 carriers compared to non-carriers (e.g., Han et al. 2007; Lind et al. 2006; Wierenga et al. 2010; Bondi et al. 2005; for review, see Wierenga and Bondi 2007). For instance, one study examined the relationship between APOE genotype and BOLD response during recall of word paired-associates in non-demented older adults (Han et al. 2007) and found that the healthy older adult ε4 carriers showed greater magnitude and extent of BOLD brain response in the right hemisphere during recall of previously studied word pairs relative to their matched ε3 counterparts (see Fig. 1). APOE genotype was also found to influence the pattern and direction of association between hippocampal activity and learning and memory performance.

More recently, Seidenberg et al. (2009) examined the effect of family history of AD and the APOE ε4 allele on whole-brain fMRI neural activity in cognitively asymptomatic older adults using a semantic memory task involving the discrimination of famous from unfamiliar names. They found that those with the ε4 allele and a family history of AD more strongly recruited bilateral posterior cingulate/precuneus, bilateral temporoparietal junction, and bilateral PFC than participants without any AD risk factors during recall of familiar versus unfamiliar names (see Fig. 2). Moreover, in comparing participants with both the ε4 allele and family history of AD to those with just a family history, ε4 carriers exhibited more activation in right-middle frontal and right- supramarginal gyri. Seidenberg et al. (2009) results suggest that the APOE ε4 allele was uniquely associated with preferential activation of right hemisphere frontal regions. Finally, Woodard et al. (2010) conducted a prospective study that evaluated genetic risk, hippocampal volume, and fMRI activation in healthy older adults. Results revealed that greater fMRI activity, absence of an APOE ε4 allele, and larger hippocampal volume were associated with reduced likelihood of cognitive decline after 18 months. The results of these fMRI studies are consistent with both the

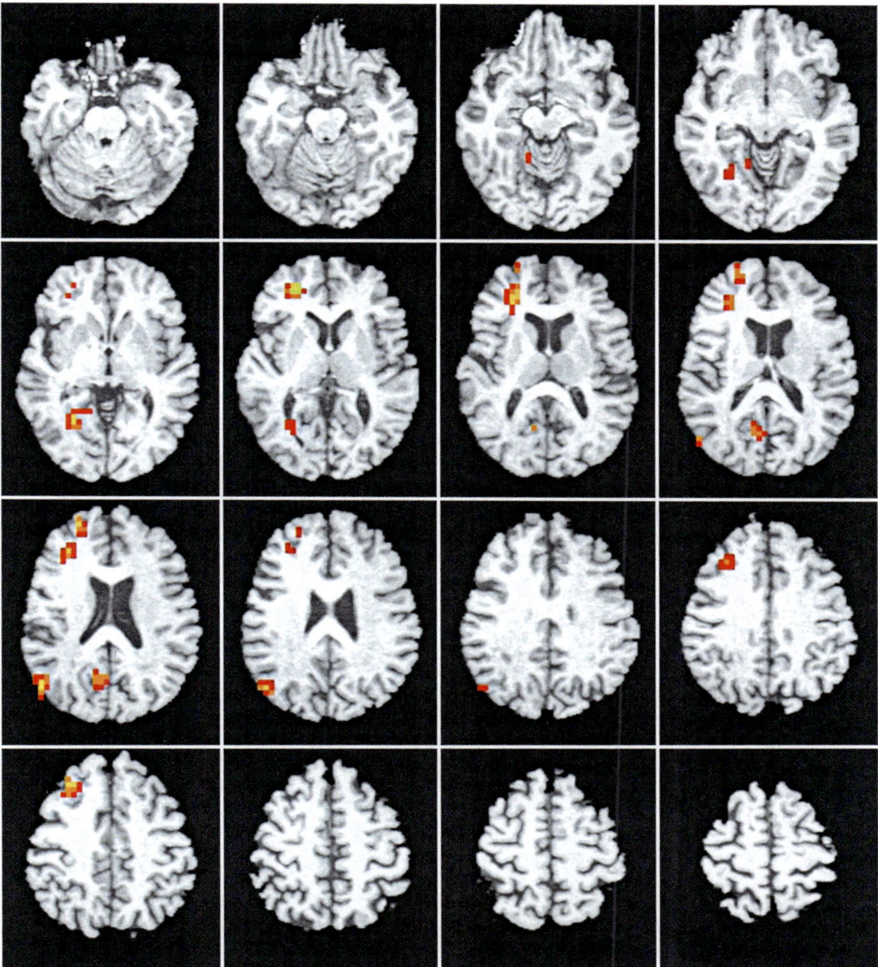

Fig. 1 Magnitude and direction of voxel-level activation to the task superimposed onto axial slices of a representative image in Taliarach space (slices span from 19 inferior to 56 superior in 4 mm increments). Activation displayed includes voxels significant at $P < 0.025$ that are contained within a cluster of 15 or more voxels. Color scale represents effect sizes for the between-subject difference between OLD items and FIXATION as measured by eta^2 (*red voxels*: $40 < \eta^2 < 60$; *orange voxels*: $60 < \eta^2 < 80$; *yellow voxels*: $80 < \eta^2 < 100$ [η^2 indexes the effect size for the magnitude of the difference between the observed response and 0]). Images are presented in radiological view (Han et al. 2007; reprinted with permission)

HAROLD model and the compensatory recruitment hypothesis, where older adult ε4 carriers appear to require additional cognitive effort to achieve comparable performance levels on tests of episodic memory encoding (Bookheimer et al. 2000; Filippini et al. 2011).

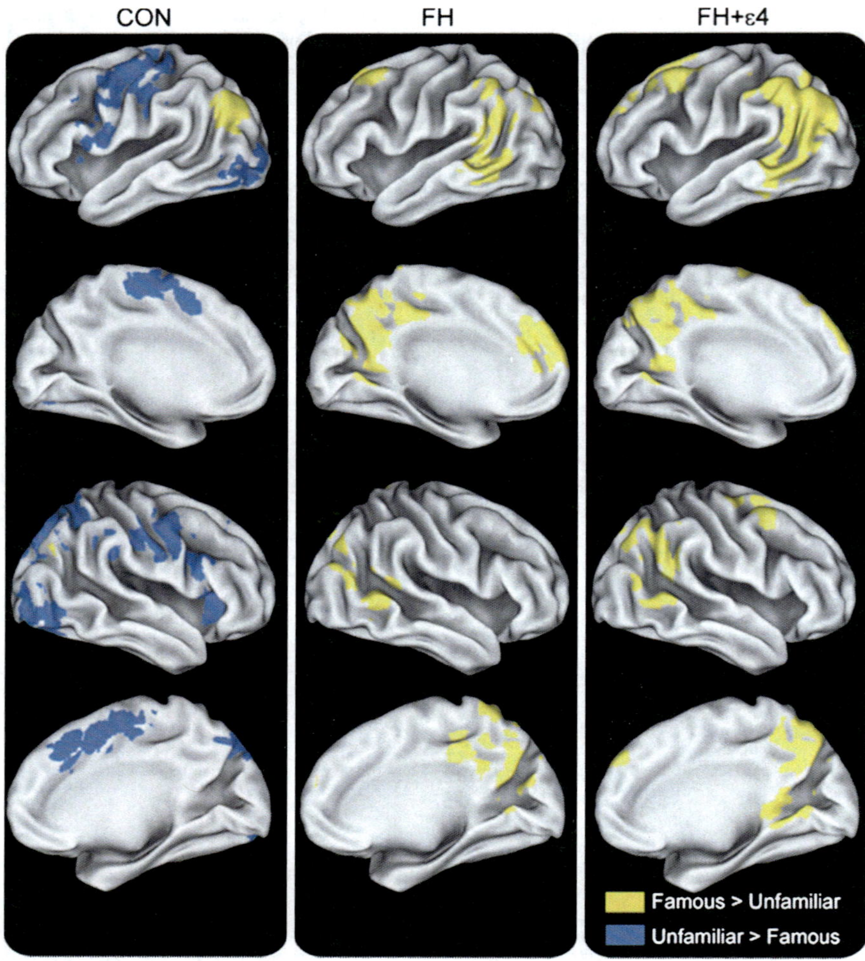

Fig. 2 Results of voxel-wise analysis demonstrating significant differences between the famous and unfamiliar name conditions, conducted separately for each group: control (*CON*), family history (*FH*), and family history and APOEε4 (*FH* + ε4) groups. *Yellow* regions showing greater activation to famous than unfamiliar names; *blue* regions showing greater activation to unfamiliar than famous names (Seidenberg et al. 2009; reprinted with permission)

While few studies have considered the effect of APOE across the lifespan, some recent research has suggested that the ε4 allele may actually confer a different effect in young carriers. Mondadori et al. (2007) studied the effect of APOE ε4 during an fMRI memory performance task in a very large group of young, healthy participants (mean age = 22.8) to find that ε4 carriers had decreased brain activation over three learning trials and retrieval, compared to ε2 and ε3 carriers. They concluded that APOE ε4 is associated with positive effects on episodic memory outside of the fMRI scanner and an economic use of memory-related neural

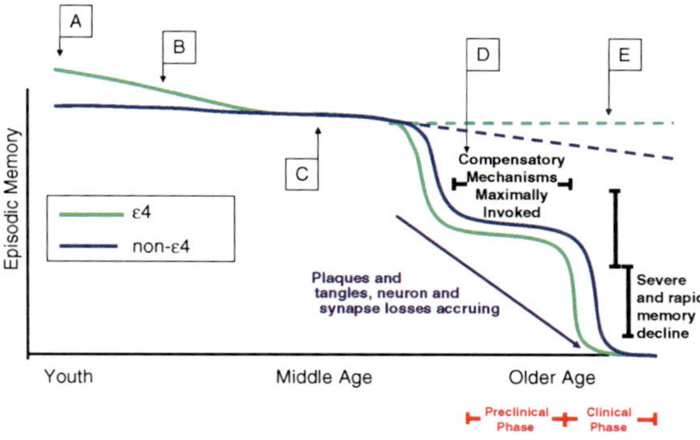

Fig. 3 The *APOE* antagonistic pleiotropy hypothesis (Han and Bondi 2008; Tuminello and Han 2011). (*A*) ε4 mediates a neurocognitive benefit very early in life. (*B*) ε4 participants recruit greater frontal-executive processes to display an advantage early in life. (*C*) Subtle cognitive genotypic differences, if at all. (*D*) ε4 subjects invoke right hemisphere frontal-executive regions to compensate for declines. (*E*) ε4 participants who do not develop dementia in old age perform better than non-ε4 participants

resources during the imaging task in young, healthy adults. Another study examined school-aged children and found that the APOE ε2 allele, normally viewed as a protective genetic trait in old age, may be associated with decreased functioning in certain cognitive domains and with atypical hemispheric dominance (i.e., left-handedness) (Bloss et al. 2010).

To reconcile these seemingly disparate findings, Han and Bondi (2008) offered a unifying hypothesis of the influence of APOE on cognitive and neural functioning under the theoretical umbrella of antagonistic pleiotropy (Williams 1957). Antagonistic pleiotropy is a concept from evolutionary biology purporting that certain genes may impact fitness (i.e., survival and reproduction) differently during different life stages. Han and Bondi (2008), along with other researchers (Alexander et al. 2007; Wright et al. 2003), argue for the notion that the APOE ε4 allele is a pleiotropic gene, such that young APOE ε4 carriers perform better on memory and other neurocognitive tasks than non-carriers, with this benefit becoming negligible by middle age (Bondi et al. 1999). By old age, Han and Bondi's (2008) model proposes that the APOE ε4 allele becomes detrimental, causing ε4 carriers to compensate for incipient cognitive declines by disproportionately invoking additional brain regions (see Fig. 3). Tuminello and Han (2011) recently re-evaluated Han and Bondi's model of the antagonistic pleiotropic effects of APOE in light of recent research, finding support for most components of the model. Furthermore, they propose that the APOE ε4 allele may again become beneficial in very-old age, based on research suggesting that oldest-old ε4 carriers show comparable or enhanced cognition compared to non-carriers (Carrion-Baralt et al. 2009; Kozauer et al. 2008; Welsh-Bohmer et al. 2009).

3 Limitations and Future Directions of fMRI Studies of Normal Aging

Despite the advancements that neuroimaging techniques have afforded us in the exploration of normal aging, there are several limitations to consider. One limitation inherent in the use of fMRI is the amount of physical noise that it produces. Studies have demonstrated that older adults are also more easily distracted than young adults. This raises the possibility that the distracting properties of fMRI scanning (i.e., the noise it produces) may disproportionately affect older adults, which may lead researchers to erroneously conclude that older adults show decrements in cognitive abilities when in less distracting environments they would not. A second limitation is that changes in the cerebrovascular system due to age or disease can significantly alter the BOLD signal and complicate its interpretation (D'Esposito et al. 2003). Only a few fMRI studies have investigated the impact of age on the neurovascular underpinnings of the BOLD signal (see Wierenga and Bondi 2007, for review). Such studies have found a significant age-related decrease in BOLD signal amplitude (Buckner et al. 2000; Tekes et al. 2005). Other studies examining the effect of age-related vascular changes on the BOLD signal have produced mixed findings. Two of three studies report that BOLD amplitude and refractory properties are similar in young and older adults (D'Esposito et al. 1999; Buckner et al. 2000), while other studies have reported that the BOLD signal has reduced signal–noise ratio in older adults (Huettel et al. 2001).

Most recently, it has been suggested that fMRI reveals an abnormal response during cognitive tasks in healthy older adults with increased vascular risk (Braskie et al. 2010). This study by Braskie et al. (2010) examined activation during a verbal memory task for healthy older adults and older adults with higher systolic blood pressure and body mass index, although these levels were still within the normal range. Results revealed activation differences in the frontal lobe, temporal lobe, precuneus, and posterior cingulate cortex in those with vascular risk factors. These results suggest that even slightly elevated cardiovascular risk is associated with changes in brain function during a memory task in the presence of comparable task performance. Such findings raise questions as to what acceptable levels of cardiovascular risk should be when recruiting older adults for studies of normal aging and suggest that this matter should be further explored. Neuroimaging applications in development such as functional arterial spin labeling (e.g., Bangen et al. 2009) may prove particularly relevant to addressing this confound.

Research on normal aging has begun to examine the impact of physical activity on cognition in older adults using fMRI technology (Smith et al. 2010). Researchers have also begun to combine imaging modalities to explore interrelationships among various indices of neural functioning (e.g., DTI, fMRI). Future work should include longitudinal examinations of the trajectory of neural and cognitive changes over time, as well as examining the impact of cerebrovascular risk factors on the BOLD response in older adults. Furthermore, in order to coherently study the effects of aging on cognition and neural activity, standardized criteria for interpreting functional neuroimaging findings as evidence for compensation are needed. Han et al. (2009) proposed such criteria, which they termed the Region-Activation-Performance model (RAP model). The RAP model proposes several guidelines for imaging researchers. Recommendations include careful consideration of (1) whole- brain patterns of activation and connectivity instead of isolated region of interest analyses, (2) the influences of structural segmentation values upon functional neuroimaging outcomes, (3) direct comparisons of contrasting activation patterns versus interpretation of isolated patterns of activation, (4) the effects of blood perfusion changes in old age upon the functional neuroimaging signal, and (5) designing a cognitive task that forces responses that are inherently compensatory in nature.

4 Conclusion

The use of functional neuroimaging techniques in aging research has proliferated since its advent and continues to expand at a rapid pace. However, there are many discrepancies in this literature and many areas that require more systematic investigation. In short, much has been learned about the impact of age on aspects of attention, executive functions, and memory; but there is still some lack of clarity in regards to how these changes are mediated by altered neural function. It also remains to be seen whether the enhanced patterns of activation seen in older adults are better accounted for by the compensatory recruitment hypothesis or the dedifferentiation hypothesis. Most importantly, there is a dearth of longitudinal studies that track cognitive and neural changes of participants as they age. As our understanding of normal aging from functional imaging research continues to expand, we can begin to explore the clinical utility of various imaging technologies with the ultimate goal of enhancing the early detection of pathological processes that disrupt normal aging (Furst and Mormino 2010) as well as intervention studies to track outcomes.

References

Alexander DM, Williams LM, Gatt JM, Dobson-Stone C, Kuan SA, Todd EG et al (2007) The contribution of apolipoprotein E alleles on cognitive performance and dynamic neural activity over six decades. Biol Psychol 75:229–238

Anderson ND, Iidaka T, McIntosh AR, Kapur S, Cabeza R, Craik FIM (2000) The effects of divided attention on encoding- and retrieval-related brain activity: a PET study of younger and older adults. J Cogn Neurosci 12(5):775–792

Andrews-Hanna JR, Snyder AZ, Vincent JL, Lustig C, Head D, Raichle ME, Buckner RL (2007) Disruption of large-scale brain systems in advanced aging. Neuron 56:924–935

Baltes PB (1987) Theoretical propositions of life-span developmental psychology: On the dynamics between growth and decline. Dev Psychol 23(5):611–626

Bangen KJ, Restom K, Liu TT, Jak AJ, Wierenga CE, Salmon DP, Bondi MW (2009) Differential age effects on cerebral blood flow and BOLD response to encoding: associations with cognition and stroke risk. Neurobiol Aging 30:1276–1287

Beason-Held LL, Kraut MA, Resnick SM (2008) Temporal patterns of longitudinal change in aging brain function. Neurobiol Aging 29:497–513

Beason-Held LL, Kraut MA, Resnick SM (2009) Stability of default-mode network activity in the aging brain. Brain Imaging Behav 3(2):123–131

Beffert U, Danik M, Krzywkowski P, Ramassamy C, Berrada F, Poirier J (1998) The neurobiology of apolipoproteins and their receptors in the CNS and Alzheimer's disease. Brain Res Rev 27(2):119–142

Bloss CS, Delis DC, Salmon DP, Bondi MW (2010) APOE genotype is associated with left-handedness and visuospatial skills in children. Neurobiol Aging 31(5):787–795

Bondi MW, Salmon DP, Galasko D, Thomas RG, Thal LJ (1999) Neuropsychological function and apolipoprotein E genotype in the preclinical detection of Alzheimer's disease. Psychol Aging 14:295–303

Bondi MW, Houston WS, Eyler LT, Brown GG (2005) fMRI evidence of compensatory mechanisms in older adults at genetic risk for Alzheimer's disease. Neurology 64:501–508

Bookheimer SY, Strojwas MH, Cohen MS, Saunders AM, Pericak-Vance MA, Mazziotta JC, Small GW (2000) Patterns of brain activation in people at risk for Alzheimer's disease. New Eng J Med 343:450–456

Braskie MN, Small GW, Bookheimer SY (2010) Vascular health risks and fMRI activation during a memory task in older adults. Neurobiol Aging 31:1532–1542

Broyd SJ, Demanuele C, Debener S, Helps SK, James CJ, Sonuga-Barke EJ (2009) Default-mode brain dysfunction in mental disorders: A systematic review. Neurosci Biobehav Rev 33: 279–296

Buckner RL, Snyder AZ, Sanders AL, Raichle ME, Morris JE (2000) Functional brain imaging of young, nondemented, and demented older adults. J Cogn Neurosci 12(suppl 2):24–34

Cabeza R (2001a) Cognitive neuroscience of aging: contributions of functional neuroimaging. Scand J Psychol 42:277–286

Cabeza R (2001b) Functional neuroimaging of cognitive aging. In: Cabeza R, Kingstone A (eds) Handbook of functional neuroimaging of cognition. MIT Press, Cambridge, pp 331–377

Cabeza R (2002) Hemispheric asymmetry reduction in old adults: The HAROLD model. Psychol Aging 17:85–100

Cabeza R, Nyberg L (1997) Imaging cognition: an empirical review of PET studies with normal subjects. J Cogn Neurosci 9:1–26

Cabeza R, Grady CL, Nyberg L, McIntosh AR, Tulving E, Kapur S, Jennings JM, Houle S, Craik FIM (1997) Age-related differences in neural activity during memory encoding and retrieval: a positron emission tomography study. J Neurosci 17(1):391–400

Cabeza R, Daselaar SM, Dolcos F, Prince SE, Budde M, Nyberg L (2004) Task- independent and task-specific age effects on brain activity during working memory, visual attention and episodic retrieval. Cereb Cortex 14(4):364–375

Carrion-Baralt JR, Melendez-Cabrero J, Rodriguez-Ubinas H, Schmeidler J, Beeri MS, Angelo G et al (2009) Impact of APOE epsilon4 on the cognitive performance of a sample of non-demented Puerto Rican nonagenarians. J Alzheimers Dis 18(3):533–540

Centers for Disease Control and Prevention and The Merck Company Foundation (2007) The State of Aging and Health in America, 2007. The Merck Company Foundation, Whitehouse Station, NJ

Chang C, Glover GH (2010) Time-frequency dynamics of resting-state brain connectivity measured with fMRI. NeuroImage 50:81–98

Chen P, Ratcliff G, Belle S, Cauley J, DeKosky S, Ganguli M (2001) Patterns of cognitive decline in presymptomatic Alzheimer disease: a prospective community study. Arch Gen Psychiatry 58(9):853–858

Chen N, Chou Y, Song AW, Madden DJ (2009) Measurement of spontaneous signal fluctuations in fMRI: Adult age differences in intrinsic functional connectivity. Brain Struct Funct 213(6):571–598

Cole DM, Smith SM, Beckmann CF (2010) Advances and pitfalls in the analysis and interpretation of resting-state fMRI data. Front Syst Neurosci 4:1–15

Corder EH, Sauders AM, Strittmatter WJ, Schmechel DE, Gaskell PC, Small GW et al (1993) Gene dose of apolipoprotein E type 4 allele and the risk of Alzheimer's disease in late-onset families. Science 261:921–923

Cordes D, Haugton VM, Arfanakis K, Wendt GJ, Turski PA, Moritz CH, Quigley MA, Meyerand ME (2000) Mapping functionally related regions of brain with functional connectivity MR imaging. Am J Neuroradiol 21:1636–1644

Cordes D, Haughton VM, Arfanakis K, Carew JD, Turski PA, Moritz CH, Quigly MA, Meyerand ME (2001) Frequencies contributing to functional connectivity in the cerebral cortex in 'resting-state' data. Am J Neuroradiol 22:1326–1333

Cummings JL (1993) Frontal-subcortical circuits and human behavior. Arch Neurol 50:873–880

D'Esposito M, Zarahn E, Aguirre GK, Rypma B (1999) The effect of normal aging on the coupling of neural activity to the BOLD hemodynamic response. NeuroImage 10:6–14

D'Esposito M, Deouell LY, Gazzaley A (2003) Alterations in the BOLD fMRI signal with ageing and disease: a challenge for neuroimaging. Nat Rev Neurosci 4:863–872

Damoiseaux JS, Greicius MD (2009) Greater than the sum of its parts: a review of studies combining structural connectivity and resting-state functional connectivity. Brain Struct Funct 213:525–533

Damoiseaux JS, Rombouts SA, Barkhof F, Scheltens P, Stam CJ, Smith SM, Beckmann CF (2006) Consistent resting-state networks across healthy subjects. Proc Natl Acad Sci U S A 103:13848–13853

Daselaar SM, Browndyke J, Cabeza R (2006) Functional neuroimaging of cognitive aging. In: Cabeza R, Kingstone A (eds) Handbook of functional neuroimaging of cognition, 2nd edn. MIT Press, Cambridge, pp 379–420

Denkova E, Botzung A, Manning L (2006) Neural correlates of remembering/knowing famous people: an event-related fMRI study. Neuropsychologia 44:2783–2791

Dennis NA, Cabeza R (2008) Neuroimaging of healthy cognitive aging. In: Craik FIM, Salthouse TA (eds) Handbook of aging and cognition, 3rd edn. Erlbaum, Mahwah

Douville K, Woodard J, Seidenberg M, Leveroni C, Nielson K, Franczak M, Antuono P, Rao S (2005) Medial temporal lobe activity for recognition of recent and remote famous names: an event-related fMRI study. Neuropsychologia 43:693–703

Erk S, Walter H, Abler B (2008) Age-related physiological responses to emotion anticipation and exposure. Neuroreport 19(4):447–452

Filippini N, Ebmeier KP, MacIntosh BJ, Trachtenberg AJ, Frisoni GB, Wilcock GK, Beckman CF, Smith SM, Matthews PM, Mackey CE (2011) Differential effects of the APOE genotype on brain function across the lifespan. NeuroImage 54:602–610

Furst AJ, Mormino EC (2010) A BOLD move: Clinical application of fMRI in aging. Neurology 74:1940–1941

Grady CL, Craik FIM (2000) Changes in memory processing with age. Curr Opin Neurobiol 10(2):224–231

Grady CL, Maisog JM, Horwitz B, Ungerleider LG, Mentis MJ, Salerno JA, Pietrini P, Wagner E, James V, Haxby JV (1994) Age-related changes in cortical blood flow activation during visual processing of faces and location. J Neurosci 14(3):1450–1462

Grady CL, McIntosh AR, Horwitz B, Maisog JM, Horwitz B, Ungerleider LG, Mentis MJ, Pietrini P, Schapiro MB, Haxby JV (1995) Age-related reductions in human recognition memory due to impaired encoding. Science 269:218–221

Grady CL, McIntosh AR, Craik FIM (2005) Task-related activity in prefrontal cortex and its relation to recognition memory performance in young and old adults. Neuropsychologia 43(10):1466–1481

Grady CL, Springer MV, Hongwanishkul D, McIntosh AR, Winocur G (2006) Age-related changes in brain activity across the adult lifespan. J Cogn Neurosci 18(2):227–241

Greicius MD, Krasnow B, Reiss AL, Menon V (2003) Functional connectivity in the resting brain: a network analysis of the default mode hypothesis. Proc Natl Acad Sci USA 100:253–258

Greicius MD, Srivastava G, Reiss AL, Menon V (2004) Default mode network activity distinguishes Alzheimer's disease from healthy aging: evidence from functional MRI. Proc Natl Acad Sci USA 101:4637–4642

Greicius MD, Supekar K, Menon V, Dougherty RF (2009) Resting state functional connectivity reflects structural connectivity in the default mode network. Cereb Cortex 19:72–78

Haist F, Bowden GJ, Mao H (2001) Consolidation of human memory over decades revealed by functional magnetic resonance imaging. Nat Neurosci 4(11):1139–1145

Han SD, Bondi MW (2008) Revision of the apolipoprotein E compensatory mechanism recruitment hypothesis. Alzheimers Dement 4:251–254

Han SD, Houston WS, Jak AJ, Eyler LT, Nagel BJ, Fleisher AS, Brown GG, Corey-Bloom J, Salmon DP, Thal LJ, Bondi MW (2007) Verbal paired-associate learning by APOE genotype in non-demented older adults: fMRI evidence of a right hemisphere compensatory response. Neurobiol Aging 28:238–247

Han SD, Delis DC, Holdnack JA (2008) Extending the WISC-IV: executive functioning. In: Prifitera A, Saklofske D, Weiss L (eds) WISC-IV: clinical assessment and intervention, 2nd edn. Elsevier Academic Press, Burlington, pp 497–512

Han SD, Bangen KJ, Bondi MW (2009) Functional magnetic resonance imaging of compensatory neural recruitment in aging and risk for alzheimer's disease: review and recommendations. Dement Geriatr Cogn Disord 27:1–10

Hedden T, Van Dijk KRA, Becker JA, Mehta A, Sperling RA, Johnson KA et al (2009) Disruption of functional connectivity in clinically normal older adults harboring amyloid burden. J Neurosci 29(40):12686–12694

Huettel SA, Singerman JD, McCarthy G (2001) The effects of aging upon the hemodynamic response measured by functional MRI. Neuroimage 13:161–175

Iidaka T, Okada T, Murata T, Omori M, Kosaka H, Sadato N et al (2002) Age-related differences in the medial temporal lobe responses to emotional faces as revealed by fMRI. Hippocampus 12(3):352–362

Jonker C, Geerlings MI, Schmand B (2000) Are memory complaints predictive for dementia? a review of clinical and population-based studies. Int J Psychiatry 15:983–991

Kastner S, Ungerleider LG (2000) Mechanisms of visual attention in the human cortex. Annu Rev Neurosci 23:315–341

Kensinger EA (2008) Emotional memory across the adult lifespan. Psychology Press, New York

Koch W, Teipel S, Mueller S, Benninghoff J, Wagner M, Bokde ALW et al (2010). Diagnostic power of default mode network resting state fMRI in the detection of Alzheimer's disease. Neurobiol Aging (in press)

Kozauer NA, Mielke MM, Chan GK, Rebok GW, Lyketsos CG (2008) Apolipoprotein E genotype and lifetime cognitive decline. Int Psychogeriatr 20(1):109–123

Lamar M, Yousem DM, Resnick SM (2004) Age differences in orbitofrontal activation: an fMRI investigation of delayed match and nonmatch to sample. NeuroImage 21(4): 1368–1376

Langenecker SA, Nielson KA, Rao SM (2004) fMRI of healthy older adults during stroop interference. Neuroimage 21(1):192–200

Leonards U, Sunaert S, Van Hecke P, Orban GA (2000) Attention mechanisms in visual search—an fMRI study. J Cogn Neurosci 12(Supp 2):61–75

Li S-C, Lindenberger U (1999) Cross-level unification: a computational exploration of the link between deterioration of neurotransmitter systems dedifferentiation of cognitive abilities in old age. In: Nilsson L-G, Markowitsch HJ (eds) Cognitive neuroscience of memory. Hogrefe & Huber, Seattle, pp 103–146

Lind J, Persson J, Ingvar M, Larsson A, Cruts M, Van Broeckhoven C, Adolfsson R, Backman L, Nilsson L, Petersson KM, Nyber L (2006) Reduced functional brain activity response in cognitively intact apolipoprotein E ε4 carriers. Brain 129:1240–1248

Logan JM, Sanders AL, Snyder AZ, Morris JC, Buckner RL (2002) Under-recruitment and nonselective recruitment: dissociable neural mechanisms associated with aging. Neuron 33:827–840

Lustig C, Synder AZ, Bhakta M, O'Brien KC, McAvoy M, Raichle ME et al (2003) Functional deactivations: change with age and dementia of the Alzheimer type. Proc Natl Acad Sci USA 100(24):14504–14509

Madden DJ, Turkington TG, Coleman RE, Provenzale JM, DeGrado TR, Hoffman JM (1996) Adult age differences in regional cerebral blood flow during visual word identification: evidence from H215O PET. Neuroimage 3:127–142

Madden DJ, Turkington TG, Provenzale JM, Hawk TC, Hoffman JM, Coleman RE (1997) Selective and divided visual attention: age-related changes in regional cerebral blood flow measured by H215O PET. Hum Brain Mapp 5:389–409

Madden DJ, Turkington TG, Provenzale JM, Denny LL, Hawk TC, Gottlob LR, Coleman RE (1999) Adult age differences in the functional neuroanatomy of verbal recognition memory. Hum Brain Mapp 7:115–135

Madden DJ, Whiting WL, Provenzale JM, Huettel SA (2004) Age-related changes in neural activity during visual target detection measured by fMRI. Cereb Cortex 14(2):143–155

Maguire EA, Frith CD (2003) Aging affects the engagement of the hippocampus during autobiographical memory retrieval. Brain 126:1511–1523

Milham MP, Erickson KI, Banich MT, Kramer AF, Webb A, Wszalek T et al (2002) Attentional control in the aging brain: insights from an fMRI study of the Stroop task. Brain Cognition 49(3):277–296

Mondadori CRA, de Quervain DJ-F, Buchmann A, Mustovic H, Wollmer MA, Schmidt CF et al (2007) Better memory and neural efficiency in young apolipoprotein E epsilon4 carriers. Cereb Cortex 17:1934–1947

Morcom AM, Fletcher PC (2007) Does the brain have a baseline? Why we should be resisting a rest. NeuroImage 37:1073–1082

Morcom AM, Good CD, Frackowiak RS, Rugg MD (2003) Age effects on the neural correlates of successful memory encoding. Brain 126:213–229

Murty VP, Sambataro F, Das S, Tan H, Callicott JH, Goldberg TE et al (2009) Age-related alterations in simple declarative memory and the effect of negative stimulus valence. J Cogn Neurosci 21(10):1920–1933

Namba Y, Tomonaga M, Kawasaki H, Otomo E, Ikeda K (1991) Apolipoprotein E immunoreactivity in cerebral amyloid deposits and neurofibrillary tangles in Alzheimer's disease and kuru plaque amyloid in Creutzfeldt-Jakob disease. Brain Res 541:163–166

Nielson KA, Langenecker SA, Garavan HP (2002) Differences in the functional neuroanatomy of inhibitory control across the adult lifespan. Psychol Aging 17:56–71

Nielson KA, Langenecker SA, Ross TJ, Garavan HP, Rao SM, Stein EA (2004) Comparability of functional MRI response in young and old during inhibition. Neuroreport 15(1):129–133

Nielson KA, Douville KL, Seidenberg M, Woodard JL, Miller SK, Antuono P, Franczak M, Rao SM (2006) Age-related functional recruitment during the recognition of famous names: an event-related fMRI study. Neurobiol Aging 27:1494–1504

Nyberg L, Cabeza R, Tulving E (1996) PET studies of encoding and retrieval: the HERA model. Psychon Bull Rev 3:135–148

Ochsner KN, Gross JJ (2005) The cognitive control of emotion. Trends Cogn Sci V 9(5):242–249

Persson J, Lustig C, Nelson JK, Reuter-Lorenz PA (2007) Age differences in deactivation: a link to cognitive control? J Cogn Neurosci 19(6):1021–1032

Petersen R, Smith G, Kokmen E, Ivnik R, Tangalos E (1992) Memory function in normal aging. Neurology 42:396–401

Piefke M, Onur OA, Fink GR (2010) Aging-related changes of neural mechanisms underlying visual-spatial working memory. Neurobiol Aging (in press)

Raichle ME, MacLeod AM, Snyder AZ, Powers WJ, Gusnard DA, Shulman GL (2001) A default mode of brain function. Proc Natl Acad Sci USA 98:676–682

Raz N (2000) Aging of the brain and its impact on cognitive performance: integration of structural and functional findings. In: Craik FIM, Salthouse TA (eds) Handbook of aging and cognition—II. Lawrence Erlbaum, Mahwah, pp 1–90

Reuter-Lorenz PA, Stanczak L, Miller A (1999) Neural recruitment and cognitive aging: two hemispheres are better than one, especially as you age. Psychol Sci 10:494–500

Salthouse T (1988) Initiating the formalization of theories of cognitive aging. Psychol Aging 3:3–16

Salthouse TA (2009) Decomposing age correlations on neuropsychological and cognitive variables. J Int Neuropsychol Soc 15:650–661

Sambataro F, Murty VP, Callicott JH, Tan H, Das S, Weinberger DR et al (2010) Age-related alterations in default mode network: impact on working memory performance. Neurobiol Aging 31(5):839–852

Saunders AM, Strittmatter WJ, Schmechel D, George-Hyslop PH, Pericak-Vance MA, Joo SH et al (1993) Association of apolipoprotein E allele epsilon 4 with late-onset familial and sporadic Alzheimer's disease. Neurology 43:1467–1472

Seidenberg M, Guidotti LM, Nielson KA, Woodard JL, Durgerian S, Antuono P, Zhang Q, Rao SM (2009) Semantic memory activation in individuals at risk for developing Alzheimer's disease. Neurology 73:612–620

Sheline YI, Raichle ME, Snyder AZ, Morris JC, Head D, Wang S et al (2010) Amyloid plaques disrupt resting state default mode network connectivity in cognitively normal elderly. Biol Psychiatry 67(6):584–587

Sliwinski M, Buschke H (1999) Cross-sectional and longitudinal relationships among age, cognition, and processing speed. Psychol Aging 14(1):18–33

Small G, Mazziotta J, Collins M, Baxter L, Phelps M, Mandelkern M et al (1995) Apolipoprotein E type 4 allele and cerebral glucose metabolism in relatives at risk for familial Alzheimer's disease. J Am Med Assoc 173(12):942–947

Smith GE, Bondi MW (2008) Normal aging, mild cognitive impairment, and Alzheimer's disease. In: Morgan JE, Ricker JH (eds) Textbook of clinical neuropsychology. Taylor & Francis, New York

Smith JC, Nielson KA, Woodard JL, Seidenberg M, Durgerian S, Antuono P, Butts A, Hantke N, Lancaster M, Rao SM (2010) Interactive effects of physical activity and APOE-e4 on BOLD semantic memory activation in healthy elders. Neuroimage 54(1):635–644

Sperling RA, LaViolette PS, O'Keefe K, O'Brien J, Rentz DM, Pihlajamaki M et al (2009) Amyloid deposition is associated with impaired default network function in older persons without dementia. Neuron 63(2):178–188

St. Jacques PL, Bessette-Symons B, Cabeza R (2009) Functional neuroimaging studies of aging and emotion: fronto-amygdalar differences during emotional perception and episodic memory. J Int Neuropsychol Soc 15(6):819–825

Stebbins GT, Carrillo MC, Dorfman J, Dirksen C, Desmond JE, Turner DA, Bennett DA, Wilson RS, Glover G, Gabrieli JDE (2002) Aging effects on memory encoding in the frontal lobes. Psychol Aging 17(1):44–55

Stevens WD, Hasher L, Chiew KS, Grady CL (2008) A neural mechanism underlying memory failure in older adults. J Neurosci 28(48):12820–12824

Tekes A, Mohamed MA, Browner NM, Calhoun VD, Yousem DM (2005) Effect of age on visuomotor functional MR imaging. Academy of Radiology 12:739–745

Tessitore A, Hariri AR, Fera F, Smith WG, Das S, Weinberger DR et al (2005) Functional changes in the activity of brain regions underlying emotion processing in the elderly. Psychiatry Res 139(1):9–18

Tulving E, Kapur S, Craik FIM, Moscovitch M, Houle S (1994) Hemispheric encoding/retrieval asymmetry in episodic memory: Positron emission tomography findings. Proc Natl Acad Sci USA 91:2016–2020

Tuminello ER, Han SD (2011) The apolipoprotein E antagonistic pleiotropy hypothesis: review and recommendations. Int J Alzheimers Dis. Article ID 726197, p 12. doi:10.4061/2011/726197.

van der Zwaag W, Francis S, Head K, Peters A, Gowland P, Morris P, Bowtell R (2009) fMRI at 1.5, 3 and 7 T: characterising BOLD signal changes. Neuroimage 47(4):1425–1434

Voss MW, Prakash RS, Erickson KI, Basak C, Chaddock L, Kim JS et al (2010) Plasticity of brain networks in a randomized intervention trial of exercise training in older adults. Front Aging Neurosci 2:1–17

Welsh-Bohmer KA, Ostbye T, Sanders L, Pieper CF, Hayden KM, Tschanz JT et al (2009) Neuropsychological performance in advanced age: influences of demographic factors and Apolipoprotein E: findings from the Cache County Memory Study. Clin Neuropsychol 23(1): 77–99

Wierenga CE, Bondi MW (2007) Use of functional magnetic resonance imaging in the early identification of Alzheimer's disease. Neuropsychol Rev 17:127–143

Wierenga CE, Stricker NH, McCauley A, Simmons A, Jak AJ, Chang Y, Delano-Wood L, Bangen KJ, Salmon DP, Bondi MW (2010) Increased functional brain response during word retrieval in cognitively intact older adults at genetic risk for Alzheimer's disease. Neuroimage 51(3):1222–1233

Williams GC (1957) Pleiotropy, natural selection, and the evolution of senescence. Evolution 11:398–411

Woodard JL, Seidenberg M, Nielson KA, Miller SK, Franczak M, Antuono P, Douville KL, Rao SM (2007) Temporally graded activation of neocortical regions in response to memories of different ages. J Cogn Neurosci 19:1113–1124

Woodard J, Seidenberg M, Nielson KA, Antuono P, Durgerian S, Guidotti LM, Zhang Q, Butts N, Hantke M, Lancaster M, Rao SM (2010) Prediction of cognitive decline in healthy older adults using fMRI. J Alzheimers Dis 21(3):871–885

Wright RO, Hu H, Silverman EK, Tsaih SW, Schwartz J, Bellinger D et al (2003) Apolipoprotein E genotype predicts 24-month Bayley scales infant development score. Pediatr Res 54: 819–825

Functional Magnetic Resonance Imaging in Aging and Dementia: Detection of Age-Related Cognitive Changes and Prediction of Cognitive Decline

John L. Woodard and Michael A. Sugarman

Abstract Functional magnetic resonance imaging (fMRI) allows for dynamic observation of the neural substrates of cognitive processing, which makes it a valuable tool for studying brain changes that may occur with both normal and pathological aging. fMRI studies have revealed that older adults frequently exhibit a greater magnitude and extent activation of the blood-oxygen-level-dependent signal compared to younger adults. This additional activation may reflect compensatory recruitment associated with functional and structural deterioration of neural resources. Increased activation has also been associated with several risk factors for Alzheimer's disease (AD), including the apolipoprotein ε4 allele. Longitudinal studies have also demonstrated that fMRI may have predictive utility in determining which individuals are at the greatest risk of developing cognitive decline. This chapter will review the results of a number of task-activated fMRI studies of older adults, focusing on both healthy aging and neuropathology associated with AD. We also discuss models that account for cognitive aging processes, including the hemispheric asymmetry reduction in older adults (HAROLD) and scaffolding theory of aging and cognition (STAC) models. Finally, we discuss methodological issues commonly associated with fMRI research in older adults.

Keywords fMRI · Aging · Cognitive decline · Alzheimer's disease

---

Both authors contributed equally to this work and share lead authorship.

J. L. Woodard (✉) · M. A. Sugarman
Department of Psychology, Wayne State University,
5057 Woodward Ave., 7th Floor, Detroit, MI 48202, USA
e-mail: john.woodard@wayne.edu

M. A. Sugarman
e-mail: msugarman5@gmail.com

Curr Topics Behav Neurosci (2012) 10: 113–136
DOI: 10.1007/7854_2011_159

Published Online: 16 September 2011

Contents

1 Age-Related Increases in Task-Related fMRI Activity 115
1.1 The HAROLD Model 115
1.2 Explanations for the HAROLD Model 115
1.3 Compensatory Recruitment in Disease and in Risk for Disease 117
1.4 A Unified Theory Accounting for Task-Related Brain Activation Across the Life span 117
2 fMRI as a Biomarker of Cognitive Decline: Differential Risk-Related Activation Patterns 118
2.1 Non-imaging Biomarkers 118
2.2 Neuroimaging Biomarkers 119
2.3 Cross-Sectional fMRI Studies of Risk of Cognitive Decline 120
2.4 Longitudinal fMRI Studies of Risk of Cognitive Decline 121
2.5 fMRI Studies of Prediction of Conversion from MCI to AD 123
3 The Default Mode Network and Pathological Aging 124
4 Methodological Issues with fMRI and Aging 125
4.1 Atrophy Correction 125
4.2 Task Performance 126
4.3 The Nature of the Control Task 126
4.4 Choice of Memory Task During fMRI 127
5 Conclusions and Future Directions 128
References 130

Over the last decade, there has been rapid growth in the application of functional magnetic resonance imaging (fMRI) to study the neural bases of age- and disease-related cognitive changes. fMRI permits a view of in vivo brain changes that may underlie age-related alterations in cognitive processing. Using this technology, these changes in cognition across the life span can now be linked to observable alterations in brain function. For example, differential patterns of neural recruitment have been observed in older compared to younger adults on a wide variety of cognitive tasks, even when performance on these tasks is equivalent. However, the nature of exactly what "increased activation" represents in older adults is still an unresolved question. fMRI has also been used to visualize the impact of degenerative conditions, such as mild cognitive impairment (MCI) and Alzheimer's disease (AD), on brain function and its relationship with cognitive decline. In addition, because it is particularly sensitive to early changes in neural function that may underlie cognitive decline, fMRI activation has recently been used as a non-invasive biomarker of the risk of cognitive decline and of converting from MCI to AD. Several recent studies have noted the potential value of fMRI in predicting development of late-life cognitive decline and dementia above and beyond other methodologies, including structural MRI (sMRI), cerebrospinal fluid (CSF) analyses, and genetic testing. If fMRI continues to show promise for early identification of those who are at the highest risk for cognitive decline, the efficacy of targeted and preventative interventions could be greatly enhanced.

In this chapter, we will review the recent behavioral neuroscience literature associated with cognitive aging in an attempt to consider interpretations of age- and disease-related increases in brain activation. We will also review the recent studies that have used fMRI to assess the likelihood of conversion from MCI into

AD as well as studies that have used fMRI with healthy older adults to predict the likelihood of cognitive decline. Next, we will cover methodological issues that are relevant to studying older individuals with functional imaging. Finally, we will conclude with the possible future directions associated with the use of this technology with older adults.

1 Age-Related Increases in Task-Related fMRI Activity

1.1 The HAROLD Model

fMRI measures neural activation using the blood-oxygen-level-dependent (BOLD) signal at high spatial resolution, and as such is an effective tool for studying age-related cerebral blood flow changes induced during cognitive processing. A main finding from early fMRI studies of memory, perception, and inhibitory control is that younger adults tend to display greater hemispheric lateralization in prefrontal activity than older adults (Cabeza 2002). For example, during a semantic encoding task (identifying single words as abstract or concrete), a left frontal lobe activation bias was observed in younger subjects, while bilateral frontal lobe activation was seen in older adults (Stebbins et al. 2002). Using a "Go/No-Go" task, predominantly right hemispheric prefrontal and parietal activation was elicited during inhibitory control trials in younger subjects, whereas older adults displayed more extensive bilateral frontal and parietal activation (Nielson et al. 2002). This age-related change in hemispheric lateralization during cognitive tasks was termed "hemispheric asymmetry reduction in older adults," or the HAROLD model (Cabeza 2002).

1.2 Explanations for the HAROLD Model

Several plausible explanations have been proposed for this differential lateralization of cerebral involvement across the life span (Cabeza 2002). One possibility is that bilateral activation in older adults reflects a compensatory mechanism for age-related declines in neurocognitive mechanisms, in which additional brain regions are recruited for tasks that younger adults are successfully able to accomplish using a single hemisphere. An alternative (but not mutually exclusive) explanation suggests that there is reduced specialization of neural modules with age (dedifferentiation), hence the decrease in hemispheric specificity in task-activated neural recruitment. Additional support for this view comes from evidence that correlations between general cognitive abilities are greater in older adults (Babcock et al. 1997; Baltes and Lindenberger 1997), since (if the assumptions of the dedifferentiation model are correct) a broader array of neural mechanisms is recruited during completion of cognitive tasks. Additional interpretations of the HAROLD model include age-related changes in cognitive strategies and task-related neural networking, as well as lateralized connectivity changes across the life span (Li et al. 2009).

Numerous studies of age-related differences on episodic memory tasks have provided considerable support for the compensatory recruitment account. In a longitudinal study investigating changes in episodic memory, older participants who demonstrated the greatest decline on tests of episodic memory at follow-up also displayed greater right ventral frontal cortex activation during episodic encoding than older individuals who remained stable (Persson et al. 2006). This finding suggests that increased frontal activation is accentuated in those at the greatest risk for cognitive decline and likely reflects compensatory recruitment. In another study (Gutchess et al. 2005), patterns of neural recruitment were examined in younger and older adults during a semantic scene-encoding task. They found greater activation in the parahippocampal gyrus (PHG) in younger adults and greater activation in middle frontal cortical regions in older adults. Additionally, PHG activation correlated inversely with interior frontal regions only in older adults, suggesting that the additional frontal activation may be a compensatory response for declining function in hippocampal areas. Finally, in a study that combined diffusion tensor imaging (DTI) with resting and task-activated fMRI, older adults demonstrated a bilateral frontal pattern of activation during a verbal working memory task, while younger adults exhibited lateralized left frontal activation (Li et al. 2009). Relative to younger participants, older adults demonstrated a decrease in left prefrontal activation with a corresponding increase in right prefrontal activation. This study suggested that age-related alterations in connectivity may underlie the functional activation changes seen in older relative to younger adults.

Increased magnitude and extent of activation in older adults relative to younger adults has been observed not only with episodic and working memory tasks, but also with semantic memory and repetition priming tasks. In one study, performance of older and younger adults was contrasted during a famous/non-famous name discrimination task (Nielson et al. 2006). Older adults demonstrated greater magnitude and extent of activation in 15 of 20 brain regions, whereas younger adults did not show greater activation magnitude in any region. Moreover, extensive left prefrontal activation was observed in the older adults, consistent with previous studies highlighting the role of prefrontal cortex in aging (DiGirolamo et al. 2001; Grady et al. 1995, 2005; Langenecker and Nielson 2003; Nielson et al. 2002, 2004). Another study using a repetition priming paradigm found that older adults exhibited less repetition-related reduction of activation in ventral and anterior left inferior prefrontal cortex than younger adults (Bergerbest et al. 2009). In addition, older adults who showed greater repetition-related activation reductions in right prefrontal cortex demonstrated better repetition priming and performance on semantic memory tests.

1.2.1 Under-Recruitment versus Compensatory Recruitment

It should be noted that increased activation in older adults is not universally seen during all cognitive tasks. Under-recruitment of frontal regions has been reported during tasks that may require self-initiated implementation of effortful organizational strategies that are typically adopted naturally by younger adults, such as verbal episodic encoding (Logan et al. 2002). Importantly, unlike increased recruitment of brain regions not typically engaged by younger adults, under-recruitment can be reversed through the adoption of explicit task strategies (e.g., clustering strategies and semantic elaboration) as well as instruction in using a specific cognitive approach (Knoke et al. 1998).

1.3 *Compensatory Recruitment in Disease and in Risk for Disease*

Compensatory recruitment of neural resources has also been reported to occur in early cognitive impairment. MCI patients who had progressed to AD after three years displayed a stronger linear association between task difficulty and neural activation during a visuospatial angle discrimination task in the left superior parietal lobules and left precuneus than patients who remained stable (Vannini et al. 2007). This increased activation in response to task demands may reflect compensatory activity resulting from reduced neural network efficiency and appears to be associated with subsequent decline. Not only can enhanced brain activity be seen in early stages of cognitive impairment, but apparent compensatory recruitment has also been reported in a 20-year-old, largely asymptomatic carrier of the presenilin I gene (Mondadori et al. 2006), a deterministic gene which leads to certain development of AD at a relatively young age. The young mutation carrier demonstrated increased activation in left frontal, temporal, and parietal cortices during detection of novel stimuli and learning and retrieval tasks relative to age-matched controls. A 45-year-old mutation carrier, who was closer to the typical age of onset of AD in this family (48 years), demonstrated significantly weaker medial temporal activation relative to age-matched controls during episodic memory tasks. Importantly, these activation differences between the younger and older mutation carriers were not observed during an fMRI task involving working memory, suggesting that the type of memory task used for preclinical detection is an important consideration.

1.4 *A Unified Theory Accounting for Task-Related Brain Activation Across the Life span*

The observation of increased brain activation during task-related fMRI in older adults contrasts with physical brain changes associated with aging, such as volume loss and ventricular dilatation (Matsumae et al. 1996), neuronal shrinkage, reductions of synapse numbers and synaptic spines, and lower numbers of synapses (Fjell and Walhovd 2010), and reductions in white matter integrity and microstructure (Barrick et al. 2010; Michielse et al. 2010). In other words, if brain volume is reduced and brain structure is disturbed, why would increased brain activation be observed? The scaffolding theory of aging cognition (STAC) was recently introduced to address this counterintuitive observation (Park and Reuter-Lorenz 2009). Briefly, STAC proposes that recruitment of additional neural circuits in response to increasing task demands occurs across the life span in order to support structures whose function has become impaired, inefficient, or both. Scaffolding may not be necessary unless task demands exceed the ability of the existing structures to respond to a challenge. A central tenet of STAC holds that the ability to rely on or recruit secondary networks may reflect an essential component of healthy cognitive aging. Such scaffolding may occur effortlessly and efficiently in response to task challenges in younger individuals. In older individuals, scaffolding may be necessary to perform even relatively basic tasks that have become more challenging due to accumulated deterioration of neural circuitry. New scaffolds can be created, or existing scaffolds that were developed early in life or in response to new learning can be recruited. This recruitment of additional neural circuits in response to challenging task demands is therefore reflected by increased regional brain activation, predominately in the frontal cortex.

2 fMRI as a Biomarker of Cognitive Decline: Differential Risk-Related Activation Patterns

AD is the most common form of dementia (Kalaria et al. 2008). Initiation of the neuropathological processes associated with AD likely occurs decades before overt symptoms appear (Kok et al. 2009). Therefore, interventions given after symptom onset are unlikely to have a meaningful impact on affecting the course of the disease because irreversible brain damage has already occurred. However, interventions given prior to the appearance of symptoms could have a greater potential to either prevent or delay symptom onset or to at least slow the progression of the disease. Detection of individuals who are at the very highest risk of developing AD using preclinical biomarkers would maximize the public health benefits of early interventions or preventative strategies.

2.1 Non-imaging Biomarkers

Given the current lack of effective treatments for AD, a growing body of research has been dedicated toward the study of promising biomarkers of AD (Clark et al. 2008; Daviglus et al. 2010; Reiman et al. 2010). Several biomarker studies have demonstrated success in predicting conversion from the acronym that follows MCI to AD using neuropsychological testing (Albert et al. 2001; De Jager et al. 2003; DeCarli et al. 2004; Nestor et al. 2004) and cerebrospinal fluid (CSF) indices, including elevated isoprostane (Brys et al. 2009; de Leon et al. 2006, 2007), elevated phosphorylated tau ($ptau_{181}$) and total tau (Buerger et al. 2002a, b; Hampel et al. 2004a), and low $A\beta_{42}$ levels (Blennow and Hampel 2003; Brys et al. 2009; Hampel et al. 2004b; Hansson et al. 2006). CSF $tau_{181}/A\beta_{42}$ and $ptau_{181}/A\beta_{42}$ ratios (Fagan et al. 2007) and absolute levels of $ptau_{181}$ and $A\beta_{42}$ (De Meyer et al. 2010) have shown promise for predicting cognitive decline in otherwise healthy older adults. However, the relative invasiveness and cost of CSF approaches may limit their utility for widespread use.

2.2 Neuroimaging Biomarkers

Less invasive neuroimaging techniques may provide more practical alternatives for identifying cognitively intact older adults at risk for future cognitive decline. Several studies have successfully identified biomarkers of dementia utilizing neuroimaging approaches, including sMRI (Cardenas et al. 2003; de Leon et al. 1989; Devanand et al. 2007; Henneman et al. 2009; Jack et al. 1999; Juottonen et al. 1998; Morra et al. 2009; Stoub et al. 2010; Wolf et al. 2003) and positron emission tomography (PET), involving regional glucose metabolism (Chetelat et al. 2003, 2005) and amyloid imaging with the ^{11}C Pittsburgh Compound B (PIB) (Rowe et al. 2007; Wolk and Klunk 2009; Wolk et al. 2009). Although these approaches are somewhat less invasive than CSF approaches, they are still costly and have disadvantages. For example, because detection of atrophy using sMRI requires the documentation of a change in volume over time, at least two sessions separated by an interval of at least several months would be needed to be able to establish such a change. Manual tracing can be time consuming, and automated and semi-automated methods of structural tracing are still subject to error (Morey et al. 2009; Tae et al. 2008). PET using ^{11}C PIB or fluorodeoxyglucose (FDG) involves exposure to ionizing radiation.

fMRI has the benefit of being minimally invasive, widely available, and potentially less labor intensive compared to other biomarker approaches. Given that it can serve as a “cognitive stress test,” fMRI has the potential to reveal possible abnormalities during cognitive performance and may be sensitive to disease-related changes. As such, fMRI has recently proven to be useful as a tool for detecting patterns of activation that may be biomarkers of subsequent cognitive decline or dementia, even in the absence of cognitive impairment, particularly in individuals who are at genetic risk of developing AD. For example, a history of dementia in first-degree relatives (Fratiglioni et al. 1993) and possession of one or more apolipoprotein E (APOE) $\varepsilon 4$ alleles (Bertram and Tanzi 2008; Corder et al. 1993; Saunders et al. 1993) are two well-known risk factors for AD. The APOE $\varepsilon 4$ allele has also been implicated in an increased risk for late-life cognitive decline (Caselli et al. 2004; Caselli et al. 2007; Swan et al. 2005). fMRI research has demonstrated that these genetic risk factors can influence neural activation patterns prior to the onset of cognitive decline and symptoms of dementia.

2.3 Cross-Sectional fMRI Studies of Risk of Cognitive Decline

APOE genotype and family history have been shown to exert separate and interacting effects on patterns of activation in healthy, cognitively intact middle-aged adults (mean age = 54 years; SD = 6.4 years) (Johnson et al. 2006). In this study, participants performed an episodic encoding task in which they had to identify a standardized set of line drawings as novel or previously learned. Individuals without a parental history of AD displayed greater activation in bilateral fusiform, hippocampus, and amygdala. However, in the hippocampus, they observed the greatest level of activation in APOE ε4 carriers without a family history of AD, and the lowest activation in APOE ε4 carriers with a family history. There was no difference in the degree of hippocampal activation between APOE ε4 non-carriers with or without a family history of dementia. This interaction indicates that family history and APOE genotype can influence brain functioning long before the onset of AD symptoms. A follow-up study that used a similar task (identification of previously viewed or novel faces instead of line drawings) found greater activation in left dorsal posterior cingulate and precuneus during recognition of previously viewed items for individuals at low risk (negative family history or APOE ε4 negative) for developing AD (Xu et al. 2009). APOE ε4 positive persons or individuals with a positive family history of AD did not show increased BOLD responses in any brain region relative to the negative risk groups. This study also found that overall recognition performance correlated strongly with BOLD signal intensity in left posterior hippocampus, parahippocampal-retrosplenial cortex, and left superior frontal cortex, regardless of risk group. Interestingly, APOE ε4 negative individuals demonstrated a stronger signal in the anterior cingulate cortex than APOE ε4 carriers, while persons without a family history of dementia showed a stronger BOLD signal in the dorsal cuneus relative to persons with a family history of dementia. This study suggests that family history and APOE ε4 status may have regionally independent effects on BOLD response during cognitive tasks.

It appears that these risk factors for AD influence brain functioning throughout life. One study (Filippini et al. 2009) demonstrated that increased activation associated with the APOE ε4 allele can be observed even in young adults (age 20–35). Eighteen ε4 carriers and 18 controls completed an episodic encoding task in which they discriminated between novel and previously learned pictures of landscapes and animals. The ε4 carriers displayed greater activation than controls in four of eight hippocampal regions. In a cross-sectional study of the role of combined risk factors on fMRI activation in healthy participants ranging in age from 18 to 84 (Trivedi et al. 2008), an interaction was observed between age and risk factors. Using the previously described line drawing encoding task (Johnson et al. 2006), the authors found that hippocampal activation increased strongly with age in APOE ε4 carriers with family history of AD, while hippocampal activation decreased significantly in participants without these risk factors. A more modest increase in activation with age was observed in participants with only one of these two risk factors. These results support the notion that the presence of risk factors may influence brain functioning in a dose-dependent fashion, even prior to the onset of clinical impairment.

One group (Bondi et al. 2005; Han et al. 2007, 2008) has proposed that the increased activation observed in asymptomatic carriers of the APOE ε4 allele represents compensatory recruitment. They propose that the compensatory mechanism includes recruitment of additional brain regions and different patterns of activation in order to maintain or improve cognitive performance. This Region-Activation-Performance model (Han et al. 2008) infers that the function of the increased signal is to combat cognitive impairment in the presence of declining neural resources, and may be indicative of an increased risk for subsequent cognitive decline. Perhaps the early, more intensive, and possibly less efficient utilization of these neural structures by APOE ε4 carriers may make them more susceptible to later degeneration and AD neuropathology. However, increased activation is not always observed in asymptomatic ε4 carriers. Decreased activation in APOE ε4 carriers compared to APOE ε4 non-carriers has been reported in young adults during an episodic memory task (Mondadori et al. 2007), and the presence of the APOE ε4 allele was actually a predictor of better memory performance in this sample. Increased regional activation was positively related to episodic memory performance in APOE ε4 carriers but was negatively related to performance in APOE ε4 non-carriers. This study suggests that the APOE ε4 allele may have beneficial effects in younger adults but may lead to deleterious effects in late-life. Some studies have even found no differences in BOLD signal between carriers and non-carriers, although this variability in study findings appears to be at least partially attributable to differences in tasks, regions studied, and whether not family history of dementia was taken into account (Trachtenberg et al. 2010). Overall, it appears that risk factors, such as the APOE ε4 allele, affect brain functioning across the life span, although they likely have multiple complex relationships with age-related cognitive performance and its underlying neurophysiology.

2.4 Longitudinal fMRI Studies of Risk of Cognitive Decline

One of the first longitudinal studies using task-activated fMRI as a predictor of cognitive decline (Bookheimer et al. 2000) reported that an increased number and spatial extent of activated brain regions at baseline can predict memory decline after a two-year retest interval. In this study of cognitively normal participants, 16 carriers of the APOE ε4 allele and 14 non-carriers, all between 47 and 82 years of age, underwent task-activated fMRI during memorization and recall of unrelated word pairs. The APOE ε4 carriers demonstrated a greater magnitude and extent of activation during learning and recall relative to rest compared to non-carriers. Participants were also administered a battery of neuropsychological measures of memory [Buschke-Fuld Selective Reminding Test (Buschke and Fuld 1974),

Logical Memory subtest from the Wechsler Memory Scale (Wechsler 1945), and the Benton Visual Retention Test (Benton and Hamsher 1976)]. Two years later, eight APOE ε4 carriers and six non-carriers were given the same battery of neuropsychological measures. The number of regions of interest showing significant activation at baseline demonstrated a −0.65 correlation with the degree of decline in verbal recall, defined as follow-up minus baseline performance on the Consistent Long-Term Retrieval index of the Buschke-Fuld selective reminding test. That is, participants (irrespective of APOE ε4 status) who demonstrated greater baseline activation during task-activated fMRI showed greater decline on consistent long-term retrieval performance relative to participants with less baseline fMRI activation. No group differences were observed on the other two memory measures or on any other summary measure from the Selective Reminding Test. Despite several limitations of this study (e.g., task performance during fMRI was not directly measured, rest was used as a control condition, and two-year follow-up was conducted on only 14/30 (46.6%) of participants), this study was the first to suggest that task-activated fMRI could have merit for prediction of memory decline.

There have been a limited number of subsequent longitudinal studies that have used fMRI to predict the risk of cognitive decline. Genetic risk in middle-aged women (family history of AD and at least one APOE ε4 allele) has been associated with decreased fMRI activation in extrastriate and posterior inferotemporal cortex at baseline, together with further decrease after four years in these regions as well as left inferior frontal and premotor cortex (Smith et al. 2005). However, there was no evidence of cognitive decline attributable to AD risk in this study. In contrast, using a word categorization task during fMRI with APOE ε4 carriers, nine older adults showing cognitive stability on episodic memory testing after five years demonstrated increased left inferior parietal activation at baseline relative to nine participants who demonstrated episodic memory decline; greater BOLD fMRI response in this region was associated with better memory performance after five years (Lind et al. 2006). However, no hippocampal volume differences were observed at baseline between stable and declining participants. Taken together, these two studies suggest that decreased baseline fMRI activation is associated with a decline in future functional brain activity (Smith et al. 2005) and/or cognitive decline (Lind et al. 2006).

Another recent longitudinal study evaluated 78 healthy, cognitively intact older adults aged 65 years and over at baseline and after 18 months (Woodard et al. 2010). All participants performed well within normal limits on baseline neuropsychological evaluation. At baseline, participants also underwent APOE ε4 genotyping, manually traced hippocampal volume measurement and task-activated fMRI during a semantic famous/non-famous name discrimination task. After 18 months, participants returned for additional neuropsychological evaluation. Approximately 35% of participants had undergone a one standard deviation or greater decline on one or more neuropsychological measures. Logistic regression analyses revealed that decreased baseline fMRI activation, smaller hippocampal volume at baseline, and presence of the APOE ε4 allele were predictive of cognitive decline over 18 months. The most effective combination for predicting future cognitive decline was fMRI activity in cortical and hippocampal regions and APOE ε4 status. That is, participants who demonstrated cognitive decline at 18-month follow-up were more likely to be APOE ε4 positive and tended to display lower levels of hippocampal and cortical activation at baseline than those who remained stable. APOE ε4 carriers displayed more activation than non-carriers at baseline, but within each APOE group, greater activation was associated with a lower probability of decline.

The results of the aforementioned studies introduce seemingly contradictory information. The previously described study (Woodard et al. 2010) observed that greater BOLD activation at baseline during a semantic memory task was *protective* against subsequent decline. However, several studies (Bondi et al. 2005; Filippini et al. 2009; Han et al. 2008; Trivedi et al. 2008; Seidenberg et al. 2009a). Semantic memory activation have found that possession of the APOE ε4 allele, a major risk factor for cognitive decline, is also associated with greater activation prior to the onset of cognitive impairment. In APOE ε4 positive persons, underlying neural structures supporting cognition may be compromised, and a greater amount of compensatory recruitment may be necessary to maintain cognitive performance. Therefore, rather than being contradictory, the finding of increased functional brain activity and maintenance of (rather than loss of) cognitive function might reflect an adaptive response to the neurobiological challenges associated with aging and disease and may represent the brain's way of protecting itself against cognitive decline using compensatory mechanisms [e.g., the previously described STAC theory (Park and Reuter-Lorenz 2009)].

2.5 fMRI Studies of Prediction of Conversion from MCI to AD

The issue of using the BOLD response as a biomarker for AD becomes further complicated in studies examining MCI. Many longitudinal studies have attempted to identify factors that may help to differentiate between MCI patients who will develop AD from those who will remain stable. Identification of factors associated with risk of further decline can provide valuable information to patients and families and facilitate protective interventions. Several of these studies have utilized fMRI and have reported that increased cortical and hippocampal activation in MCI patients may be predictive of further cognitive decline and dementia. For the purposes of succinctly illustrating the utility of fMRI as a biomarker, this discussion will center on hippocampal activation, a main focus for MCI and AD research.

One study (Woodard et al. 2009) observed that MCI patients display greater hippocampal activation during a semantic memory task involving discrimination of famous names from unfamiliar names relative to age-matched, cognitively intact controls without family history of dementia or the APOE ε4 allele. Another study (Dickerson et al. 2005) compared MCI patients to both healthy controls and patients with probable AD during a face/name associative encoding task. They found that the MCI patients had the greatest hippocampal activation out of the three groups. Other studies have suggested that elevated hippocampal activation in the early phases of dementia may sometimes be associated with a poorer prognosis. In a longitudinal study of conversion from MCI to AD (Miller et al. 2008), MCI patients with greater hippocampal activation during episodic encoding were at a higher risk of developing further cognitive decline at 4-year follow-up. Another longitudinal study (O'Brien et al. 2010) followed participants in the prodromal phase of dementia (displaying minor clinical symptoms but not meeting diagnostic criteria for MCI) and cognitively intact controls. fMRI scans and neuropsychological testing were performed at baseline and two-year follow-up. Interestingly, participants with the highest activation at baseline were most likely to demonstrate decreases in hippocampal activation during a face-name associative memory task and neuropsychological test performance. Finally, a recent study suggested that hippocampal activation in MCI patients is compensatory in nature and is needed to support successful memory encoding (Kircher et al. 2007). Overall, these studies suggest that increased hippocampal activation may occur in MCI, and this hyperactivation appears to be compensatory in nature, but may also be a harbinger of subsequent decline in BOLD activation and cognitive functioning that are associated with progression to AD (Furst and Mormino 2010).

3 The Default Mode Network and Pathological Aging

When not engaged in any specific task, the brain is far from "at rest". In fact, several structures typically display fMRI BOLD activation during passive states, including the hippocampus and medial temporal lobes, regions in the frontal and parietal lobes, and posterior cingulate cortex. However, this "default mode" network (also known as "resting state") is typically deactivated once a person becomes involved in a specific task (Buckner et al. 2008; Buckner and Vincent 2007) (see also chapter in this volume by Breting, Tuminello and Han, for review of normal age-related changes in the default mode network).

However, lack of an expected suppression of the default mode network during task activity may be indicative of abnormalities with the default mode network and associated with pathological aging. Research has substantiated the notion that AD may be characterized by irregularities in the default mode network. One study (Rombouts et al. 2005) observed more default mode deactivation during active cognitive processing in healthy older adults than in persons with MCI, who in turn displayed more deactivation than participants with AD. Another group (Greicius et al. 2004) observed that AD patients display less default mode activation in the posterior cingulate and hippocampus than healthy controls. In addition, default mode activity has elucidated disruptions in functional connectivity associated with AD (Wang et al. 2006). Decreased connectivity (as measured by correlations in activation) was observed between hippocampus and several cortical structures in

AD patients compared to healthy controls. A longitudinal study (Petrella et al. 2007) demonstrated that default mode activity also has prognostic utility in the early detection of dementia pathology. They found that MCI patients who converted to AD at 3.5-year follow-up had less task-activated suppression of posteromedial cortex than MCI patients who remained stable. Finally, some studies have observed a relationship between genetic risk for AD and the default mode network (Filippini et al. 2009), although others have not (Koch et al. 2010).

One group (Buckner et al. 2005) went so far as to speculate that the continuous activity of the default mode network may play a role in the formation of the amyloid-beta plaques that are characteristic of AD pathology. They noticed that the regions of activation of the default mode network bear a striking resemblance to the distribution of amyloid-beta plaques that are commonly seen throughout the brain in the early phases of AD. This "metabolism hypothesis" might imply that the default overuse (or inefficient use) of these regions throughout life may make them more susceptible to late-life degeneration from the accumulation of amyloid-beta neuropathology, and the increased activation observed in ε4 carriers (Filippini et al. 2009) may contribute to this risk.

4 Methodological Issues with fMRI and Aging

4.1 Atrophy Correction

When analyzing a group of older participants, differential atrophy needs to be taken into account when interpreting BOLD fMRI activation. For example, if a decreased BOLD activation is observed in a specific structure, it may not necessarily be due to poor functioning but rather to a reduction in tissue volume in that region. This issue is especially important in gerontological research because many structures display atrophy across the life span (Raz et al. 2005), and accelerated atrophy is observed with pathological aging (Agosta et al. 2009). Additionally, with tissue loss and/or ventricular enlargement, it is difficult to compare area activation within and between groups. To account for between-subject volume differences, a spatial normalization is typically applied to warp structural data into a common coordinate system such as Talairach space (Talairach and Tournoux 1988) prior to activation analysis. However, if the warping process utilizes a rigid body transformation, differences in anatomical and ventricular shape and size will not be matched between subjects. A common process to compensate for this variation is to smooth the images using a Gaussian filter.

The process of normalization to standard space is predicated under the assumption that there is a positive linear relationship between total volume and activation. Atrophic structures are expanded to match to template, and the activation is increased accordingly. However, one study (Johnson et al. 2000) observed a *positive* correlation between atrophy and fMRI activation in AD patients in the left inferior frontal gyrus, potentially representing compensatory recruitment from the remaining tissue. Thus, spatial warping may not be an optimal procedure for managing between-subject variation in atrophy.

4.2 Task Performance

One issue in fMRI interpretation of the task-activated BOLD signal concerns whether comparisons in activation can be made between participants that have differential task performance. For example, is the decreased hippocampal activation observed in AD patients compared to healthy controls a product of neuropathology, poorer task performance, or a combination of both? From a research standpoint, the issue that is of concern is the former, and functional imaging is not necessary to assess the latter. A way to address this potential confound is to match performance between groups, either by controlling for covariates or selecting subsamples of the experimental groups with equivalent performance, although both these methods come with substantial problems. Matching one or more variables between subjects assumes a linear relationship between performance and activation, and may systematically create a mismatch on another variable. Selecting subsamples reduces statistical power, which is problematic because sample size is often a limitation due to the expense of fMRI studies. Additionally, when analyzing groups with differences in cognitive abilities, the subsamples would most likely be composed of the lower end of the control group and the higher end of the patient group. These subsamples most likely will not be representative of the entire group (Brown and Eyler 2006).

Another way to address the task performance issue is the utilization of a task that all participants can perform satisfactorily, thereby minimizing group differences. An example of such a paradigm is a famous/non-famous name discrimination task (Douville et al. 2005). Differences in activation with equivalent performance on this task have successfully distinguished healthy controls from MCI patients (Woodard et al. 2009) and older adults with the APOE ε4 allele and/or a family history of dementia (Seidenberg et al. 2009a). Additionally, in a longitudinal study, functional activation was a significant predictor of which participants would develop cognitive decline at 18-month follow-up (Woodard et al. 2010).

4.3 The Nature of the Control Task

A major concern when choosing a task for the scanner is deciding on a proper control task. Comparing task activation to resting state is often not appropriate because of the aforementioned systematic differences that exist in the default mode network between healthy controls and those with memory impairment and dementia. Additionally, using resting state as a control task hinges on the assumption that task performance does not alter the default mode network. However, this assumption is clearly not true, as the default mode network is typically deactivated during task performance (Buckner et al. 2008; Buckner and Vincent 2007), and an inability to deactivate the default mode network during cognitive tasks has been associated with AD (Lustig et al. 2003). The comparison of two active states is typically more meaningful.

An issue with utilizing a specific task as a control is that it assumes an additive relationship between the experimental and comparison tasks in terms of activation. Subtracting the control from experimental task implies that what remains is the "task-activated" BOLD signal. By demonstrating that task activation varies as a function of the comparison, it has been argued that this "pure insertion" assumption is not appropriate for fMRI studies (Friston et al. 1996). These authors contend that the brain does not function in a linear, additive manner. Therefore, a factorial design with multiple baselines and the use of an interaction term between the two tasks may be a more valid design and interpretation of the data. When analyzing between-groups differences in older adults, this issue is especially relevant. For example, it is possible that additive activation could be observed between experimental and comparison conditions for the control group while the experimental group displays an interactive pattern. These between-groups differences would go unnoticed in fMRI designs predicated on pure insertion (Brown and Eyler 2006).

An additional strategy for combating the assumption of pure insertion is to compare activation for control over experimental conditions, in addition to experimental over control conditions. For example, using a famous/non-famous name discrimination task, one study (Seidenberg et al. 2009a) observed that healthy older adults with risk factors for AD (APOE ε4 allele and/or a family history of dementia) displayed greater cortical activation during famous name recognition, whereas controls without these risk factors displayed greater activation during identification of unfamiliar names. These differing activation patterns may reflect different cognitive strategies used by each group to complete the task. This example illustrates that pure insertion may not be the most efficient design, and contrasts of both activation minus control and control minus activation should be considered when elucidating between-groups differences in fMRI studies.

4.4 Choice of Memory Task During fMRI

A final methodological consideration involved when implementing fMRI with older adults relates to the task used in the scanner. A number of prior studies have used episodic memory tasks to investigate the effects of risk and/or disease on activation pattern. However, if the goal is to use fMRI for prediction of cognitive decline, episodic memory paradigms may present particular challenges. First, episodic memory performance declines as symptoms appear in MCI or AD (Bondi and Kaszniak 1991; Irle et al. 1990; Petersen et al. 1994, 1999, 2001). Therefore, performance reductions on an episodic memory task may suggest that the individual has already experienced a significant cognitive decline and has become symptomatic, and discrepancies in task performance may confound the interpretation of fMRI data. Episodic memory tasks are also challenging for older adults, especially if cognitively impaired. Therefore, they may be more likely to evoke frustration and greater effort, and may paradoxically produce greater activation in persons who are most susceptible to imminent decline [e.g., (Persson et al. 2006)]. Finally, because episodic memory is known to decline with normal aging as well as with degenerative conditions (Nilsson 2003; Petersen et al. 1992), the activation resulting from performance on this type of task may not be able to make accurate distinctions between the effects of aging and cognitive impairment.

Semantic memory paradigms may represent a useful alternative to episodic memory tasks when used in conjunction with fMRI. Semantic memory tasks are often considerably easier than episodic memory tasks because presentation of the material to be recognized evokes an almost immediate familiarity of the information that has previously been learned. There are relatively few semantic memory changes associated with age (Nilsson 2003), although semantic memory is commonly affected in degenerative dementias (Hodges et al. 1990, 1992; Nebes 1989). Therefore, semantic memory tasks may be more likely to be able to differentiate disease-related changes from performance that would be characteristic of healthy aging. For instance, temporally graded remote memory loss is typically seen in conditions such as Alzheimer's disease in which recently learned information is more effortful to recall than remotely learned information (Butters et al. 1987). Specifically, a breakdown of semantic knowledge has been demonstrated in both AD and MCI (Seidenberg et al. 2009b). Finally, although semantic and episodic memory may engage slightly different brain regions (Binder et al. 2009), they also engage a common set of brain regions, including hippocampus, posterior cingulate, and precuneus (Desgranges et al. 1998a, b; Fletcher et al. 1997; Moscovitch et al. 2005). Thus, semantic memory paradigms may offer a more effective way to assess the functional state of neural circuitry that is vulnerable to the effects of degenerative conditions, such as AD.

5 Conclusions and Future Directions

In the past decade, fMRI has emerged as a valuable, non-invasive tool for understanding the neural bases of both normal and pathological aging. Regional elevations in BOLD signal and reduced hemispheric asymmetry despite equivalent task performance have been associated with aging. These patterns may be representative of compensatory recruitment of additional circuitry for declining neural resources [e.g., scaffolding (Park and Reuter-Lorenz 2009)] and/or dedifferentiation of specialized modules. fMRI has also demonstrated differential patterns of activation associated with pathological aging, including hippocampal hyperactivation in MCI and hypoactivation in AD. Task-related deactivation of the default mode network declines normally with aging and may be accentuated in pathological aging. Additionally, fMRI has made significant contributions to the understanding of the neural basis of risk factors for cognitive decline and dementia, and how these factors may impact brain function prior to the onset of symptoms. Recently, longitudinal studies have demonstrated the effectiveness of fMRI for prediction of cognitive decline, with more accurate prediction being observed when using BOLD activation from fMRI than when using hippocampal volume or demographic information (Woodard et al. 2010).

Future gerontological fMRI research must continue to address the "increased activation" phenomenon that is observed with older adults and persons with risk factors for AD. It is still not clear whether the additional activation and recruitment are universally indicative of adaptive aging, a biomarker of future decline as a byproduct of neuropathology, or both. The implications of increased activation may also differ by region. Increased task-related engagement of cortical structures may have different meanings and effects compared to subcortical regions. Finally, longitudinal studies can also determine whether evidence of cognitive scaffolding at mid- and late-life may be predictive of cognitive decline and dementia, as well as how risk factors for AD, such as family history and the APOE ε4 allele, may influence neural functioning and confer susceptibility to neuropathology.

Future research should also determine whether fMRI can be used to evaluate the effectiveness of pharmacological and non-pharmacological interventions by leading to changes in the BOLD signal during certain types of cognitive tasks. A related question involves whether fMRI can be used to assess whether these interventional strategies may be beneficial in delaying or preventing the occurrence of neurodegeneration. It has been demonstrated that the BOLD signal during memorization can be enhanced in older adults by the implementation of a semantic encoding technique (Logan et al. 2002), suggesting that the intentional use of strategic *cognitive* interventions may be able to influence brain functioning. Lifestyle factors such as engagement in physical (Rolland et al. 2008), cognitive (Wilson et al. 2003, 2007), and social (Saczynski et al. 2006) activities have also been associated with a reduced risk of cognitive decline, MCI, and AD. It is possible that these activities could protect against degeneration by enhancing neurogenesis, optimizing neurotransmission, or improving tissue oxygenation and metabolism, thereby influencing patterns of functional activation.

Physical activity has specifically been identified as a promising intervention for enhancing late-life cognitive and neural functioning. One recent study (Erickson et al. 2011) demonstrated that a six-month exercise intervention can increase hippocampal volume in older adults. Another study (Smith et al. 2011) found increased fMRI activation in cortical areas during a famous name discrimination task in physically active, cognitively intact older adults with the APOE ε4 allele. A longitudinal follow-up study demonstrated that these participants were at a lower risk of cognitive decline after 18 months compared to physically inactive APOE ε4 carriers (Woodard et al., Manuscript under review). Further research might include controlled intervention studies observing the impact of lifestyle factors (including physical activity) on the fMRI BOLD signal, and whether any potential alterations in physical activity frequency or intensity may confer resistance to subsequent cognitive decline and/or dementia.

Acknowledgments This work was supported in part by NIH grant R01 AG022304. The content is solely the responsibility of the authors and does not necessarily represent the official views of the National Institute on Aging or the National Institutes of Health.

References

Agosta F, Vossel KA, Miller BL, Migliaccio R, Bonasera SJ, Filippi M et al (2009) Apolipoprotein E epsilon4 is associated with disease-specific effects on brain atrophy in Alzheimer's disease and frontotemporal dementia. Proc Natl Acad Sci USA 106(6):2018–2022

Albert MS, Moss MB, Tanzi R, Jones K (2001) Preclinical prediction of AD using neuropsychological tests. J Int Neuropsychol Soc 7(5):631–639

Babcock RL, Laguna KD, Roesch SC (1997) A comparison of the factor structure of processing speed for younger and older adults: testing the assumption of measurement equivalence across age groups. Psychol Aging 12(2):268–276

Baltes PB, Lindenberger U (1997) Emergence of a powerful connection between sensory and cognitive functions across the adult life span: a new window to the study of cognitive aging? Psychol Aging 12(1):12–21

Barrick TR, Charlton RA, Clark CA, Markus HS (2010) White matter structural decline in normal ageing: a prospective longitudinal study using tract-based spatial statistics. NeuroImage 51(2):565–577

Benton AL, Hamsher Kd (1976) Multilingual Aphasia examination. University of Iowa, Iowa City

Bergerbest D, Gabrieli JD, Whitfield-Gabrieli S, Kim H, Stebbins GT, Bennett DA et al (2009) Age-associated reduction of asymmetry in prefrontal function and preservation of conceptual repetition priming. NeuroImage 45(1):237–246

Bertram L, Tanzi RE (2008) Thirty years of Alzheimer's disease genetics: the implications of systematic meta-analyses. Nat Rev Neurosci 9(10):768–778

Binder JR, Desai RH, Graves WW, Conant LL (2009) Where is the semantic system? A critical review and meta-analysis of 120 functional neuroimaging studies. Cereb Cortex 19(12): 2767–2796

Blennow K, Hampel H (2003) CSF markers for incipient Alzheimer's disease. Lancet Neurol 2(10):605–613

Bondi MW, Kaszniak AW (1991) Implicit and explicit memory in Alzheimer's disease and Parkinson's disease. J Clin Exp Neuropsychol 13:339–358

Bondi MW, Houston WS, Eyler LT, Brown GG (2005) fMRI evidence of compensatory mechanisms in older adults at genetic risk for Alzheimer disease. Neurol 64:501–508

Bookheimer SY, Strojwas MH, Cohen MS, Saunders AM, Pericak-Vance MA, Mazziotta JC et al (2000) Patterns of brain activation in people at risk for Alzheimer's Disease. N Engl J Med 343(7):450–456

Brown GG, Eyler LT (2006) Methodological and conceptual issues in functional magnetic resonance imaging: applications to schizophrenia research. Annu Rev Clin Psychol 2:51–81

Brys M, Pirraglia E, Rich K, Rolstad S, Mosconi L, Switalski R et al (2009) Prediction and longitudinal study of CSF biomarkers in mild cognitive impairment. Neurobiol Aging 30(5): 682–690

Buckner RL, Vincent JL (2007) Unrest at rest: default activity and spontaneous network correlations. Neuroimage 37(4):1091–1096, discussion 1097–1099

Buckner RL, Andrews-Hanna JR, Schacter DL (2008) The brain's default network: anatomy, function, and relevance to disease. Ann N Y Acad Sci 1124:1–38

Buckner RL, Snyder AZ, Shannon BJ, LaRossa G, Sachs R, Fotenos AF et al (2005) Molecular, structural, and functional characterization of Alzheimer's disease: evidence for a relationship between default activity, amyloid, and memory. J Neurosci 25(34):7709–7717

Buerger K, Teipel SJ, Zinkowski R, Blennow K, Arai H, Engel R et al (2002a) CSF tau protein phosphorylated at threonine 231 correlates with cognitive decline in MCI subjects. Neurol 59(4):627–629

Buerger K, Zinkowski R, Teipel SJ, Tapiola T, Arai H, Blennow K et al (2002b) Differential diagnosis of Alzheimer disease with cerebrospinal fluid levels of tau protein phosphorylated at threonine 231. Arch Neurol 59(8):1267–1272

Buschke H, Fuld PA (1974) Evaluating storage, retention, and retrieval in disordered memory and learning. Neurol 24:1019–1025

Butters N, Grandholm E, Salmon DP, Grant I, Wolfe J (1987) Episodic and semantic memory: a comparison of amnesic and demented patients. J Clin Exp Neuropsychol 9:479–497

Cabeza R (2002) Hemispheric asymmetry reduction in older adults: the HAROLD model. Psychol Aging 17(1):85–100

Cardenas VA, Du AT, Hardin D, Ezekiel F, Weber P, Jagust WJ et al (2003) Comparison of methods for measuring longitudinal brain change in cognitive impairment and dementia. Neurobiol Aging 24(4):537–544

Caselli RJ, Reiman E, Osborne D, Hentz JG, Baxter LC, Hernandez JL et al (2004) Longitudinal changes in cognition and behavior in asymptomatic carriers of the APOE ε4 allele. Neurol 62:1990–1995

Caselli RJ, Reiman EM, Locke DE, Hutton ML, Hentz JG, Hoffman-Snyder C et al (2007) Cognitive domain decline in healthy apolipoprotein E epsilon4 homozygotes before the diagnosis of mild cognitive impairment. Arch Neurol 64(9):1306–1311

Chetelat G, Desgranges B, De La Sayette V, Viader F, Eustache F, Baron JC (2003) Mild cognitive impairment: Can FDG-PET predict who is to rapidly convert to Alzheimer's disease? Neurol 60:1374–1377

Chetelat G, Eustache F, Viader F, De la Sayette V, Pelerin A, Mezenge F et al (2005) FDG-PET measurement is more accurate than neuropsychological assessments to predict global cognitive deterioration in patients with mild cognitive impairment. Neurocase 11:14–25

Clark CM, Davatzikos C, Borthakur A, Newberg A, Leight S, Lee VM et al (2008) Biomarkers for early detection of Alzheimer pathology. Neurosignals 16(1):11–18

Corder EH, Saunders AM, Strittmatter WJ, Schmechel DE, Gaskell PC, Small GW et al (1993) Gene dose of apolipoprotein E type 4 allele and the risk of Alzheimer's disease in late onset families. Sci 261:921–923

Daviglus ML, Bell CC, Berrettini W, Bowen PE, Connolly ES, Cox NJ et al (2010) National Institutes of Health State-of-the-Science Conference Statement: Preventing Alzheimer's disease and cognitive decline. NIH Consens State Sci Statements 27(4):1–30

De Jager CA, Hogervorst E, Combrinck M, Budge MM (2003) Sensitivity and specificity of neuropsychological tests for mild cognitive impairment, vascular cognitive impairment and Alzheimer's disease. Psychol Med 33(6):1039–1050

de Leon MJ, George AE, Stylopoulos LA, Smith G, Miller DC (1989) Early marker for Alzheimer's disease: the atrophic hippocampus. Lancet 2(8664):672–673

de Leon MJ, DeSanti S, Zinkowski R, Mehta PD, Pratico D, Segal S et al (2006) Longitudinal CSF and MRI biomarkers improve the diagnosis of mild cognitive impairment. Neurobiol Aging 27(3):394–401

de Leon MJ, Mosconi L, Li J, De Santi S, Yao Y, Tsui WH et al (2007) Longitudinal CSF isoprostane and MRI atrophy in the progression to AD. J Neurol 254(12):1666–1675

De Meyer G, Shapiro F, Vanderstichele H, Vanmechelen E, Engelborghs S, De Deyn PP et al (2010) Diagnosis-independent Alzheimer disease biomarker signature in cognitively normal elderly people. Arch Neurol 67(8):949–956

DeCarli C, Mungas D, Harvey D, Reed B, Weiner M, Chui H et al (2004) Memory impairment, but not cerebrovascular disease, predicts progression of MCI to dementia. Neurol 63(2): 220–227

Desgranges B, Baron JC, de la Sayette V, Petit-Taboue MC, Benali K, Landeau B et al (1998a) The neural substrates of memory systems impairment in Alzheimer's disease. A PET study of resting brain glucose utilization [In Process Citation]. Brain 121(Pt 4):611–631

Desgranges B, Baron JC, Eustache F (1998b) The functional neuroanatomy of episodic memory: the role of the frontal lobes, the hippocampal formation, and other areas. [Review] [120 refs]. Neuroimage 8(2):198–213

Devanand DP, Pradhaban G, Liu X, Khandji A, De Santi S, Segal S et al (2007) Hippocampal and entorhinal atrophy in mild cognitive impairment: prediction of Alzheimer disease. Neurol 68(11):828–836

Dickerson BC, Salat DH, Greve DN, Chua EF, Rand-Giovannetti E, Rentz DM et al (2005) Increased hippocampal activation in mild cognitive impairment compared to normal aging and AD. Neurol 65(3):404–411

DiGirolamo GJ, Kramer AF, Barad V, Cepeda NJ, Weissman DH, Milham MP et al (2001) General and task-specific frontal lobe recruitment in older adults during executive processes: a fMRI investigation of task-switching. Neuroreport 12(9):2065–2071

Douville K, Woodard JL, Seidenberg M, Miller SK, Leveroni CL, Nielson KA et al (2005) Medial temporal lobe activity for recognition of recent and remote famous names: an event-related fMRI study. Neuropsychologia 43(5):693–703

Erickson KI, Voss MW, Prakash RS, Basak C, Szabo A, Chaddock L et al (2011) Exercise training increases size of hippocampus and improves memory. Proc Natl Acad Sci USA 108(7):3017–3022

Fagan AM, Roe CM, Xiong C, Mintun MA, Morris JC, Holtzman DM (2007) Cerebrospinal fluid tau/beta-amyloid(42) ratio as a prediction of cognitive decline in nondemented older adults. Arch Neurol 64(3):343–349

Filippini N, MacIntosh BJ, Hough MG, Goodwin GM, Frisoni GB, Smith SM et al (2009) Distinct patterns of brain activity in young carriers of the APOE-epsilon4 allele. Proc Natl Acad Sci USA 106(17):7209–7214

Fjell AM, Walhovd KB (2010) Structural brain changes in aging: courses, causes and cognitive consequences. Rev Neurosci 21(3):187–221

Fletcher PC, Frith CD, Rugg MD (1997) The functional neuroanatomy of episodic memory. Trends Neurosci 20(5):213–218

Fratiglioni L, Ahlbom A, Viitanen M, Winblad B (1993) Risk factors for late-onset Alzheimer's disease: a population-based, case-control study. Ann Neurol 33(3):258–266

Friston KJ, Price CJ, Fletcher P, Moore C, Frackowiak RS, Dolan RJ (1996) The trouble with cognitive subtraction. Neuroimage 4(2):97–104

Furst AJ, Mormino EC (2010) A BOLD move: clinical application of fMRI in aging. Neurol 74(24):1940–1941

Grady CL, McIntosh AR, Horwitz B, Maisog JM, Ungerleider LG, Mentis MJ et al (1995) Age-related reductions in human recognition memory due to impaired encoding. Sci 269:218–221

Grady CL, McIntosh AR, Craik FI (2005) Task-related activity in prefrontal cortex and its relation to recognition memory performance in young and old adults. Neuropsychologia 43(10):1466–1481

Greicius MD, Srivastava G, Reiss AL, Menon V (2004) Default-mode network activity distinguishes Alzheimer's disease from healthy aging: evidence from functional MRI. Proc Natl Acad Sci USA 101(13):4637–4642

Gutchess AH, Welsh RC, Hedden T, Bangert A, Minear M, Liu LL et al (2005) Aging and the neural correlates of successful picture encoding: frontal activations compensate for decreased medial-temporal activity. J Cogn Neurosci 17(1):84–96

Hampel H, Buerger K, Zinkowski R, Teipel SJ, Goernitz A, Andreasen N et al (2004a) Measurement of phosphorylated tau epitopes in the differential diagnosis of Alzheimer disease: a comparative cerebrospinal fluid study. Arch Gen Psychiatry 61(1):95–102

Hampel H, Teipel SJ, Fuchsberger T, Andreasen N, Wiltfang J, Otto M et al (2004b) Value of CSF beta-amyloid1–42 and tau as predictors of Alzheimer's disease in patients with mild cognitive impairment. Mol Psychiatry 9(7):705–710

Han SD, Houston WS, Jak AJ, Eyler LT, Nagel BJ, Fleisher AS et al (2007) Verbal paired-associate learning by APOE genotype in non-demented older adults: fMRI evidence of a right hemispheric compensatory response. Neurobiol Aging 28(2):238–247

Han SD, Bangen KJ, Bondi MW (2008) Functional magnetic resonance imaging of compensatory neural recruitment in aging and risk for Alzheimer's disease: review and recommendations. Dement Geriatr Cogn Disord 27(1):1–10

Hansson O, Zetterberg H, Buchhave P, Londos E, Blennow K, Minthon L (2006) Association between CSF biomarkers and incipient Alzheimer's disease in patients with mild cognitive impairment: a follow-up study. Lancet Neurol 5(3):228–234

Henneman WJ, Sluimer JD, Barnes J, van der Flier WM, Sluimer IC, Fox NC et al (2009) Hippocampal atrophy rates in Alzheimer disease: added value over whole brain volume measures. Neurol 72(11):999–1007

Hodges JR, Salmon DP, Butters N (1990) Differential impairment of semantic and episodic memory in Alzheimer's and Huntington's diseases: a controlled prospective study. J Neurol Neurosurg Psychiatry 53:1089–1095

Hodges JR, Salmon DP, Butters N (1992) Semantic memory impairment in Alzheimer's disease: failure of access or degraded knowledge? Neuropsychologia 30(4):301–314

Irle E, Kaiser P, Naumann-Stoll G (1990) Differential patterns of memory loss in patients with Alzheimer's disease and Korsakoff's disease. Int J Neurosci 52(1–2):67–77

Jack CR, Petersen RC, Xu YC, O'Brien PC, Smith GE, Ivnik RJ et al (1999) Prediction of AD with MRI-based hippocampal volume in mild cognitive impairment. Neurol 52:1397–1403

Johnson SC, Saykin AJ, Baxter LC, Flashman LA, Santulli RB, McAllister TW et al (2000) The relationship between fMRI activation and cerebral atrophy: comparison of normal aging and alzheimer disease. Neuroimage 11(3):179–187

Johnson SC, Schmitz TW, Trivedi MA, Ries ML, Torgerson BM, Carlsson CM et al (2006) The influence of Alzheimer disease family history and apolipoprotein E epsilon4 on mesial temporal lobe activation. J Neurosci 26(22):6069–6076

Juottonen K, Lehtovirta M, Helisalmi S, Riekkinen PJ Sr, Soininen H (1998) Major decrease in the volume of the entorhinal cortex in patients with Alzheimer's disease carrying the apolipoprotein E epsilon4 allele. J Neurol Neurosurg Psychiatry 65(3):322–327

Kalaria RN, Maestre GE, Arizaga R, Friedland RP, Galasko D, Hall K et al (2008) Alzheimer's disease and vascular dementia in developing countries: prevalence, management, and risk factors. Lancet Neurol 7(9):812–826

Kircher TT, Weis S, Freymann K, Erb M, Jessen F, Grodd W et al (2007) Hippocampal activation in patients with mild cognitive impairment is necessary for successful memory encoding. J Neurol Neurosurg Psychiatry 78(8):812–818

Knoke D, Taylor AE, Saint-Cyr JA (1998) The differential effects of cueing on recall in Parkinson's disease and normal subjects. Brain Cogn 38(2):261–274

Koch W, Teipel S, Mueller S, Benninghoff J, Wagner M, Bokde AL et al (2010) Diagnostic power of default mode network resting state fMRI in the detection of Alzheimer's disease. Neurobiol Aging

Kok E, Haikonen S, Luoto T, Huhtala H, Goebeler S, Haapasalo H et al (2009) Apolipoprotein E-dependent accumulation of Alzheimer disease-related lesions begins in middle age. Ann Neurol 65(6):650–657

Langenecker SA, Nielson KA (2003) Frontal recruitment during response inhibition in older adults replicated with fMRI. Neuroimage 20(2):1384–1392

Li Z, Moore AB, Tyner C, Hu X (2009) Asymmetric connectivity reduction and its relationship to "HAROLD" in aging brain. Brain Res 1295:149–158

Lind J, Persson J, Ingvar M, Larsson A, Cruts M, Van Broeckhoven C et al (2006) Reduced functional brain activity response in cognitively intact apolipoprotein E epsilon4 carriers. Brain 129(Pt 5):1240–1248

Logan JM, Sanders AL, Snyder AZ, Morris JC, Buckner RL (2002) Under-recruitment and nonselective recruitment: dissociable neural mechanisms associated with aging. Neuron 33(5): 827–840

Lustig C, Snyder AZ, Bhakta M, O'Brien KC, McAvoy M, Raichle ME et al (2003) Functional deactivations: change with age and dementia of the Alzheimer type. Proc Natl Acad Sci USA 100(24):14504–14509

Matsumae M, Kikinis R, Morocz IA, Lorenzo AV, Sandor T, Albert MS et al (1996) Age-related changes in intracranial compartment volumes in normal adults assessed by magnetic resonance imaging. J Neurosurg 84(6):982–991

Michielse S, Coupland N, Camicioli R, Carter R, Seres P, Sabino J et al (2010) Selective effects of aging on brain white matter microstructure: a diffusion tensor imaging tractography study. NeuroImage 52(4):1190–1201

Miller SL, Fenstermacher E, Bates J, Blacker D, Sperling RA, Dickerson BC (2008) Hippocampal activation in adults with mild cognitive impairment predicts subsequent cognitive decline. J Neurol Neurosurg Psychiatry 79(6):630–635

Mondadori CR, Buchmann A, Mustovic H, Schmidt CF, Boesiger P, Nitsch RM et al (2006) Enhanced brain activity may precede the diagnosis of Alzheimer's disease by 30 years. Brain 129(Pt 11):2908–2922

Mondadori CR, de Quervain DJ, Buchmann A, Mustovic H, Wollmer MA, Schmidt CF et al (2007) Better memory and neural efficiency in young apolipoprotein E epsilon4 carriers. Cereb Cortex 17(8):1934–1947

Morey RA, Petty CM, Xu Y, Hayes JP, Wagner HR 2nd, Lewis DV et al (2009) A comparison of automated segmentation and manual tracing for quantifying hippocampal and amygdala volumes. Neuroimage 45(3):855–866

Morra JH, Tu Z, Apostolova LG, Green AE, Avedissian C, Madsen SK et al (2009) Automated 3D mapping of hippocampal atrophy and its clinical correlates in 400 subjects with Alzheimer's disease, mild cognitive impairment, and elderly controls. Hum Brain Mapp 30(9):2766–2788

Moscovitch M, Rosenbaum RS, Gilboa A, Addis DR, Westmacott R, Grady C et al (2005) Functional neuroanatomy of remote episodic, semantic and spatial memory: a unified account based on multiple trace theory. J Anat 207(1):35–66

Nebes RD (1989) Semantic memory in Alzheimer's disease. Psychol Bull 106(3):377–394

Nestor PJ, Scheltens P, Hodges JR (2004) Advances in the early detection of Alzheimer's disease. Nat Med 10:S34–S41

Nielson KA, Langenecker SA, Garavan H (2002) Differences in the functional neuroanatomy of inhibitory control across the adult lifespan. Psychol Aging 17(1):56–57

Nielson KA, Langenecker SA, Ross TJ, Garavan H, Rao SM, Stein EA (2004) Comparability of functional MRI response in young and old during inhibition. Neuroreport 15(1):129–133

Nielson KA, Douville KL, Seidenberg M, Woodard JL, Miller SK, Franczak M et al (2006) Age-related functional recruitment for famous name recognition: an event-related fMRI study. Neurobiol Aging 27(10):1494–1504

Nilsson LG (2003) Memory function in normal aging. Acta Neurol Scand 179:7–13

O'Brien JL, O'Keefe KM, LaViolette PS, DeLuca AN, Blacker D, Dickerson BC et al (2010) Longitudinal fMRI in elderly reveals loss of hippocampal activation with clinical decline. Neurol 74(24):1969–1976

Park DC, Reuter-Lorenz P (2009) The adaptive brain: aging and neurocognitive scaffolding. Annu Rev Psychol 60:173–196

Persson J, Nyberg L, Lind J, Larsson A, Nilsson LG, Ingvar M et al (2006) Structure-function correlates of cognitive decline in aging. Cereb Cortex 16(7):907–915

Petersen R, Smith G, Kokmen E, Ivnik R, Tangalos E (1992) Memory function in normal aging. Neurol 42:396–401

Petersen RC, Smith GE, Ivnik RJ, Kokmen E, Tangalos EG (1994) Memory function in very early Alzheimer's disease. Neurology 44:867–872

Petersen RC, Smith GE, Waring SC, Ivnik RJ, Tangalos EG, Kokmen E (1999) Mild cognitive impairment: clinical characterization and outcome [In Process Citation]. Arch Neurol 56(3): 303–308

Petersen RC, Stevens JC, Ganguli M, Tangalos EG, Cummings JL, DeKosky ST (2001) Practice parameter: early detection of dementia: mild cognitive impairment (an evidence-based review). Report of the Quality Standards Subcommittee of the American Academy of Neurology. Neurol 56(9):1133–1142

Petrella JR, Prince SE, Wang L, Hellegers C, Doraiswamy PM (2007) Prognostic value of posteromedial cortex deactivation in mild cognitive impairment. PLoS One 2(10):e1104

Raz N, Lindenberger U, Rodrigue KM, Kennedy KM, Head D, Williamson A et al (2005) Regional brain changes in aging healthy adults: general trends, individual differences and modifiers. Cereb Cortex 15(11):1676–1689

Reiman EM, Langbaum JB, Tariot PN (2010) Alzheimer's prevention initiative: a proposal to evaluate presymptomatic treatments as quickly as possible. Biomark Med 4(1):3–14

Rolland Y, Abellan van Kan G, Vellas B (2008) Physical activity and Alzheimer's disease: from prevention to therapeutic perspectives. J Am Med Dir Assoc 9(6):390–405

Rombouts SA, Barkhof F, Goekoop R, Stam CJ, Scheltens P (2005) Altered resting state networks in mild cognitive impairment and mild Alzheimer's disease: an fMRI study. Hum Brain Mapp 26(4):231–239

Rowe CC, Ng S, Ackermann U, Gong SJ, Pike K, Savage G et al (2007) Imaging beta-amyloid burden in aging and dementia. Neurol 68(20):1718–1725

Saczynski JS, Pfeifer LA, Masaki K, Korf ES, Laurin D, White L et al (2006) The effect of social engagement on incident dementia: the Honolulu-Asia Aging Study. Am J Epidemiol 163(5):433–440

Saunders AM, Strittmatter WJ, Schmechel D, St. George-Hyslop PH, Pericak-Vance MA, Joo SH et al (1993) Association of apolipoprotein E allele ε4 with late-onset familial and sporadic Alzheimer's disease. Neurol 43:1467–1472

Seidenberg M, Guidotti L, Nielson KA, Woodard JL, Durgerian S, Antuono P et al (2009a) Semantic memory activation in individuals at risk for developing Alzheimer disease. Neurology 73(8):612–620

Seidenberg M, Guidotti L, Nielson KA, Woodard JL, Durgerian S, Zhang Q et al (2009b) Semantic knowledge for famous names in mild cognitive impairment. J Int Neuropsychol Soc 15(1):9–18

Smith CD, Kryscio RJ, Schmitt FA, Lovell MA, Blonder LX, Rayens WS et al (2005) Longitudinal functional alterations in asymptomatic women at risk for Alzheimer's disease. J Neuroimaging 15(3):271–277

Smith JC, Nielson KA, Woodard JL, Seidenberg M, Durgerian S, Antuono P et al (2011) Interactive effects of physical activity and APOE-epsilon4 on BOLD semantic memory activation in healthy elders. Neuroimage 54(1):635–644

Stebbins GT, Carrillo MC, Dorfman J, Dirksen C, Desmond JE, Turner DA et al (2002) Aging effects on memory encoding in the frontal lobes. Psychol Aging 17(1):44–55

Stoub TR, Rogalski EJ, Leurgans S, Bennett DA, de Tolego-Morrell L (2010) Rate of entorhinal and hippocampal atrophy in incipient and mild AD: relation to memory functioning. Neurobiol Aging 31(7):1089–1098

Swan GE, Lessov-Schlaggar CN, Carmelli D, Schellenberg GD, La Rue A (2005) Apolipoprotein E epsilon4 and change in cognitive functioning in community-dwelling older adults. J Geriatr Psychiatry Neurol 18(4):196–201

Tae WS, Kim SS, Lee KU, Nam EC, Kim KW (2008) Validation of hippocampal volumes measured using a manual method and two automated methods (FreeSurfer and IBASPM) in chronic major depressive disorder. Neuroradiol 50(7):569–581

Talairach J, Tournoux P (1988) Co-planar stereotaxic atlas of the human brain. Thieme, New York

Trachtenberg AJ, Filippini N, Mackay CE (2010) The effects of APOE-epsilon4 on the BOLD response. Neurobiol Aging

Trivedi MA, Schmitz TW, Ries ML, Hess TM, Fitzgerald ME, Atwood CS et al (2008) fMRI activation during episodic encoding and metacognitive appraisal across the lifespan: risk factors for Alzheimer's disease. Neuropsychologia 46(6):1667–1678

Vannini P, Almkvist O, Dierks T, Lehmann C, Wahlund LO (2007) Reduced neuronal efficacy in progressive mild cognitive impairment: a prospective fMRI study on visuospatial processing. Psychiatry Res 156(1):43–57

Wang L, Zang Y, He Y, Liang M, Zhang X, Tian L et al (2006) Changes in hippocampal connectivity in the early stages of Alzheimer's disease: evidence from resting state fMRI. Neuroimage 31(2):496–504

Wechsler D (1945) A standardized memory scale for clinical use. J Psychol 19:87–95

Wilson RS, Bennett DA, Bienias JL, Mendes de Leon CF, Morris MC, Evans DA (2003) Cognitive activity and cognitive decline in a biracial community population. Neurol 61(6):812–816

Wilson RS, Scherr PA, Schneider JA, Tang Y, Bennett DA (2007) Relation of cognitive activity to risk of developing Alzheimer disease. Neurol 69(20):1911–1920

Wolf H, Jelic V, Gertz HJ, Nordberg A, Julin P, Wahlund LO (2003) A critical discussion of the role of neuroimaging in mild cognitive impairment. Acta Neurol Scand 179:52–76

Wolk DA, Klunk W (2009) Update on amyloid imaging: from healthy aging to Alzheimer's disease. Curr Neurol Neurosci Rep 9(5):345–352

Wolk DA, Price JC, Saxton JA, Snitz BE, James JA, Lopez OL et al (2009) Amyloid imaging in mild cognitive impairment subtypes. Ann Neurol 65(5):557–568

Woodard JL, Seidenberg M, Nielson KA, Antuono P, Guidotti L, Durgerian S et al (2009) Semantic memory activation in amnestic mild cognitive impairment. Brain 132(Pt 8): 2068–2078

Woodard JL, Seidenberg M, Nielson KA, Smith JC, Antuono P, Durgerian S et al (2010) Prediction of cognitive decline in healthy older adults using fMRI. J Alzheimers Dis 21(3): 871–885

Woodard JL, Nielson KA, Sugarman MA, Smith JC, Seidenberg M, Durgerian S et al (Manuscript under review). Lifestyle and genetic contributions to cognitive decline and hippocampal integrity and healthy aging

Xu G, McLaren DG, Ries ML, Fitzgerald ME, Bendlin BB, Rowley HA et al (2009) The influence of parental history of Alzheimer's disease and apolipoprotein E epsilon4 on the BOLD signal during recognition memory. Brain 132(Pt 2):383–391

Neuroanatomical Changes Associated with Cognitive Aging

Janice M. Juraska and Nioka C. Lowry

Abstract The literature on the neuroanatomical changes that occur during normal, non-demented aging is reviewed here with an emphasis on the improved accuracy of studies that use stereological techniques. Loss of neural tissue involved in cognition occurs during aging of humans as well as the other mammals that have been examined. There is considerable regional specificity within the cerebral cortex and the hippocampus in both the degree and cellular basis for loss. The anatomy of the prefrontal cortex is especially vulnerable to the effects of aging while the major subfields of the hippocampus are not. A loss of neurons, dendrites and synapses has been documented, as well as changes in neurotransmitter systems, in some regions of the cortex and hippocampus but not others. Species differences are also apparent in the cortical white matter and the corpus callosum where there are indications of loss of myelin in humans, but most evidence favors preservation in rats. The examination of whether the course of neuroanatomical aging is altered by hormone replacement in females is just beginning. When hormone replacement is started close to the time of cycle cessation, there are indications in humans and rats that replacement can preserve neural tissue but there is some variability due to the type of hormones and regimen of administration.

J. M. Juraska (✉)
Department of Psychology and Program in Neuroscience,
University of Illinois, 603 E Daniel, Champaign,
IL 61820, USA
e-mail: jjuraska@illinois.edu

N. C. Lowry
Department of Psychology, University of Illinois,
603 E Daniel, Champaign, IL 61820, USA
e-mail: nchisho2@illinois.edu

Curr Topics Behav Neurosci (2012) 10: 137–162
DOI: 10.1007/7854_2011_137

Published Online: 14 June 2011

Keywords Stereology · Neuron number · Dendrites · Synapses · Synaptophysin · Dopamine · Estrogen · Progesterone · Medroxyprogesterone acetate · Hormone treatment · Menopause · Hippocampus · Prefrontal cortex · Acetylcholine · Glutamate · White matter · Myelination · Basolateral amygdala · Corpus callosum

Abbreviations

MRI	Magnetic resonance imaging
PET	Positron emission tomography
SPECT	Single-photon emission computed tomography
DAT	Dopamine transporter
MAO-A	Monoamine oxidase A
MAO-B	Monoamine oxidase B
mRNA	Messenger ribonucleic acid
mPFC	Medial prefrontal cortex
NMDA R	N-methyl-D-aspartate receptor
MPA	Medroxyprogesterone acetate
E	17-β estradiol
P	Progesterone
CEE	Conjugated equine estrogens
ChAT	Choline acetyltransferase
AChE	Acetylcholinesterase

Contents

1 Human Imaging Studies ... 139
2 Neuron Loss ... 140
  2.1 The Cerebral Cortex ... 140
  2.2 The Hippocampus ... 143
3 White Matter ... 144
4 Dendrites and Synapses ... 145
  4.1 The Cerebral Cortex ... 145
  4.2 Hippocampus ... 147
5 Neurotransmission ... 148
  5.1 Monoamines ... 148
  5.2 Acetylcholine ... 149
  5.3 Glutamate ... 149
6 Hormone Treatment ... 150
  6.1 Humans ... 150
  6.2 Animal Models ... 150
7 Conclusions ... 153
References ... 153

As life expectancy rises in industrialized countries, there is a growing need to understand the cognitive and neural declines that accompany normal aging. Many neural changes are evident even at the gross size level and can be observed with magnetic resonance imaging (MRI). These in turn are a reflection of underlying cellular changes that are harder to assess in humans. Understanding these cellular changes associated with aging might provide insight in understanding the causes of aging and how they can be modified.

The current chapter will review what is known about normal brain aging in mammals. It does not include Alzheimer's disease, which is a separate and specialized literature, or other forms of dementia. Cellular changes are necessary for our ultimate understanding of aging; therefore, work from non-human animals will be emphasized here. All of the literature cited used stereological techniques (both unbiased counting and volume estimation) unless otherwise noted because studies of cell density alone are difficult to interpret at best and at worst, misleading. The little that is known about the neuroanatomical effects of ovarian hormone replacement during aging will also be reviewed. This portion of the chapter is complementary to the chapter in this volume by Boulware, Kent and Frick on behavioral effects of hormone treatment during aging.

1 Human Imaging Studies

The advent of structural MRI has meant that aging of the brain can be studied in healthy individuals. This technique has revealed a clear concordance across both cross-sectional and longitudinal studies that neural tissue is lost during aging, most prominently seen as a decrease in the volume of the gray matter in the cerebral cortex (Bartzokis et al. 2001; Raz et al. 1997, 2010; Resnick et al. 2003). There is also a general consensus that the frontal cortex, especially the prefrontal cortex is the most vulnerable, while the occipital cortex shows the least loss in volume with increasing age (Resnick et al. 2003; Sowell et al. 2003; Allen et al. 2005; Raz et al. 2005). In fact, Fjell et al. (2009) found that prefrontal cortex volume decreases could be detected in a 1-year interval for healthy non-demented people at an average age of 76 years. Fjell et al. (2009) argued that the pattern of loss is different than that seen in Alzheimer's patients, so the losses are not primarily attributable to preclinical Alzheimer's disease.

The hippocampus is another neural area of obvious cognitive importance. There is general agreement that the hippocampus loses volume with age (e.g., Scahill et al. 2003; Raz et al. 2005; Walhovd et al. 2009). Like the cortex, the volume of the hippocampus shows detectable decreases in the aged at observation intervals of 2.5 years or less (Fjell et al. 2009; Raz et al. 2010).

It should be noted that most areas of the brain, with the exception of the brain stem, show varying degrees of shrinkage with aging (Raz and Rodrigue 2006; Fjell et al. 2009). This includes other structures associated with cognition such as the cerebellum and striatum that are less often studied but can contribute to cognitive changes.

How the loss of neural tissue relates to function is not well-understood. There is evidence for increased recruitment of frontal, as opposed to posterior, areas of the cortex with age even as executive function declines (Phillips and Andres 2010), which indicates that there are basic organizational changes. At an even more basic level, the cellular changes that are the basis for the loss of volume have not been fully investigated. There is an assumption that dendrites and synapses, but not neurons, are lost. We will explore the known evidence for this assumption in subsequent sections.

The other major discernable neural compartment in structural MRI is white matter, and there is a consensus that the cortical white matter and the corpus callosum decrease in the aged (e.g., Bartzokis et al. 2001; Sullivan et al. 2002; Sowell et al. 2003; Salat et al. 2009) with the white matter associated with frontal regions being the most susceptible (Resnick et al. 2003; Gunning-Dixon et al. 2009; Raz et al. 2010). Furthermore an even more refined technique is diffusion tensor imaging, which is a measure of the integrity of white matter structure through the assessment of water diffusion. There should be more diffusion along the length of an axon than perpendicular to it which results in high anisotropy, and this indicates greater axon packing and myelination. There is a decrease in anisotropy in the cortical white matter of the aged (O'Sullivan et al. 2001; Ota et al. 2006a) and again the white matter within the prefrontal areas is the most vulnerable (Salat et al. 2005). Diffusion tensor imaging has also revealed degenerative aging effects in the perforant path associated with hippocampal function (Yassa et al. 2010). White matter loss has been correlated with a wide array of behavioral functions from gait and the mini mental exam (Ryberg et al. 2007) to task switching (Gratton et al. 2009). The cellular basis for white matter loss will also be reviewed in a later section.

2 Neuron Loss

2.1 The Cerebral Cortex

2.1.1 Humans

Historically, there was a report from human autopsy tissue that aged humans have large decreases in cell number (20–50%) (Brody 1955). This study had many shortcomings: it looked at postmortem brains from birth to 95 years, so that only 9 brains were from individuals at 70 or more years of age; dementia status was not noted and the methods were inadequate even apart from stereology. Yet the reaction to this study seems to continue, so that other indications of neuron loss are often ignored because they are minimal in comparison to this early study.

For example, Terry et al. (1987) did not use stereology but they did carefully note that the lack of neuron density change combined with a decrease in cortical size led to an estimate of approximately 10% loss in temporal and 15% loss of neurons in frontal cortices. This observation was made in the Discussion of the manuscript but has not been noted in the literature, perhaps because it was not included in the Abstract. These estimates were confirmed in a stereologically correct study by Pakkenberg and Gundersen (1997) who showed that between ages 20 and 93 years, 10% of cortical neurons are lost across all of the hemispheres in both genders. They did not separate lobes or other areas within the cortex so that this decrease may be unevenly spaced across the cortex as the volume data suggests. To date, there do not appear to be any studies of cortical subregions that have combined stereological counting techniques with consideration of reference volume (either through reconstruction of reference volume or through the optical fractionators) so that the total number of neurons can be calculated during aging.

2.1.2 Rhesus Monkeys

Like the literature on age-related neuron loss in humans, surprisingly little work has looked at this variable in laboratory primates or rodents. The small extant literature does indicate that loss of neurons is variable across cortical and hippocampal regions. In aging rhesus monkeys, Hof et al. (2000) examined two neuronal populations within the visual cortex and did not detect a difference; however, only four young adult and four aged monkeys were assessed which is insufficient to detect small differences. Also looking at rhesus monkeys, Smith et al. (2004) found a 32% decrease in the number of neurons in area 8A, a prefrontal region, and a 50% reduction in cholinergic projection neurons to this region from the nucleus basalis. Interestingly in the same animals, there was no loss of neurons in neighboring prefrontal area 46 or in the projecting neurons within the nucleus basalis to area 46 (Smith et al. 2004), which does not change in volume in aged rhesus monkeys (O'Donnell et al. 1999). It should be noted that both of these prefrontal regions are involved in working memory, which declines in both human and non-human primates (Walker et al. 1988).

2.1.3 Rodents

There is also a localized pattern of neuron loss in the rodent cerebral cortex. Heumann and Leuba (1983) reconstructed the volume of the layers in the entire cortex of the Swiss mouse and counted neurons (with a split nuclei technique) in four cortical areas to calculate density. They found decreases of 7% in the upper layers (II–IV) by 12 months of age and 15% by 24 months compared to young adults. No overall changes were seen in the lower layers. By examining specific cortical areas, Yates et al. (2008) found a loss of neurons (15%) in the lower layers (V and VI) of the ventral (areas IL and PL) medial prefrontal

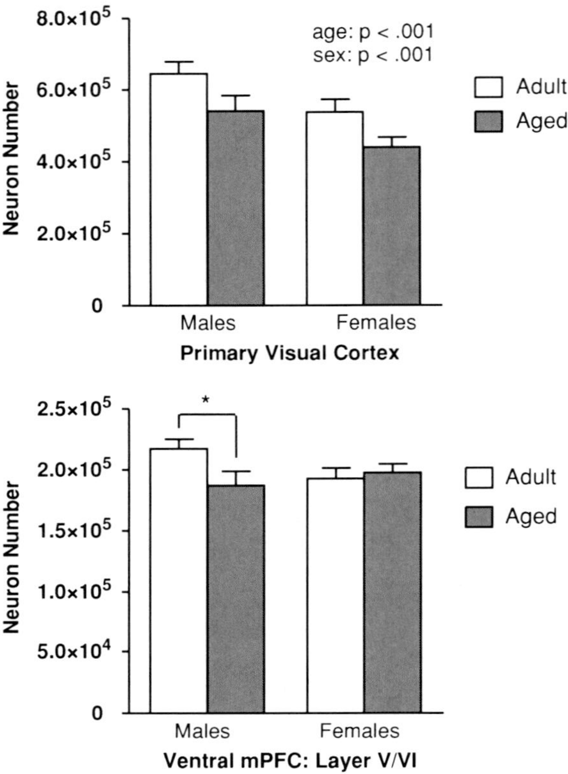

Fig. 1 Mean number ± SEM of neurons per animal in the primary visual cortex (*upper*) and in the ventral mPFC, layers 5/6 (*lower*) of adult and aged male and female rats. *$p<.04$. Modified from Yates et al. (2008)

cortex of aged male rats (Fig. 1). This loss was not found in female rats of the same age or in the upper layers. The specificity of this loss is shown by the lack of neuronal loss in either sex in the neighboring dorsal region (ACd and ACv) of the medial prefrontal cortex. Curcio and Coleman (1982) also found no neuronal loss in the cortical barrels (layer IV) of aging mice. In contrast, Yates et al. (2008) found a loss of neurons (18–20%) from all of the layers of the primary visual cortex (OC1) except for layer IV in both males and females (Fig. 1). There have been reports of loss of photoreceptors in aged albino rats that affect water-maze performance (O'Steen et al. 1995; Spencer et al. 1995) but pigmented (Long Evans hooded) rats were examined in Yates et al., and the principal layer that receives thalamic input, layer IV, did not lose neurons. This makes it unlikely that aging changes in the retina are responsible. It is also of note that Peters et al. (1983) found no aging changes in neuronal density or cortical thickness in rat primary visual cortex which would seem to indicate that there were no neuronal losses in the region. However, we found that the neuronal loss was dependent on decreases in the length and width of the visual cortex, not the thickness, and we also did not find a change in neuronal density, yet we found a relatively large loss of neurons (Yates et al. 2008). This illustrates the importance of taking the volume of a structure into account when examining aging and the number of neurons.

Another example of the specificity of neuronal loss during aging is illustrated by Shi et al. (2006). They found that there was no overall loss of neurons or inhibitory interneurons in the somatosensory cortex in aged F344 × brown Norway rats, but there was a decrease in the parvalbumin subtype of inhibitory interneuron by 25–29 months of age.

2.1.4 Summary

Both monkey and rodent data indicate that there are losses of neurons in the cortex during aging but they are localized to particular cortical areas while other parts of the cortex do not show detectable losses. The regional losses of volume in MRI human studies support, but do not prove, this generalization. This means that any overall measure of neuronal loss in the cortex will show only subtle differences at best. Also, where there is a loss of neurons, the volume of the area often decreases so that density measures alone, even when accompanied by cortical thickness, are not sufficient to establish whether there was a loss of neurons in the region.

2.2 The Hippocampus

Performance on hippocampal-dependent tasks often shows age-related declines (Gallagher and Rapp 1997) and neurogenesis in the dentate gyrus declines with age (Klempin and Kempermann 2007), yet stereological studies have found only occasional signs of neuronal loss in the major subfields that comprise the hippocampal tri-synaptic circuit (dentate gyrus, CA 2/3, CA1) in any species. In humans, the major subfields do not detectably lose neurons with age but there are sizable losses in both the hilus (31%) and subiculum (52%) (West 1993). Simic et al. (1997) also used stereological methods and found decreases in neuron number during normal human aging in the hilus and CA1 but not the other subfields. In contrast to the human data, no loss of neurons has been found in any subfield of the rhesus monkey (Keuker et al. 2003) or tree shrew (Keuker et al. 2004). Likewise, no neuronal losses were found in any portion of the major hippocampal subfields in rats (Rasmussen et al. 1996; Rapp and Gallagher 1996) or C57BL/6 mice (Calhoun et al. 1998). However, there is a report of a decrease of neurons in the rat hilus at an age (24 months) that is in range of the previous studies in humans (Azcoitia et al. 2005). In sum, the major subfields generally do not have a detectable loss of neurons in the aged, but other portions of the hippocampus have been documented to have losses.

3 White Matter

White matter is composed of axons, both myelinated and unmyelinated, and glial cells, and all of these cellular elements need to be quantified to understand the basis for the decreases in cortical white matter found in human aging. This quantification is not always possible. Axons, especially unmyelinated ones, can be very small ($< 1\ \mu$) and cannot be discerned in light microscopy so that electron microscopy must be used. The tightly packed axons and large amount of myelination in the white matter that result in the high anisotropy, which is used in diffusion tensor imaging, also make it impossible for fixative to reach the whole structure as quickly as is required for optimal electron microscopy. This is even more of a problem in aging where there is a loss of neural vasculature (reviewed by Riddle et al. 2003). It has been our experience that as myelin is added to the axons of the rat corpus callosum during development, the fixation of the unmyelinated fibers becomes increasingly poor to the point that it is not possible to unequivocally quantify the number of unmyelinated axons by adulthood. The human cortical white matter is even more heavily myelinated than in the rat so that accurate counts of the number of unmyelinated axons within it have never been performed. All of this means that we do not know if axons are lost in the cortical white matter, including the corpus callosum, during aging in any species.

In contrast, myelin sheaths are easily preserved and can be quantified even in poorly preserved human tissue. There is evidence of loss of myelin in the corpus callosum and white matter under the cortex in human autopsy tissue (Meier-Ruge et al. 1992; Marner et al. 2003) that corroborates the findings of decreased anisotropy in diffusion tensor imaging. In the rhesus monkey, disruptions in the myelin sheath in the corpus callosum have been found (Peters et al. 2000; Peters and Sethares 2003) and more recently a decrease in the number of myelinated fibers per unit area has been documented (Bowley et al. 2010). If the corpus callosum does not increase in size in the aged monkey, this represents a decrease in the number of myelinated axons per se. It is not known whether there is a decrease in the number of unmyelinated axons.

The laboratory rat does not present the same clear picture as the human/nonhuman primates. We have found no decrease in the size of the genu and splenium of the corpus callosum in aged male and female rats (Yates and Juraska 2007). Also using a light microscopic counting technique that correlates with the counts of myelinated axons in electron microscopy (Markham et al. 2009), we found that the area of myelinated profiles did not change in the aged compared to middle aged rats (Yates and Juraska 2007). Although the average age of these rats was 21 months, there were rats in the 24–26 month range that showed no evidence of loss. It is of note that both the area and amount of myelin tended to increase between young adults (4 months) and middle aged (13 months), especially in female rats. Our results were corroborated by Peiffer et al. (2010) using diffusion tensor imaging that indicated an increase in myelin between young adulthood and old age. In contrast, Yang et al. (2009) found a decrease in cortical white matter volume and in myelination in aged (27 months) rats. Disruptions in myelination (that may or may not indicate a change in the number of myelinated axons) have also been seen in electron microscopy of the aged rat corpus callosum (Sargon et al. 2007). One might speculate that the disparities between studies could be due to environmental factors.

4 Dendrites and Synapses

4.1 The Cerebral Cortex

There are numerous studies showing dendritic and synaptic changes, most often decreases, in parts of the cortex and hippocampus during aging of several species. There is surprising concordance on these measures given how readily they can change with both enriching (Green et al. 1983; Greenough et al. 1986) and stressful (Fuchs et al. 2006; Holmes and Wellman 2009) environments. Work has been concentrated on the prefrontal cortex which appears to be particularly vulnerable to aging.

4.1.1 Humans

No decrease in the density of synapses was detected in the temporal cortex (Gibson 1983) but without knowing volume, it is not possible to determine if synapses were lost during aging in this region. However, a decrease in synaptic density has been found in the aging prefrontal cortex (Huttenlocher et al. 1979; Gibson 1983). Although the volume of the prefrontal cortex was not measured in either study, current MRI studies (reviewed above) indicate that it is likely the volume decreased (or at least stayed the same). Thus this density change probably represents a decrease in synaptic number unless the aged synapses decreased in length which would make them less likely to appear in an electron micrograph. There is support for a decrease in synapses from studies using the Golgi method. A loss of dendritic spines, and to a lesser degree dendrites, has been found in portions of the prefrontal and visual cortex (Jacobs et al. 1997). Also, de Brabander et al. (1998) found a loss of dendrites in layer V in two portions (areas 9 and 46) of the aged prefrontal/frontal cortex, although no losses were detected in layer IIIc.

4.1.2 Rhesus Monkeys

There have been several studies showing dendritic and synapse loss in the prefrontal cortex of aging rhesus monkeys. These studies are consistent with the findings of loss in the prefrontal cortex of humans. Cupp and Uemura (1980) found a loss of dendrites using the Golgi method, and similarly Duan et al. (2003) found a loss of dendrites and spines in intracellularly filled prefrontal neurons that projected to the temporal cortex. Furthermore, Dumitriu et al. (2010) have shown that the loss of spines tended to be the thin, more plastic spines. They also found a comparable loss of axospinal synapses in layer III of the prefrontal cortex using electron microscopy and this loss was correlated with performance on a delayed non-match to sample task.

Soghomonian et al. (2010) found that the size of inhibitory synapses (axosomatic and axodendritic) in the upper layers of the aging prefrontal cortex increases as does the number of synaptic vesicles in axosomatic synapses, all of which suggests increased inhibition in aged monkeys. On the other hand, an analysis of the cable properties of dendrites and spines in the aged indicates that neuronal excitability of neurons should be increasing (Kabaso et al. 2009). There is an increased action potential firing rate in the aged prefrontal cortex of monkeys that correlates with impaired performance on several delay tasks (Chang et al. 2005) confirming the altered functional properties of aging neurons in the prefrontal cortex.

4.1.3 Rodents

The studies looking at the aging rat cortex have examined more cortical areas than in the rhesus monkey. As early as 1975, Feldman and Dowd reported a decrease in dendritic spines in the visual cortex of the aged rat, and Vaughan (1977) found that there was atrophy of the dendritic tree in the aged rat auditory cortex. There are age-related losses of dendrites and spines in the upper (Grill and Riddle 2002; Wallace et al. 2007) and lower (Markham and Juraska 2002) layers of the prefrontal cortex. Wong et al. (1998) found a decrease in synaptophysin, a protein found in presynaptic vesicles and thus a biochemical marker for synapses, in the aging parietal cortex. In examining the electrophysiological properties of the parietal neurons, Wong et al. (2000) found that both excitatory and inhibitory spontaneous post-synaptic potentials were decreased during aging, but the aged animals with impairments in the Morris water maze had a bias toward more inhibitory potentials (Wong et al. 2006).

Although most of the cortical areas that have been examined in the rat show losses of dendrites, not all of the closely connected subcortical areas mirror these effects. For example, the principle neurons of the basolateral amygdala project to and receive projections from the medial prefrontal cortex (Krettek and Price 1977; Neafsey et al. 1993). However in contrast to the prefrontal neurons, the principle neurons of the basolateral amygdala have larger dendritic trees in aged rats compared to young adults (Rubinow et al. 2009). This is not a compensation for a loss of neurons because no change in the number of neurons during aging was found in a stereological study (Rubinow and Juraska 2009). It is not obvious why some neural areas are vulnerable to neuroanatomical losses during aging and others not.

4.2 Hippocampus

4.2.1 Humans

Unlike the cerebral cortex, the literature is mixed on whether dendrites and synapses are lost in the hippocampus. In humans, Eastwood et al. (2006) did not detect a change in synaptophysin protein between adulthood and old age in any of the hippocampal subfields. There are limits to the sensitivity of the immunoautoradiographic technique used and there were only 4 aged brains in the study but it can be concluded that massive synaptic loss does not occur in the aging human hippocampus.

4.2.2 Rhesus Monkeys

The studies examining aged rhesus monkeys are mixed. Haley et al. (2010) found a decrease in synaptophysin in all subfields of the hippocampus and the entorhinal cortex. On the other hand, Tigges et al. (1996), using a stereological method for electron microscopy, found no decrease in the number of synapses in aged monkey hippocampus. Uemura (1985) found a decrease in the dendritic tree of the subiculum, an area that is broadly part of the hippocampal formation but outside of the major subfields. This is a neural region that merits more study.

4.2.3 Rodents

There is more research in the rodent but not more clarity. Calhoun et al. (1998) did not find a decrease in the number of synaptophysin positive boutons in CA1 of the aging C57BL/6 mouse. Likewise, Nicolle et al. (1999) found no age-related decrease in synaptophysin in Western blots. This was corroborated by Smith et al. (2000) who found no changes in synaptophysin with aging in any of the hippocampal subfields but did find that synaptophysin levels correlated with performance on the water maze in the dentate and CA3 fields of aged rats. However, Geinisman et al. (2004) found no decrease in synapses in CA1 using electron microscopy and stereological techniques in either behaviorally impaired or unimpaired aged rats. In contrast, Markham et al. (2005) found small decreases in the apical dendritic tree in CA1, and this was corroborated by Shi et al. (2005), who did stereological electron microscopy in this portion of CA1 and found a decrease in the number of synapses. Rat strain does not account for any the differences between studies and one can only speculate that environmental factors, even if experienced when the rat is developing might alter the course of later aging (Black et al. 1991).

5 Neurotransmission

5.1 *Monoamines*

5.1.1 Humans

While age-related changes occur in the dopaminergic, serotoninergic and noradrenergic systems, the focus here will be the dopaminergic system because of its role in maintaining cognitive function (reviewed in Bäckman et al. 2010). Dopamine is also involved in many pathologies associated with aging and as a result has been extensively studied during aging. Human autopsy studies have found decreases in both striatal D_1 (Rinne et al. 1990) and D_2 (Seeman et al. 1987; Severson et al. 1982) receptor densities across the life span. In addition, positron emission tomography (PET) and single-photon emission computed tomography (SPECT) studies have found similar decreases in D_1 (Suhara et al. 1991; Wang et al. 1998) and D_2 (Antonini et al. 1993; Ichise et al. 1998) receptor densities in the striatum. Furthermore, PET and autopsy studies have found decreases in D_1 (Suhara et al. 1991; De Keyser et al. 1990) and D_2 receptors in both the frontal cortex and hippocampus, with the fastest rate of decline found in the frontal cortex (Kaasinen et al. 2000, 2002; Inoue et al. 2001). Studies have also found a decrease in dopamine synthesis during aging and this decrease was the greatest in the dorsal lateral prefrontal cortex (Ota et al. 2006b). In addition to the changes found in receptors densities, age-related declines in the dopamine transporter (DAT) have been found in both postmortem (Allard and Marcusson, 1989; Bannon and Whitty 1997) and PET and SPECT studies (Van Dyck et al. 1995; Rinne et al. 1998). Furthermore there is evidence that the enzymes that degrade monoamines are altered during aging. Monoamines are degraded by two monoamine oxidases, MAO-A and MAO-B. Although the activity of MAO-A was not found to increase during aging, an age-related increase in MAO-B was found in the cerebral cortex and hippocampus starting in the fifth decade of life (Saura et al. 1997).

5.1.2 Rhesus Monkeys

Decreases in D_2 receptors have also been found in the aged monkey brain (Lai et al. 1987; Morris et al. 1999) as well as reductions in dopamine synthesis and DAT availability (Harada et al. 2002). Similar to the human research, endogenous levels of dopamine decrease in the cerebral cortex of rhesus monkeys with the prefrontal cortex experiencing the greatest decline (Goldman-Rakic and Brown 1981).

5.1.3 Rodents

Research in rodents has produced findings similar to those in humans and other primates. A decrease in D_1 and D_2 receptors was found in aged rats as well as a significant decrease in striatal levels of dopamine (Gozlan et al. 1990; Hyttel 1987).

Studies have also found decreased levels of D_2 receptor mRNA during aging in rats (Mesco et al. 1993). In addition, MAO-B is decreased in the frontal cortex of aged male rats (Amenta et al. 1994) Furthermore, mRNA levels for both the dopamine transporter and tyrosine hydroxylase, a rate limiting enzyme in dopamine synthesis, are significantly reduced in the aged rat substantia nigra which projects to both the hippocampus and prefrontal cortex (Himi et al. 1995). Indeed, a recent study found that activity of tyrosine hydroxylase was decreased in the medial prefrontal cortex during aging and this resulted in impaired performance on a mPFC mediated task (Mizoguchi et al. 2009).

5.2 *Acetylcholine*

The prefrontal cortex and hippocampus also receive cholinergic projections from the basal forebrain and these projections are known to play a role in learning and memory (Baxter and Chiba 1999). Cholinergic degeneration in the hippocampus and neocortex is commonly associated with Alzheimer's disease (Schliebs and Arendt 2006), but these changes are not consistently found during normal aging. Although the concentration of choline or acetylcholine does not appear to change, the synthesis of acetylcholine is decreased during aging in two strains of mice (Gibson et al. 1981) and measures of cholinergic functioning such as evoked acetylcholine release (Moore et al. 1996; Takei et al. 1989) and cholinergic receptor plasticity (Pedigo and Polk 1985) are reduced in the aged rodent brain. In addition, studies have found a loss of cholinergic neurons in the basal forebrain of aged rats (Fischer et al. 1991) and rhesus monkeys (Stroessner-Johnson et al. 1992) and this is in agreement with a recent study that found a decrease in cholinergic appositions on pyramidal neurons in aged rats (Casu et al. 2002). Importantly, there is evidence that the cholinergic system in rodents can be altered by estradiol (Gibbs et al. 2009) and this will be discussed in the section below, Hormone Treatment.

5.3 *Glutamate*

The glutamatergic system is known to be involved in learning and memory and is also altered with aging (McEntee and Crook 1993); however, studies have produced contradictory results that may be explained by differences between species and strains. One of the most consistent findings during aging is a decrease in *N*-methyl-D-aspartate receptor (NMDA R) density in the frontal cortex and hippocampus of rodents (Castorina et al. 1994; Magnusson and Cotman 1993; Miyoshi et al. 1990) and non-human primates (Hof et al. 2002; Gazzaley et al. 1996). Furthermore, decreases in glutamate uptake during aging have been reported (Vatassery et al. 1998; Saransaari and Oja 1995; Wheeler and Ondo

1986), although glutamate release in the hippocampus and cortex does not appear to be altered (Palmer et al. 1994; Sanchez-Prieto et al. 1994; Dawson et al. 1989).

6 Hormone Treatment

6.1 Humans

One question of practical importance is what happens to the course of neural aging when gonadal hormone levels decrease, as in human menopause, or when they are kept from such a precipitous decrease with hormone replacement. Males also have lower gonadal hormones during aging but the hormone decline is a slow and steady one over several decades and not marked by an event that occurs in a couple of years. In contrast, females have a relatively fast decline in gonadal steroids during perimenopause that is associated with negative symptoms (e.g., hot flashes, sleep disturbance, etc.). Still, females debate whether to take hormone replacement to relieve symptoms associated with menopause, in part because the long-term consequences of hormone replacement on neural function are relatively unexplored. For these reasons, this review will concentrate on the effects of hormones on the aging female brain.

The Women's Health Initiative gave a negative report on the cognitive (including dementia) effects of estrogen replacement with or without MPA, the synthetic progestin (Rapp et al. 2003; Shumaker et al. 2003, 2004; Espeland et al. 2004). However, this cannot be considered the final word on hormone therapy especially given the long interval between the onset of menopause and the age at first replacement which animal studies indicate is critical for the behavioral outcome of estrogen exposure (Gibbs 2000).

A number of structural MRI studies have shown that hormone replacement of various types (estrogen alone or estrogen with a progestagen) decreases the shrinkage associated with aging of both the cortex and hippocampus (Boccardi et al. 2006; Resnick et al. 2009; Robertson et al. 2009; Lord et al. 2010) including a longitudinal study with a 5-year interval (Raz et al. 2004). While one might classify this as "neuroprotective", this term generally refers to protection from cell death, especially following an ischemic episode (McCullough and Hurn 2003). Given that most of the neural shrinkage is not due to cell loss, it is not clear that neuroprotection, as specifically defined, applies.

6.2 Animal Models

The cellular bases for the effects of hormone replacement have only been partially investigated in animal models during aging. In aging rhesus monkeys, a periodic dose (every 3 weeks) of estrogen increased spine numbers on layer III pyramidal neurons (Hao et al. 2006), especially the thin spines that are associated with plasticity (Hao et al. 2007). There were no effects on dendritic branching.

In rats, there are indications that gonadal steroids may be "neuroprotective" for the dendritic tree in intact females. Female rats, unlike human females, do not have ovarian failure at estropause, but rather a loss of cyclicity so that moderate levels of estrogen and progesterone are secreted at relatively constant levels (Clemens and Meites 1971; Wise and Ratner 1980). We have found that intact females have less age related atrophy of the dendritic tree and loss of spine density in the medial prefrontal cortex than males (Markham and Juraska 2002). Also male rats, but not female rats, have a loss of neurons in the upper layers of the medial prefrontal cortex (Yates et al. 2008) (Fig. 1) and a small decrease in the dendritic tree of hippocampal CA1 pyramidal neurons (Markham et al. 2005).

Whether this moderation of loss during aging is due to estrogen has not been thoroughly tested. Estrogen does not increase dendritic spines in hippocampal CA1 as it does in young adult rats. However, similar to young adult rats, it does increase NMDA receptor subunits (Adams et al. 2001). In mice, Fernandez and Frick (2004) administered estrogen for 2 months to middle aged ovariectomized female mice and measured synaptophysin using Western blots. They found that synaptophysin increased in the hippocampus with high doses of estrogen but their results in the frontoparietal tissue were equivocal in that a medium dose of estrogen decreased synaptophysin while low and high doses were without effect.

We have ovariectomized middle aged female rats and immediately treated them with estrogen (17-β estradiol; E) alone or in combination with either progesterone (P) or the synthetic progestin, medroxyprogesterone acetate (MPA), for 6 months. Estrogen treatment occurred through the drinking water either chronically or cyclically (3 days E water and 1 day regular water). Progesterone and MPA were either present chronically through pellet implants or cyclically (1 day every 4 days) by mixing with food. After 6 months of hormone treatment, we quantified synaptophysin-labeled boutons in the medial prefrontal cortex with stereological techniques (density X volume) as an indication of the number of synapses in this area.

We currently have preliminary data (Lowry and Juraska, unpublished data) that females receiving chronic E plus P had fewer synaptophysin boutons in layers 2/3 of the medial prefrontal cortex than females receiving E treatment alone (Fig. 2). Interestingly, the groups receiving cyclic E plus P, as well as those given chronic and cyclic E plus MPA, were not different than chronic E alone. In fact compared to the no replacement group, cyclic E plus MPA had more synaptophysin boutons in layers 2/3, and in layers 5/6, females given chronic E plus MPA had more synaptophysin boutons than those receiving chronic E plus P.

These results are counter to our work on the negative effects of MPA on performance on the water maze when administered in combination with estrogen (Lowry et al. 2010) or in another laboratory when administered alone (Braden et al. 2010). However, we have found that both chronic E plus P and E plus MPA enhance performance on a T maze alternation task (Lowry et al. 2008), which indicates that simple generalizations from structural to behavioral effects are complicated and undoubtedly involve more than one neural area.

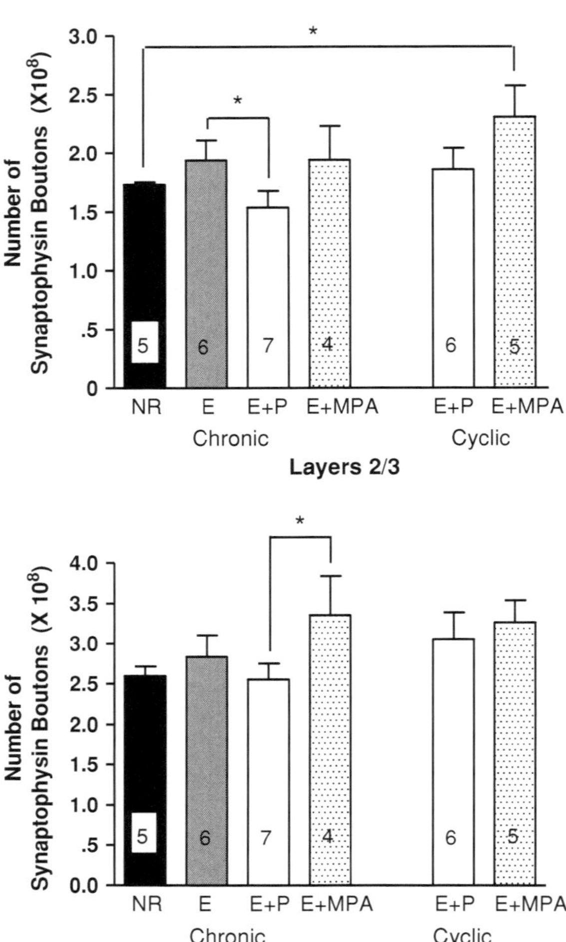

Fig. 2 Mean number ± SEM per animal of synaptophysin boutons in Layers 2/3 (*upper*) and Layers 5/6 (*lower*) of the ventral medial prefrontal cortex. *$p<.05$ No replacement (NR), 17 β-estradiol (E), E and progesterone (E + P), E and MPA (E + MPA)

There is little basis for speculation about the types of synapses that were altered with that long-term hormone treatment. Because studies in young animals have indicated that estrogen alters the cholinergic system, the neurotransmitter that is most often studied in the context of estrogen replacement in aging is acetylcholine (reviewed in Gibbs 2010). Importantly, the neural mechanisms underlying the behavioral effects of hormone treatment may be different in young and aged animals. Although a SPECT study in humans found more labeling of vesicular acetylcholine transporters in the posterior cingulate of women receiving estrogen alone compared with women receiving estrogen and progestin, there was no overall effect of estrogen treatment on this measure as compared to non-estrogen users (Smith et al. 2001). In non-human primates, long-term treatment with either conjugated equine estrogens (CEE) alone or in combination with MPA did not significantly increase choline acetyltransferase (ChAT) or acetylcholinesterase (AChE) in any brain region studied, however CEE in combination with MPA significantly decreased both ChAT and AChE in the medial septum (Gibbs et al. 2002), a brain region containing cholinergic neurons that project to the hippocampus (Mesulam et al. 1983). Long-term hormone treatment in rats failed to prevent age-related reductions in ChAT in the medial septum (Gibbs 2003) and estradiol benzoate did not alter ChAT activity in the hippocampus or frontal cortex of aged female mice (Frick et al. 2002). Interestingly, short-term estrogen treatment increased ChAT in the hippocampus of middle aged female rats (Gibbs et al. 2009; Bohacek et al. 2008) but not aged animals (Gibbs et al. 2009). Few studies have investigated the effects of hormone treatment during aging on other neurotransmitter systems. Similar to the cholinergic system no differences were found in dopamine levels after long-term treatment with CEE in non-human primates (Gibbs et al. 2006). However, acute estrogen treatment increased basal dopamine levels in the striatum of aged rats (McDermott 1993). There are many factors that may alter the neural effects of hormone replacement during aging, including the length of treatment and it is imperative that future studies examine the effects of long-term exposure to hormones in aged animals to better model their effects on the course of neural aging.

7 Conclusions

There is ample evidence that neural areas associated with cognition, such as the cerebral cortex, decrease in size during normal, non-demented aging, but there is only partial understanding of the cellular basis for the loss. Decreases in the number of neurons, dendrites and synapses have all been detected, as have changes in neurotransmitter systems. There is, however, considerable regional specificity to aging effects and some species differences as well. The picture is not yet complete, but it would seem to be a necessary step for understanding the effects of aging on cognitive behavior and what factors could ameliorate the effects. Among these factors is hormone replacement which may have the capacity to modify the course of neural aging.

Acknowledgements Our work was supported by grants from the National Institute of Aging, AG18046 and AG022499. We thank Wendy Koss and Renee Sadowski for comments on the manuscript.

References

Adams MM, Shah RA, Janssen WG, Morrison JH (2001) Different modes of hippocampal plasticity in response to estrogenestrogen in young and aged female rats. Proc Natl Acad Sci USA 98:8071–8076. doi:10.1073/pnas.141215898

Allard P, Marcusson JO (1989) Age-correlated loss of dopamine uptake sites labeled with [3H]GBR-12935 in human putamen. Neurobiol Aging 10:661–664

Allen JS, Bruss J, Brown CK, Damasio H (2005) Normal neuroanatomical variation due to age: the major lobes and a parcellation of the temporal region. Neurobiol Aging 26:1245–1260; discussion 1279–1282. doi:10.1016/j.neurobiolaging.2005.05.023

Amenta F, Bograni S, Cadel S, Ferrante F, Valsecchi B, Vega JA (1994) Microanatomical changes in the frontal cortex of aged rats: effect of L-deprenyl treatment. Brain Res Bull 34:125–131

Antonini A, Leenders KL, Reist H, Thomann R, Beer HF, Locher J (1993) Effect of age on D2 dopamine receptors in normal human brain measured by positron emission tomography and 11C-raclopride. Arch Neurol 50:474–480

Azcoitia I, Perez-Martin M, Salazar V, Castillo C, Ariznavarreta C, Garcia-Segura LM, Tresguerres JA (2005) Growth hormone prevents neuronal loss in the aged rat hippocampus. Neurobiol Aging 26:697–703. doi:10.1016/j.neurobiolaging.2004.06.007

Bäckman L, Lindenberger U, Li SC, Nyberg L (2010) Linking cognitive aging to alterations in dopamine neurotransmitter functioning: recent data and future avenues. Neurosci Biobehav Rev 34:670–677. doi:10.1016/j.neubiorev.2009.12.008

Bannon MJ, Whitty CJ (1997) Age-related and regional differences in dopamine transporter mRNA expression in human midbrain. Neurology 48:969–977

Bartzokis G, Beckson M, Lu PH, Nuechterlein KH, Edwards N, Mintz J (2001) Age-related changes in frontal and temporal lobe volumes in men: a magnetic resonance imaging study. Arch Gen Psychiatry 58:461–465.

Baxter MG, Chiba AA (1999) Cognitive functions of the basal forebrain. Curr Opin Neurobiol 9:178–183

Black JE, Isaacs KR, Greenough WT (1991) Usual vs successful aging: some notes on experiential factors. Neurobiol Aging 12:325–328; discussion 352–355

Boccardi M, Ghidoni R, Govoni S, Testa C, Benussi L, Bonetti M, Binetti G, Frisoni GB (2006) Effects of hormone therapy on brain morphology of healthy postmenopausal women: a Voxel-based morphometry study. Menopause 13:584–591. doi:10.1097/01.gme.0000196811.88505.10

Bohacek J, Bearl AM, Daniel JM (2008) Long-term ovarian hormone deprivation alters the ability of subsequent oestradiol replacement to regulate choline acetyltransferase protein levels in the hippocampus and prefrontal cortex of middle-aged rats. J Neuroendocrinol 20:1023–1027. doi:10.1111/j.1365-2826.2008.01752.x

Bowley MP, Cabral H, Rosene DL, Peters A (2010) Age changes in myelinated nerve fibers of the cingulate bundle and corpus callosum in the rhesus monkey. J Comp Neurol 518: 3046–3064. doi:10.1002/cne.22379

Braden BB, Talboom JS, Crain ID, Simard AR, Lukas RJ, Prokai L, Scheldrup MR, Bowman BL, Bimonte-Nelson HA (2010) Medroxyprogesterone acetate impairs memory and alters the GABAergic system in aged surgically menopausal rats. Neurobiol Learn Mem 93:444–453. doi:10.1016/j.nlm.2010.01.002

Brody H (1955) Organization of the cerebral cortex. III. A study of aging in the human cerebral cortex. J Comp Neurol 102:511–516

Calhoun ME, Kurth D, Phinney AL, Long JM, Hengemihle J, Mouton PR, Ingram DK, Jucker M (1998) Hippocampal neuron and synaptophysin-positive bouton number in aging C57BL/6 mice. Neurobiol Aging 19:599–606

Castorina M, Ambrosini AM, Pacific L, Ramacci MT, Angelucci L (1994) Age-dependent loss of NMDA receptors in hippocampus, striatum, and frontal cortex of the rat: prevention by acetyl-L-carnitine. Neurochem Res 19:795–798

Casu MA, Wong TP, De Koninck Y, Ribeiro-da-Silva A, Cuello AC (2002) Aging causes a preferential loss of cholinergic innervation of characterized neocortical pyramidal neurons. Cereb Cortex 12:329–337

Chang YM, Rosene DL, Killiany RJ, Mangiamele LA, Luebke JI (2005) Increased action potential firing rates of layer 2/3 pyramidal cells in the prefrontal cortex are significantly

related to cognitive performance in aged monkeys. Cereb Cortex 15:409–418. doi:10.1093/cercor/bhh144
Clemens JA, Meites J (1971) Neuroendocrine status of old constant-estrous rats. Neuroendocrinology 7:249–256
Cupp CJ, Uemura E (1980) Age-related changes in prefrontal cortex of Macaca mulatta: quantitative analysis of dendritic branching patterns. Exp Neurol 69:143–163. doi:10.1016/0197-4580(91)90077
Curcio CA, Coleman PD (1982) Stability of neuron number in cortical barrels of aging mice. J Comp Neurol 212:158–172. doi:10.1002/cne.902120206
Dawson R Jr, Wallace DR, Meldrum MJ (1989) Endogenous glutamate release from frontal cortex of adult and aged rats. Neurobiol Aging 10:665–668
de Brabander JM, Kramers RJ, Uylings HB (1998) Layer-specific dendritic regression of pyramidal cells with ageing in the human prefrontal cortex. Eur J Neurosci 10:1261–1269
de Keyser J, De Backer JP, Vauquelin G, Ebinger G (1990) The effect of aging on the D1 dopamine receptors in human frontal cortex. Brain Res 528:308–310
Duan H, Wearne SL, Rocher AB, Macedo A, Morrison JH, Hof PR (2003) Age-related dendritic and spine changes in corticocortically projecting neurons in macaque monkeys. Cereb Cortex 13:950–961
Dumitriu D, Hao J, Hara Y, Kaufmann J, Janssen WG, Lou W, Rapp PR, Morrison JH (2010) Selective changes in thin spine density and morphology in monkey prefrontal cortex correlate with aging-related cognitive impairment. J Neurosci 30:7507–7515. doi:10.1523/JNEUROSCI.6410-09.2010
Eastwood SL, Weickert CS, Webster MJ, Herman MM, Kleinman JE, Harrison PJ (2006) Synaptophysin protein and mRNA expression in the human hippocampal formation from birth to old age. Hippocampus 16:645–654. doi:10.1002/hipo.20194
Espeland MA, Rapp SR, Shumaker SA, Brunner R, Manson JE, Sherwin BB, Hsia J, Margolis KL, Hogan PE, Wallace R, Dailey M, Freeman R, Hays J, Women's Health Initiative Memory Study (2004) Conjugated equine estrogens and global cognitive function in postmenopausal women: Women's Health Initiative Memory Study. JAMA 291:2959–2968. doi:10.1001/jama.291.24.2959
Feldman ML, Dowd C (1975) Loss of dendritic spines in aging cerebral cortex. Anat Embryol (Berl) 148:279–301
Fernandez SM, Frick KM (2004) Chronic oral estrogen affects memory and neurochemistry in middle-aged female mice. Behav Neurosci 118:1340–1351. doi:10.1037/0735-7044.118.6.1340
Fischer W, Bjorklund A, Chen K, Gage FH (1991) NGF improves spatial memory in aged rodents as a function of age. J Neurosci 11:1889–1906
Fjell AM, Walhovd KB, Fennema-Notestine C, McEvoy LK, Hagler DJ, Holland D, Brewer JB, Dale AM (2009) One-year brain atrophy evident in healthy aging. J Neurosci 29:15223–15231. doi:10.1523/JNEUROSCI.3252-09.2009
Frick KM, Fernandez SM, Bulinski SC (2002) Estrogen replacement improves spatial reference memory and increases hippocampal synaptophysin in aged female mice. Neuroscience 115:547–558
Fuchs E, Flugge G, Czeh B (2006) Remodeling of neuronal networks by stress. Front Biosci 11:2746–2758
Gallagher M, Rapp PR (1997) The use of animal models to study the effects of aging on cognition. Annu Rev Psychol 48:339–370. doi:10.1146/annurev.psych.48.1.339
Gazzaley AH, Siegel SJ, Kordower JH, Mufson EJ, Morrison JH (1996) Circuit-specific alterations of N-methyl-D-aspartate receptor subunit 1 in the dentate gyrus of aged monkeys. Proc Natl Acad Sci USA 93:3121–3125
Geinisman Y, Ganeshina O, Yoshida R, Berry RW, Disterhoft JF, Gallagher M (2004) Aging, spatial learning, and total synapse number in the rat CA1 stratum radiatum. Neurobiol Aging 25:407–416. doi:10.1016/j.neurobiolaging.2003.12.001
Gibbs RB (2010) Estrogen therapy and cognition: a review of the cholinergic hypothesis. Endocr Rev 31:224–253. doi:10.1210/er.2009-0036

Gibbs RB, Mauk R, Nelson D, Johnson DA (2009) Donepezil treatment restores the ability of estradiol to enhance cognitive performance in aged rats: evidence for the cholinergic basis of the critical period hypothesis. Horm Behav 56:73–83. doi:10.1016/j.yhbeh.2009.03.003

Gibbs RB, Edwards D, Lazar N, Nelson D, Talameh J (2006) Effects of long-term hormone treatment and of tibolone on monoamines and monoamine metabolites in the brains of ovariectomised, Cynomologous monkeys. J Neuroendocrinol 18:643–654. doi:10.1111/j.1365-2826.2006.01463.x

Gibbs RB (2003) Effects of ageing and long-term hormone replacement on cholinergic neurones in the medial septum and nucleus basalis magnocellularis of ovariectomized rats. J Neuroendocrinol 15:477–485

Gibbs RB, Nelson D, Anthony MS, Clarkson TB (2002) Effects of long-term hormone replacement and of tibolone on choline acetyltransferase and acetylcholinesterase activities in the brains of ovariectomized, cynomologus monkeys. Neuroscience 113:907–914

Gibbs RB (2000) Long-term treatment with estrogen and progesteroneprogesterone enhances acquisition of a spatial memory task by ovariectomized aged rats. Neurobiol Aging 21: 107–116. doi:10.1016/S0197-4580(00)00103-2

Gibson GE, Peterson C, Jenden DJ (1981) Brain acetylcholine synthesis declines with senescence. Science 213:674–676

Gibson PH (1983) EM study of the numbers of cortical synapses in the brains of ageing people and people with Alzheimer-type dementia. Acta Neuropathol 62:127–133

Goldman-Rakic PS, Brown RM (1981) Regional changes of monoamines in cerebral cortex and subcortical structures of aging rhesus monkeys. Neuroscience 6:177–187

Gozlan H, Daval G, Verge D, Spampinato U, Fattaccini CM, Gallissot MC, el Mestikawy S, Hamon M (1990) Aging associated changes in serotoninergic and dopaminergic pre- and postsynaptic neurochemical markers in the rat brain. Neurobiol Aging 11:437–449

Gratton G, Wee E, Rykhlevskaia EI, Leaver EE, Fabiani M (2009) Does white matter matter? Spatio-temporal dynamics of task switching in aging. J Cogn Neurosci 21:1380–1395. doi: 10.1162/jocn.2009.21093

Green EJ, Greenough WT, Schlumpf BE (1983) Effects of complex or isolated environments on cortical dendrites of middle-aged rats. Brain Res 264:233–240

Greenough WT, McDonald JW, Parnisari RM, Camel JE (1986) Environmental conditions modulate degeneration and new dendrite growth in cerebellum of senescent rats. Brain Res 380:136–143

Grill JD, Riddle DR (2002) Age-related and laminar-specific dendritic changes in the medial frontal cortex of the rat. Brain Res 937:8–21

Gunning-Dixon FM, Brickman AM, Cheng JC, Alexopoulos GS (2009) Aging of cerebral white matter: a review of MRI findings. Int J Geriatr Psychiatry 24:109–117. doi:10.1002/gps.2087

Haley GE, Kohama SG, Urbanski HF, Raber J (2010) Age-related decreases in SYN levels associated with increases in MAP-2, apoE, and GFAP levels in the rhesus macaque prefrontal cortex and hippocampus Age (Dordr) 32:283–296. doi:10.1007/s11357-010-9137-9

Hao J, Rapp PR, Janssen WG, Lou W, Lasley BL, Hof PR, Morrison JH (2007) Interactive effects of age and estrogen on cognition and pyramidal neurons in monkey prefrontal cortex. Proc Natl Acad Sci USA 104:11465–11470. doi:10.1073/pnas.0704757104

Hao J, Rapp PR, Leffler AE, Leffler SR, Janssen WG, Lou W, McKay H, Roberts JA, Wearne SL, Hof PR, Morrison JH (2006) Estrogen alters spine number and morphology in prefrontal cortex of aged female rhesus monkeys. J Neurosci 26:2571–2578. doi:10.1523/JNEUROSCI. 3440-05.2006

Harada N, Nishiyama S, Satoh K, Fukumoto D, Kakiuchi T, Tsukada H (2002) Age-related changes in the striatal dopaminergic system in the living brain: a multiparametric PET study in conscious monkeys. Synapse 45:38–45. doi:10.1002/syn.10082

Heumann D, Leuba G (1983) Neuronal death in the development and aging of the cerebral cortex of the mouse. Neuropathol Appl Neurobiol 9:297–311

Himi T, Cao M, Mori N (1995) Reduced expression of the molecular markers of dopaminergic neuronal atrophy in the aging rat brain. J Gerontol A Biol Sci Med Sci 50:B193–200

Hof PR, Duan H, Page TL, Einstein M, Wicinski B, He Y, Erwin JM, Morrison JH (2002) Age-related changes in GluR2 and NMDAR1 glutamate receptor subunit protein immunoreactivity in corticocortically projecting neurons in macaque and patas monkeys. Brain Res 928:175–186

Hof PR, Nimchinsky EA, Young WG, Morrison JH (2000) Numbers of meynert and layer IVB cells in area V1: a stereologic analysis in young and aged macaque monkeys. J Comp Neurol 420:113–126

Holmes A, Wellman CL (2009) Stress-induced prefrontal reorganization and executive dysfunction in rodents. Neurosci Biobehav Rev 33:773–783. doi:10.1016/j.neubiorev.2008.11.005

Huttenlocher PR (1979) Synaptic density in human frontal cortex–developmental changes and effects of aging. Brain Res 163:195–205

Hyttel J (1987) Age related decrease in the density of dopamine D1 and D2 receptors in corpus striatum of rats. Pharmacol Toxicol 61:126–129

Ichise M, Ballinger JR, Tanaka F, Moscovitch M, St George-Hyslop PH, Raphael D, Freedman M (1998) Age-related changes in D2 receptor binding with iodine-123-iodobenzofuran SPECT. J Nucl Med 39:1511–1518

Inoue M, Suhara T, Sudo Y, Okubo Y, Yasuno F, Kishimoto T, Yoshikawa K, Tanada S (2001) Age-related reduction of extrastriatal dopamine D2 receptor measured by PET. Life Sci 69:1079–1084

Jacobs B, Driscoll L, Schall M (1997) Life-span dendritic and spine changes in areas 10 and 18 of human cortex: a quantitative Golgi study. J Comp Neurol 386:661–680

Kaasinen V, Kemppainen N, Nagren K, Helenius H, Kurki T, Rinne JO (2002) Age-related loss of extrastriatal dopamine D(2) -like receptors in women. J Neurochem 81:1005–1010

Kaasinen V, Vilkman H, Hietala J, Nagren K, Helenius H, Olsson H, Farde L, Rinne J (2000) Age-related dopamine D2/D3 receptor loss in extrastriatal regions of the human brain. Neurobiol Aging 21:683–688

Kabaso D, Coskren PJ, Henry BI, Hof PR, Wearne SL (2009) The electrotonic structure of pyramidal neurons contributing to prefrontal cortical circuits in macaque monkeys is significantly altered in aging. Cereb Cortex 19:2248–2268. doi:10.1093/cercor/bhn242

Keuker JI, de Biurrun G, Luiten PG, Fuchs E (2004) Preservation of hippocampal neuron numbers and hippocampal subfield volumes in behaviorally characterized aged tree shrews. J Comp Neurol 468:509–517. doi:10.1002/cne.10996

Keuker JI, Luiten PG, Fuchs E (2003) Preservation of hippocampal neuron numbers in aged rhesus monkeys. Neurobiol Aging 24:157–165

Klempin F, Kempermann G (2007) Adult hippocampal neurogenesis and aging. Eur Arch Psychiatry Clin Neurosci 257:271–280. doi:10.1007/s00406-007-0731-5

Krettek JE, Price JL (1977) Projections from the amygdaloid complex to the cerebral cortex and thalamus in the rat and cat. J Comp Neurol 172:687–722. doi:10.1002/cne.901720408

Lai H, Bowden DM, Horita A (1987) Age-related decreases in dopamine receptors in the caudate nucleus and putamen of the rhesus monkey (Macaca mulatta). Neurobiol Aging 8:45–49

Lord C, Engert V, Lupien SJ, Pruessner JC (2010) Effect of sex and estrogen therapy on the aging brain: a voxel-based morphometry study. Menopause 17:846–851. doi:10.1097/gme.0b013e3181e06b83

Lowry NC, Yates MA, Juraska JM (2008) Ovarian hormones in aged female rats benefit acquisistion of a spatial alternation task, but do not improve performance during delayed alternation. Soc Neurosci Abs Online

Lowry NC, Pardon LP, Yates MA, Juraska JM (2010) Effects of long-term treatment with 17 beta-estradiol and medroxyprogesterone acetate on water maze performance in middle aged female rats. Horm Behav 58:200–207. doi:10.1016/j.yhbeh.2010.03.018

Magnusson KR, Cotman CW (1993) Age-related changes in excitatory amino acid receptors in two mouse strains. Neurobiol Aging 14:197–206

Markham JA, Herting MM, Luszpak AE, Juraska JM, Greenough WT (2009) Myelination of the corpus callosum in male and female rats following complex environment housing during adulthood. Brain Res 1288:9–17. doi:10.1016/j.brainres.2009.06.087

Markham JA, McKian KP, Stroup TS, Juraska JM (2005) Sexually dimorphic aging of dendritic morphology in CA1 of hippocampus. Hippocampus 15:97–103. doi:10.1002/hipo.20034

Markham JA, Juraska JM (2002) Aging and sex influence the anatomy of the rat anterior cingulate cortex. Neurobiol Aging 23:579–588

Marner L, Nyengaard JR, Tang Y, Pakkenberg B (2003) Marked loss of myelinated nerve fibers in the human brain with age. J Comp Neurol 462:144–152. doi:10.1002/cne.10714

McCullough LD, Hurn PD (2003) Estrogen and ischemic neuroprotection: an integrated view. Trends Endocrinol Metab 14:228–235

McDermott JL (1993) Effects of estrogen upon dopamine release from the corpus striatum of young and aged female rats. Brain Res 606:118–125

McEntee WJ, Crook TH (1993) Glutamate: its role in learning, memory, and the aging brain. Psychopharmacology (Berl) 111:391–401

Meier-Ruge W, Ulrich J, Bruhlmann M, Meier E (1992) Age-related white matter atrophy in the human brain. Ann N Y Acad Sci 673:260–269

Mesco ER, Carlson SG, Joseph JA, Roth GS (1993) Decreased striatal D2 dopamine receptor mRNA synthesis during aging. Brain Res Mol Brain Res 17:160–162

Mesulam MM, Mufson EJ, Levey AI, Wainer BH (1983) Cholinergic innervation of cortex by the basal forebrain: cytochemistry and cortical connections of the septal area, diagonal band nuclei, nucleus basalis (substantia innominata), and hypothalamus in the rhesus monkey. J Comp Neurol 214:170–197. doi:10.1002/cne.902140206

Miyoshi R, Kito S, Doudou N, Nomoto T (1990) Age-related changes of strychnine-insensitive glycine receptors in rat brain as studied by in vitro autoradiography. Synapse 6:338–343. doi:10.1002/syn.890060405

Mizoguchi K, Shoji H, Tanaka Y, Maruyama W, Tabira T (2009) Age-related spatial working memory impairment is caused by prefrontal cortical dopaminergic dysfunction in rats. Neuroscience 162:1192–1201. doi:10.1016/j.neuroscience.2009.05.023

Moore H, Stuckman S, Sarter M, Bruno JP (1996) Potassium, but not atropine-stimulated cortical acetylcholine efflux, is reduced in aged rats. Neurobiol Aging 17:565–571

Morris ED, Chefer SI, Lane MA, Muzic RF Jr, Wong DF, Dannals RF, Matochik JA, Bonab AA, Villemagne VL, Grant SJ, Ingram DK, Roth GS, London ED (1999) Loss of D2 receptor binding with age in rhesus monkeys: importance of correction for differences in striatal size. J Cereb Blood Flow Metab 19:218–229. doi:10.1097/00004647-199902000-00013

Neafsey EJ, Terreberry RR, Hurley KM, Ruit KG, Frysztak RJ (1993) Anterior cingulate cortex in rodents: connections, visceral control functions, and implications for emotion. In: Vogt BA, Gabriel M (eds) Neurobiology of cingulate cortex and limbic thalamus. Birkhauser, Boston, pp 206–223

Nicolle MM, Gallagher M, McKinney M (1999) No loss of synaptic proteins in the hippocampus of aged, behaviorally impaired rats. Neurobiol Aging 20:343–348

O'Donnell KA, Rapp PR, Hof PR (1999) Preservation of prefrontal cortical volume in behaviorally characterized aged macaque monkeys. Exp Neurol 160:300–310. doi:10.1006/exnr.1999.7192

O'Steen WK, Spencer RL, Bare DJ, McEwen BS (1995) Analysis of severe photoreceptor loss and Morris water-maze performance in aged rats. Behav Brain Res 68:151–158

O'Sullivan M, Jones DK, Summers PE, Morris RG, Williams SC, Markus HS (2001) Evidence for cortical "disconnection" as a mechanism of age-related cognitive decline. Neurology 57:632–638

Ota M, Obata T, Akine Y, Ito H, Ikehira H, Asada T, Suhara T (2006) Age-related degeneration of corpus callosum measured with diffusion tensor imaging. Neuroimage 31:1445–1452. doi:10.1016/j.neuroimage.2006.02.008

Ota M, Yasuno F, Ito H, Seki C, Nozaki S, Asada T, Suhara T (2006) Age-related decline of dopamine synthesis in the living human brain measured by positron emission tomography with L-[beta-11C]DOPA. Life Sci 79:730–736. doi:10.1016/j.lfs.2006.02.017

Pakkenberg B, Gundersen HJ (1997) Neocortical neuron number in humans: effect of sex and age. J Comp Neurol 384:312–320

Palmer AM, Robichaud PJ, Reiter CT (1994) The release and uptake of excitatory amino acids in rat brain: effect of aging and oxidative stress. Neurobiol Aging 15:103–111

Pedigo NW Jr, Polk DM (1985) Reduced muscarinic receptor plasticity in frontal cortex of aged rats after chronic administration of cholinergic drugs. Life Sci 37:1443–1449

Peiffer AM, Shi L, Olson J, Brunso-Bechtold JK (2010) Differential effects of radiation and age on diffusion tensor imaging in rats. Brain Res 1351:23–31. doi:10.1016/j.brainres.2010.06.049

Peters A, Sethares C (2003) Is there remyelination during aging of the primate central nervous system? J Comp Neurol 460:238–254. doi:10.1002/cne.10639

Peters A, Moss MB, Sethares C (2000) Effects of aging on myelinated nerve fibers in monkey primary visual cortex. J Comp Neurol 419:364–376

Peters A, Feldman ML, Vaughan DW (1983) The effect of aging on the neuronal population within area 17 of adult rat cerebral cortex. Neurobiol Aging 4:273–282

Phillips LH, Andres P (2010) The cognitive neuroscience of aging: new findings on compensation and connectivity. Cortex 46:421–424. doi:10.1016/j.cortex.2010.01.005

Rapp PR, Gallagher M (1996) Preserved neuron number in the hippocampus of aged rats with spatial learning deficits. Proc Natl Acad Sci USA 93:9926–9930

Rapp SR, Espeland MA, Shumaker SA, Henderson VW, Brunner RL, Manson JE, Gass ML, Stefanick ML, Lane DS, Hays J, Johnson KC, Coker LH, Dailey M, Bowen D, WHIMS Investigators (2003) Effect of estrogen plus progestin on global cognitive function in postmenopausal women: the Women's Health Initiative Memory Study: a randomized controlled trial. JAMA 289:2663–2672. doi:10.1001/jama.289.20.2663

Rasmussen T, Schliemann T, Sorensen JC, Zimmer J, West MJ (1996) Memory impaired aged rats: no loss of principal hippocampal and subicular neurons. Neurobiol Aging 17:143–147

Raz N, Ghisletta P, Rodrigue KM, Kennedy KM, Lindenberger U (2010) Trajectories of brain aging in middle-aged and older adults: regional and individual differences. Neuroimage 51:501–511. doi:10.1016/j.neuroimage.2010.03.020

Raz N, Rodrigue KM (2006) Differential aging of the brain: patterns, cognitive correlates and modifiers. Neurosci Biobehav Rev 30:730–748. doi:10.1016/j.neubiorev.2006.07.001

Raz N, Lindenberger U, Rodrigue KM, Kennedy KM, Head D, Williamson A, Dahle C, Gerstorf D, Acker JD (2005) Regional brain changes in aging healthy adults: general trends, individual differences and modifiers. Cereb Cortex 15:1676–1689. doi:10.1093/cercor/bhi044

Raz N, Rodrigue KM, Kennedy KM, Acker JD (2004) Hormone replacement therapy and age-related brain shrinkage: regional effects. Neuroreport 15:2531–2534

Raz N, Gunning FM, Head D, Dupuis JH, McQuain J, Briggs SD, Loken WJ, Thornton AE, Acker JD (1997) Selective aging of the human cerebral cortex observed in vivo: differential vulnerability of the prefrontal gray matter. Cereb Cortex 7:268–282

Resnick SM, Espeland MA, Jaramillo SA, Hirsch C, Stefanick ML, Murray AM, Ockene J, Davatzikos C (2009) Postmenopausal hormone therapy and regional brain volumes: the WHIMS-MRI Study. Neurology 72:135–142. doi:10.1212/01.wnl.0000339037.76336.cf

Resnick SM, Pham DL, Kraut MA, Zonderman AB, Davatzikos C (2003) Longitudinal magnetic resonance imaging studies of older adults: a shrinking brain. J Neurosci 23:3295–3301

Riddle DR, Sonntag WE, Lichtenwalner RJ (2003) Microvascular plasticity in aging. Ageing Res Rev 2:149–168

Rinne JO, Sahlberg N, Ruottinen H, Nagren K, Lehikoinen P (1998) Striatal uptake of the dopamine reuptake ligand [11C]beta-CFT is reduced in Alzheimer's disease assessed by positron emission tomography. Neurology 50:152–156

Rinne JO, Lonnberg P, Marjamaki P (1990) Age-dependent decline in human brain dopamine D1 and D2 receptors. Brain Res 508:349–352

Robertson D, Craig M, van Amelsvoort T, Daly E, Moore C, Simmons A, Whitehead M, Morris R, Murphy D (2009) Effects of estrogen therapy on age-related differences in gray matter concentration. Climacteric 12:301–309. doi:10.1080/13697130902730742

Rubinow MJ, Hagerbaumer DA, Juraska JM (2009) The food-conditioned place preference task in adolescent, adult and aged rats of both sexes. Behav Brain Res 198:263–266. doi:10.1016/j.bbr.2008.11.024

Rubinow MJ, Juraska JM (2009) Neuron and glia numbers in the basolateral nucleus of the amygdala from preweaning through old age in male and female rats: a stereological study. J Comp Neurol 512:717–725. doi:10.1002/cne.21924

Ryberg C, Rostrup E, Stegmann MB, Barkhof F, Scheltens P, van Straaten EC, Fazekas F, Schmidt R, Ferro JM, Baezner H, Erkinjuntti T, Jokinen H, Wahlund LO, O'brien J, Basile AM, Pantoni L, Inzitari D, Waldemar G, LADIS study group (2007) Clinical significance of corpus callosum atrophy in a mixed elderly population. Neurobiol Aging 28:955–963. doi:10.1016/j.neurobiolaging.2006.04.008

Salat DH, Greve DN, Pacheco JL, Quinn BT, Helmer KG, Buckner RL, Fischl B (2009) Regional white matter volume differences in nondemented aging and Alzheimer's disease. Neuroimage 44:1247–1258. doi:10.1016/j.neuroimage.2008.10.030

Salat DH, Tuch DS, Greve DN, van der Kouwe AJ, Hevelone ND, Zaleta AK, Rosen BR, Fischl B, Corkin S, Rosas HD, Dale AM (2005) Age-related alterations in white matter microstructure measured by diffusion tensor imaging. Neurobiol Aging 26:1215–1227. doi:10.1016/j.neurobiolaging.2004.09.017

Sanchez-Prieto J, Herrero I, Miras-Portugal MT, Mora F (1994) Unchanged exocytotic release of glutamic acid in cortex and neostriatum of the rat during aging. Brain Res Bull 33:357–359

Saransaari P, Oja SS (1995) Age-related changes in the uptake and release of glutamate and aspartate in the mouse brain. Mech Ageing Dev 81:61–71

Sargon MF, Denk CC, Celik HH, Surucu HS, Aldur MM (2007) Electron microscopic examination of the myelinated axons of corpus callosum. in perfused young and old rats. Int J Neurosci 117:999–1010. doi:10.1080/00207450600934382

Saura J, Andres N, Andrade C, Ojuel J, Eriksson K, Mahy N (1997) Biphasic and region-specific MAO-B response to aging in normal human brain. Neurobiol Aging 18:497–507

Scahill RI, Frost C, Jenkins R, Whitwell JL, Rossor MN, Fox NC (2003) A longitudinal study of brain volume changes in normal aging using serial registered magnetic resonance imaging. Arch Neurol 60:989–994. doi:10.1001/archneur.60.7.989

Schliebs R, Arendt T (2006) The significance of the cholinergic system in the brain during aging and in Alzheimer's disease. J Neural Transm 113:1625–1644. doi:10.1007/s00702-006-0579-2

Seeman P, Bzowej NH, Guan HC, Bergeron C, Becker LE, Reynolds GP, Bird ED, Riederer P, Jellinger K, Watanabe S (1987) Human brain dopamine receptors in children and aging adults. Synapse 1:399–404. doi:10.1002/syn.890010503

Severson JA, Marcusson J, Winblad B, Finch CE (1982) Age-correlated loss of dopaminergic binding sites in human basal ganglia. J Neurochem 39:1623–1631

Shi L, Pang H, Linville MC, Bartley AN, Argenta AE, Brunso-Bechtold JK (2006) Maintenance of inhibitory interneurons and boutons in sensorimotor cortex between middle and old age in Fischer 344 X Brown Norway rats. J Chem Neuroanat 32:46–53. doi:10.1016/j.jchemneu.2006.04.001

Shi L, Linville MC, Tucker EW, Sonntag WE, Brunso-Bechtold JK (2005) Differential effects of aging and insulin-like growth factor-1 on synapses in CA1 of rat hippocampus. Cereb Cortex 15:571–577. doi:10.1093/cercor/bhh158

Shumaker SA, Legault C, Kuller L, Rapp SR, Thal L, Lane DS, Fillit H, Stefanick ML, Hendrix SL, Lewis CE, Masaki K, Coker LH, Women's Health Initiative Memory Study (2004) Conjugated equine estrogens and incidence of probable dementia and mild cognitive impairment in postmenopausal women: Women's Health Initiative Memory Study. JAMA 291:2947–2958. doi:10.1001/jama.291.24.2947

Shumaker SA, Legault C, Rapp SR, Thal L, Wallace RB, Ockene JK, Hendrix SL, Jones BN 3rd, Assaf AR, Jackson RD, Kotchen JM, Wassertheil-Smoller S, Wactawski-Wende J, WHIMS

Investigators (2003) Estrogen plus progestin and the incidence of dementia and mild cognitive impairment in postmenopausal women: the Women's Health Initiative Memory Study: a randomized controlled trial. JAMA 289:2651–2662. doi:10.1001/jama.289.20.2651

Simic G, Kostovic I, Winblad B, Bogdanovic N (1997) Volume and number of neurons of the human hippocampal formation in normal aging and Alzheimer's disease. J Comp Neurol 379:482–494

Smith DE, Rapp PR, McKay HM, Roberts JA, Tuszynski MH (2004) Memory impairment in aged primates is associated with focal death of cortical neurons and atrophy of subcortical neurons. J Neurosci 24:4373–4381. doi:10.1523/JNEUROSCI.4289-03.2004

Smith TD, Adams MM, Gallagher M, Morrison JH, Rapp PR (2000) Circuit-specific alterations in hippocampal synaptophysin immunoreactivity predict spatial learning impairment in aged rats. J Neurosci 20:6587–6593

Smith YR, Minoshima S, Kuhl DE, Zubieta JK (2001) Effects of long-term hormone therapy on cholinergic synaptic concentrations in healthy postmenopausal women. J Clin Endocrinol Metab 86:679–684

Soghomonian JJ, Sethares C, Peters A (2010) Effects of age on axon terminals forming axosomatic and axodendritic inhibitory synapses in prefrontal cortex. Neuroscience 168:74–81. doi:10.1016/j.neuroscience.2010.03.020

Sowell ER, Peterson BS, Thompson PM, Welcome SE, Henkenius AL, Toga AW (2003) Mapping cortical change across the human life span. Nat Neurosci 6:309–315. doi:10.1038/nn1008

Spencer RL, O'Steen WK, McEwen BS (1995) Water maze performance of aged Sprague-Dawley rats in relation to retinal morphologic measures. Behav Brain Res 68:139–150

Stroessner-Johnson HM, Rapp PR, Amaral DG (1992) Cholinergic cell loss and hypertrophy in the medial septal nucleus of the behaviorally characterized aged rhesus monkey. J Neurosci 12:1936–1944

Suhara T, Fukuda H, Inoue O, Itoh T, Suzuki K, Yamasaki T, Tateno Y (1991) Age-related changes in human D1 dopamine receptors measured by positron emission tomography. Psychopharmacology (Berl) 103:41–45

Sullivan EV, Pfefferbaum A, Adalsteinsson E, Swan GE, Carmelli D (2002) Differential rates of regional brain change in callosal and ventricular size: a 4-year longitudinal MRI study of elderly men. Cereb Cortex 12:438–445

Takei N, Nihonmatsu I, Kawamura H (1989) Age-related decline of acetylcholine release evoked by depolarizing stimulation. Neurosci Lett 101:182–186

Terry RD, DeTeresa R, Hansen LA (1987) Neocortical cell counts in normal human adult aging. Ann Neurol 21:530–539. doi:10.1002/ana.410210603

Tigges J, Herndon JG, Rosene DL (1996) Preservation into old age of synaptic number and size in the supragranular layer of the dentate gyrus in rhesus monkeys. Acta Anat (Basel) 157:63–72

Uemura E (1985) Age-related changes in the subiculum of Macaca mulatta: dendritic branching pattern. Exp Neurol 87:412–427

van Dyck CH, Seibyl JP, Malison RT, Laruelle M, Wallace E, Zoghbi SS, Zea-Ponce Y, Baldwin RM, Charney DS, Hoffer PB (1995) Age-related decline in striatal dopamine transporter binding with iodine-123-beta-CITSPECT. J Nucl Med 36:1175–1181

Vatassery GT, Lai JC, Smith WE, Quach HT (1998) Aging is associated with a decrease in synaptosomal glutamate uptake and an increase in the susceptibility of synaptosomal vitamin E to oxidative stress. Neurochem Res 23:121–125

Vaughan DW (1977) Age-related deterioration of pyramidal cell basal dendrites in rat auditory cortex. J Comp Neurol 171:501–515. doi:10.1002/cne.901710406

Walhovd KB, Westlye LT, Amlien I, Espeseth T, Reinvang I, Raz N, Agartz I, Salat DH, Greve DN, Fischl B, Dale AM, Fjell AM (2009) Consistent neuroanatomical age-related volume differences across multiple samples. Neurobiol Aging. doi:10.1016/j.neurobiolaging.2009.05.013

Walker LC, Kitt CA, Struble RG, Wagster MV, Price DL, Cork LC (1988) The neural basis of memory decline in aged monkeys. Neurobiol Aging 9:657–666

Wallace M, Frankfurt M, Arellanos A, Inagaki T, Luine V (2007) Impaired recognition memory and decreased prefrontal cortex spine density in aged female rats. Ann N Y Acad Sci 1097: 54–57. doi:10.1196/annals.1379.026

Wang Y, Chan GL, Holden JE, Dobko T, Mak E, Schulzer M, Huser JM, Snow BJ, Ruth TJ, Calne DB, Stoessl AJ (1998) Age-dependent decline of dopamine D1 receptors in human brain: a PET study. Synapse 30:56–61

West MJ (1993) Regionally specific loss of neurons in the aging human hippocampus. Neurobiol Aging 14:287–293

Wheeler DD, Ondo JG (1986) Endogenous GABA concentration in cortical synaptosomes from young and aged rats. Exp Gerontol 21:79–85

Wise PM, Ratner A (1980) Effect of ovariectomy on plasma LH, FSH, estradiol, and progesterone and medial basal hypothalamic LHRH concentrations old and young rats. Neuroendocrinology 30:15–19

Wong TP, Marchese G, Casu MA, Ribeiro-da-Silva A, Cuello AC, De Koninck Y (2006) Imbalance towards inhibition as a substrate of aging-associated cognitive impairment. Neurosci Lett 397:64–68. doi:10.1016/j.neulet.2005.11.055

Wong TP, Marchese G, Casu MA, Ribeiro-da-Silva A, Cuello AC, De Koninck Y (2000) Loss of presynaptic and postsynaptic structures is accompanied by compensatory increase in action potential-dependent synaptic input to layer V neocortical pyramidal neurons in aged rats. J Neurosci 20:8596–8606

Wong TP, Campbell PM, Ribeiro-da-Silva A, Cuello AC (1998) Synaptic numbers across cortical laminae and cognitive performance of the rat during ageing. Neuroscience 84:403–412

Yang S, Li C, Lu W, Zhang W, Wang W, Tang Y (2009) The myelinated fiber changes in the white matter of aged female Long-Evans rats. J Neurosci Res 87:1582–1590. doi:10.1002/jnr.21986

Yassa MA, Muftuler LT, Stark CE (2010) Ultrahigh-resolution microstructural diffusion tensor imaging reveals perforant path degradation in aged humans in vivo. Proc Natl Acad Sci USA 107:12687–12691. doi:10.1073/pnas.1002113107

Yates MA, Markham JA, Anderson SE, Morris JR, Juraska JM (2008) Regional variability in age-related loss of neurons from the primary visual cortex and medial prefrontal cortex of male and female rats. Brain Res 1218:1–12. doi:10.1016/j.brainres.2008.04.055

Yates MA, Juraska JM (2007) Increases in size and myelination of the rat corpus callosum during adulthood are maintained into old age. Brain Res 1142:13–18. doi:10.1016/j.brainres.2007.01.043

Part III
Reproductive Aging

The Impact of Age-Related Ovarian Hormone Loss on Cognitive and Neural Function

Marissa I. Boulware, Brianne A. Kent and Karyn M. Frick

Abstract On average, women now live one-third of their lives after menopause. Because menopause has been associated with an elevated risk of dementia, an increasing body of research has studied the effects of reproductive senescence on cognitive function. Compelling evidence from humans, nonhuman primates, and rodents suggests that ovarian sex-steroid hormones can have rapid and profound effects on memory, attention, and executive function, and on regions of the brain that mediate these processes, such as the hippocampus and prefrontal cortex. This chapter will provide an overview of studies in humans, nonhuman primates, and rodents that examine the effects of ovarian hormone loss and hormone replacement on cognitive functions mediated by the hippocampus and prefrontal cortex. For humans and each animal model, we outline the effects of aging on reproductive function, describe how ovarian hormones (primarily estrogens) modulate hippocampal and prefrontal physiology, and discuss the effects of both reproductive aging and hormone treatment on cognitive function. Although this review will show that much has been learned about the effects of reproductive senescence on cognition, many critical questions remain for future investigation.

Keywords Estradiol · Hippocampus · Prefrontal cortex · Memory · Menopause · Aging

M. I. Boulware · K. M. Frick (✉)
Department of Psychology, University of Wisconsin-Milwaukee,
2441 E. Hartford Ave, Milwaukee, WI 53211, USA
e-mail: frickk@uwm.edu

B. A. Kent
Department of Psychology, Yale University, New Haven, CT 06520, USA

Curr Topics Behav Neurosci (2012) 10: 165–184
DOI: 10.1007/7854_2011_122

Published Online: 30 April 2011

Contents

1 Background ... 167
2 Humans ... 168
3 Nonhuman Primates ... 171
4 Rodents ... 173
5 Conclusions ... 176
References ... 177

The female reproductive system is a complex network in which interactions between ovarian and neural processes are crucial for the expression of behaviors related to sexual maturation, procreation, mood, and cognition. Like other organ systems in the body, the reproductive system is vulnerable to the ravages of aging. However, the impact of aging on the reproductive system of females is unique due to its inevitable failure during middle age. Understanding the influence of reproductive senescence in females on biological and psychological processes is of mounting importance given the ever-increasing amount of time that women now live beyond menopause. In 2006, the average life expectancy of women in the United States was 80.2 years and is estimated to increase to nearly 82 years in the next decade (United States Census Bureau 2008; United States National Center for Health Statistics 2009). Yet the average age of menopause onset has remained stable, and thus, women are spending significantly more of their lifetimes in a state of reproductive senescence.

The impact of this prolonged period of ovarian hormone deprivation on peripheral organs and tissues (e.g., heart, breast, uterus, and bone) has long been studied. However, research in the past two decades has revealed that ovarian sex-steroid hormones, such as estrogens and progestagens, can also rapidly and profoundly affect parts of the brain critical for cognitive function, such as the hippocampus and prefrontal cortex [for review, see (Sherwin and Henry 2008)]. As such, a growing literature has assessed the effects of ovarian hormone loss and replacement on cognitive processes such as learning and memory. Much of this literature stems from the two most common animal models of reproductive senescence, nonhuman primates and rodents (rats and mice). These species are attractive model systems because of their short life-spans, similarities to humans in the effects of aging and sex-steroid hormones on cognitive function, and the ability to conduct invasive studies that permit examination of neural function at the cellular and molecular levels. Therefore, the goal of this chapter is to review the literature on reproductive senescence as it pertains to humans, nonhuman primates, and rodents. Because extensive reviews have been published recently for these species (e.g., see Dumitriu et al. 2010; Frick 2009; Lacreuse 2006; Sherwin and Henry 2008; Voytko et al. 2009), this chapter will provide a brief overview of the most seminal findings published in recent years. The chapter will conclude by discussing future avenues for research.

1 Background

Estrogens, such as estrone, estriol, and the potent 17β-estradiol (termed "estradiol" or "E_2"), are synthesized and secreted in both males and females, albeit, at higher levels in females. In females, estrogen synthesis begins in the theca interna cells of the ovaries, where cholesterol is first converted into pregnenolone (Farkash et al. 1986). Acting mostly as a prohormone, pregnenolone is the precursor for both progesterone and androstenedione. The enzyme aromatase then converts androstenedione and testosterone into estrogens including estrone and E_2 (Brodie et al. 1976). In addition to the ovaries, recent evidence demonstrates that estrogens and progesterone are also synthesized in small quantities in the brain (Hojo et al. 2004; Kretz et al. 2004).

Once synthesized, estrogens are released into the bloodstream where they can bind to intracellular ligand-activated transcription factors, termed estrogen receptor α (ERα) and β (ERβ) (Koike et al. 1987; Spreafico et al. 1992; Tremblay et al. 1997). Binding at or near the cell nucleus to ERα and ERβ initiates the traditional "genomic" actions of estrogens, whereby the hormone–receptor complex binds to an estrogen response element on the DNA and serves as a nuclear transcription factor. Both ERs are expressed in brain regions critical for cognitive function, thereby providing an opportunity for estrogens to modulate multiple cognitive processes. For example, ERα and ERβ are both expressed in the dorsal and ventral hippocampus, where they are primarily found in CA1 and CA3 pyramidal neurons (Shughrue and Merchenthaler 2000). Both ERs are also expressed in the cerebral cortex, basal forebrain, and amygdala (Milner et al. 2005, 2001; Osterlund et al. 2000; Shughrue et al. 1997). In the cortex, ERβ is expressed in greater abundance than ERα, especially within frontal, parietal, and entorhinal cortices (Osterlund et al. 2000; Shughrue et al. 1997). Basal forebrain cholinergic neurons projecting to the hippocampus and neocortex also express both ERα and ERβ, although ERα is predominant (Shughrue et al. 2000). Although their nuclear localization suggests a relatively slow genomic mechanism of action, both ERs have been identified at extranuclear sites within the hippocampus, including dendritic spines, axons and axon terminals (Milner et al. 2005, 2001), where they may be involved in rapid effects of estrogens on cell signaling and epigenetic mechanisms (Fernandez et al. 2008; Zhao et al. 2010).

The vulnerability of the hippocampus and prefrontal cortex to aging and Alzheimer's disease (deToledo-Morrell et al. 2007; Driscoll and Sutherland 2005) has driven the fledgling field of hormones and cognition to focus primarily on these brain regions. The hippocampus is a bilateral medial temporal lobe structure critical for memories involving spatial, relational, and contextual information, and is necessary only for consolidation of such memories, not their long-term storage (Eichenbaum 1997, 2002; Squire 1992). As detailed in the sections below, E_2-induced alterations in the hippocampus have been most often observed in the CA1 subregion, the dentate gyrus, and to a lesser extent, the CA3 subregion. The prefrontal cortex, particularly the dorsolateral prefrontal cortex, is also critically necessary for memory, particularly a form of short-term memory called working memory (Goldman-Rakic 1992). However, the prefrontal cortex is thought to subserve a broader array of cognitive processes than the hippocampus, including attention, executive function (e.g., planning, judgment, mental flexibility, and verbal fluency), source memory, and episodic memory (Kandel et al. 2000). Despite the prevalence of deficits in both memory and executive functioning in the elderly (Woodruff-Pak 1997), the majority of studies on the neurobiological effects of hormone loss in rodents have focused on the hippocampus, in part because the preponderance of early studies were conducted in the hippocampus. As such, the rodent section below will focus primarily on hippocampal morphology and hippocampal-dependent memory. However, prefrontal function has long been of interest to human and nonhuman primate researchers, and so considerably more information on prefrontal function is available for primates than for rodents. Thus, hormonal effects on both prefrontal and hippocampal function will be discussed in the human and nonhuman primate sections.

2 Humans

As the human female ages, the reproductive system undergoes a plethora of changes that eventually lead to the cessation of reproductive abilities known as the menopause. The gradual transition to menopause occurs at approximately age 51 and can last anywhere from 2 to 7 years, with the most notable changes being amenorrhea, significant decreases in ovarian hormone levels, and ultimate ovarian failure (Bellantoni and Blackman 1996). The menopausal transition requires a highly orchestrated series of events within the Hypothalamic–Pituitary–Ovarian axis that includes changes in both the brain (e.g., elevated follicle stimulating hormone (FSH) and luteinizing hormone (LH) levels secreted by the pituitary) and ovaries (e.g., depletion of ovarian follicles and resultant drop in circulating ovarian estrogen concentrations) (Bellantoni and Blackman 1996). Because these events occur over the course of several years, the menopausal transition is generally divided into three phases based on the regularity and occurrence of the menses (Soules et al. 2001). The late reproductive years (often referred to as "premenopause") are characterized by regular menstrual cycles and an enduring increase in FSH that lasts for the remainder of a woman's life (Soules et al. 2001). Perimenopause, which can be divided into early and late stages, is defined by variable cycle lengths, skipped cycles, and at least one period of ammenorhea lasting over 60 days (Soules et al. 2001). Finally, postmenopause is characterized by the complete cessation of menses for at least 12 months (Gold et al. 2000; Soules et al. 2001). The dramatic decline of circulating ovarian hormones, which is a defining characteristic of reproductive senescence, is theorized to be the driving factor behind the apparent decline in cognitive functioning that is observed in postmenopausal women.

Anecdotally and in the laboratory, perimenopausal and postmenopausal women report more difficulties with memory and concentration than premenopausal women (Amore et al. 2007; Gold et al. 2000). Women are also three times more likely to develop Alzheimer's disease than men (Yaffe et al. 1998). Nevertheless, relatively few studies have examined the effects of reproductive senescence on brain and cognitive function; these studies are far outnumbered by those testing the effects of hormone replacement in postmenopausal women. The comparatively slim literature associating menopausal status or E_2 levels with cognition and brain function in women may stem from methodological challenges, including difficulties in accurately assessing menopausal status and measuring circulating E_2 in older women (Barrett-Connor and Laughlin 2009). With regard to brain function, a recent functional neuroimaging study reported that perimenopausal women (mean age 47.5 yrs) with moderate to severe menopausal symptoms (e.g., hot flashes) have reduced cerebral blood flow in the medial prefrontal cortex compared to age-matched asymptomatic controls (Abe et al. 2006). Also of note are post-mortem data showing that nuclear ERα expression, aromatase expression, and neuronal metabolic activity in the hippocampus are significantly higher in postmenopausal women (mean age 72.8 yrs) relative to pre- and perimenopausal women (Ishunina and Swaab 2007). This increase could arise if the drop in ovarian hormone levels triggers an increase in de novo estrogen production that up-regulates nuclear ERα expression (Ishunina et al. 2007). The possibility that increased E_2 levels negatively impact cognitive function in aging is supported by another study of postmenopausal women (mean age 70 yrs) in which higher total E_2 levels were associated with smaller hippocampal volumes and worse verbal memory (den Heijer et al. 2003). This result suggests that in older women, an up-regulation of E_2 synthesis and ERα expression may be detrimental to cognitive functioning.

Cross-sectional data on menopausal status suggest detrimental effects of menopause on executive functioning tasks such as mental flexibility, planning and reaction time (Elsabagh et al. 2007; Halbreich et al. 1995). However, no effects on verbal fluency, spatial ability, and episodic memory were reported in these studies (Elsabagh et al. 2007; Halbreich et al. 1995; Herlitz et al. 2007; Thilers et al. 2010). In contrast, longitudinal studies indicate that perimenopause is associated with reduced verbal fluency (Fuh et al. 2006) and postmenopause with impaired verbal fluency and visuospatial abilities (Thilers et al. 2010). Interestingly, this last study reported a significant interaction between menopausal status and body mass index (BMI), such that overweight postmenopausal women exhibited less cognitive decline than those with normal BMIs (Thilers et al. 2010). Serum E_2 levels were also positively correlated with BMI (Thilers et al. 2010), suggesting an association between lower E_2 levels and a faster rate of verbal and visuospatial decline. This relationship is consistent with findings showing that women with higher total or bioavailable E_2 levels exhibit better global cognitive function, verbal memory, executive function, and a lower risk of mild cognitive impairment and Alzheimer's disease than those with lower levels (Drake et al. 2000; Lebrun et al. 2005; Wolf and Kirschbaum 2002; Yaffe et al. 2000; Zandi et al. 2002); (Barrett-Connor et al. 1999; den Heijer et al. 2003; Herlitz et al. 2007; Laughlin et al. 2010). Notably, the only study to examine a substantially biracial population found that both African-American and Caucasian women with low bioavailable E_2

levels were 2–3 times more likely to exhibit verbal memory impairments and global cognitive decline than those with high E_2 levels (Yaffe et al. 2007). Collectively, these findings provide some support for the notion that the menopausal transition is associated with cognitive dysfunction.

Perhaps the most convincing argument for the role of ovarian hormones in maintaining cognitive function (particularly verbal memory) in women comes from studies of women who underwent bilateral oophorectomy for benign disease prior to menopause. In an influential series of studies, women who were treated with E_2 immediately after surgery maintained performance on tests of verbal memory, whereas those receiving placebo experienced significant verbal memory decline (Phillips and Sherwin 1992; Sherwin 1988). In contrast, visuospatial abilities were not affected by treatment, suggesting an effect specific to verbal learning and memory. Despite this specificity, these findings, as well as data from animal models demonstrating potent effects of E_2 on hippocampal synaptic plasticity, neuroproliferation, and neuroprotection [reviewed in (Spencer et al. 2008; Wise et al. 2001)], initially provided robust support for the notion that the loss of ovarian hormones at menopause renders cognitive regions of the brain more vulnerable to the detrimental effects of aging. This work subsequently stimulated numerous investigations into the effects of hormone therapy (both estrogens alone or estrogens plus a progestin) on cognitive function in menopausal women.

Observational and longitudinal studies in postmenopausal women generally suggest that estrogen use after surgical or natural menopause is beneficial for cognitive function. In observational studies, estrogen use has been associated with better verbal fluency (Hogervorst et al. 1999), verbal memory (Kampen and Sherwin 1994; Maki et al. 2001), working memory (Duff and Hampson 2000), and visuospatial function (Duka et al. 2000). Similar findings have been reported in longitudinal studies (Grodstein et al. 2000; Matthews et al. 1999; Steffens et al. 1999). Hormone therapy has also been associated with a lower risk of dementia (LeBlanc et al. 2001), although this decreased risk is most evident among women who initiated hormone therapy during or soon after the menopause (Yaffe et al. 1998).

Data from randomized clinical trials have been mixed, with those testing the effects of E_2 in recently menopausal women reporting beneficial effects of treatment on verbal and working memory [e.g., (Joffe et al. 2006; Phillips and Sherwin 1992; Viscoli et al. 2005; Wolf et al. 1999)], and those testing effects of conjugated equine estrogens in older postmenopausal women generally reporting no effect or detrimental effects of treatment on global cognitive decline and verbal memory [e.g., (Barrett-Connor and Kritz-Silverstein 1993; Binder et al. 2001; Grady et al. 2002; Janowsky et al. 2000; LeBlanc et al. 2007); reviewed in (Maki 2005; Sherwin and Henry 2008)]. In particular, data from the Women's Health Initiative Memory Study (WHIMS), the largest randomized clinical trial of the commonly prescribed conjugated equine estrogens, demonstrate that estrogen (with or without a synthetic progestin) significantly increases the risk of global cognitive decline and dementia in postmenopausal women over age 65 (Espeland et al. 2004; Rapp et al. 2003b; Shumaker et al. 2004). A follow-up study from the WHI Study of Cognitive Aging (WHISCA) reported that treatment impaired verbal memory and had no effects on tests of attention, working memory, spatial ability, affect, or depression (Resnick et al. 2009a, 2006). A subsequent neuroimaging study of WHIMS subjects found that estrogen treatment was associated with smaller hippocampal, frontal cortex, and total brain volumes (Resnick et al. 2009b). This finding is inconsistent with several smaller studies that show a positive association between estrogen use and volumes of the hippocampus and cortical regions (Berent-Spillson et al. 2010; Boccardi et al. 2006; Eberling et al. 2003; Lord et al. 2008), but are in keeping with some findings indicating that postmenopausal women with naturally higher levels of E_2 had smaller hippocampi and worse verbal memory (den Heijer et al. 2003). Although the large sample size of the WHI provides greater statistical power than the smaller studies demonstrating a positive relationship between estrogen use and hippocampal volumes, a number of design flaws limit the generalizability of the WHI findings as has been discussed elsewhere (Maki 2006; Sherwin and Henry 2008), including the type of hormone treatment used and an older subject population at high risk for cardiovascular and cerebrovascular disease.

Despite the apparent inconsistencies in the clinical literature, several important principles about hormone treatment have begun to emerge from these studies. First, treatment is most effective for younger women. For both cognitive and neural function, the data support a limited window of opportunity in which treatment during a "critical period" at or near the onset of menopause protects against cognitive decline, whereas treatment several years after menopause is ineffective or detrimental to cognitive health (Erickson et al. 2010; Maki 2006; Sherwin and Henry 2008). Data from nonhuman primates and rodents also support the existence of a critical period for estrogen treatment (Frick 2009; Sherwin and Henry 2008). Second, E_2 may be a more effective treatment than conjugated equine estrogens, as suggested by several randomized clinical trials (Joffe et al. 2006; Phillips and Sherwin 1992; Viscoli et al. 2005; Wolf et al. 1999). Third, clinical trials of hormone therapies are susceptible to a "healthy user bias" due to the fact that women who initiate hormone therapy are generally healthier and more educated than women who do not elect treatment (Keating et al. 1999; Matthews et al. 1996). As such, this bias must be considered when interpreting data and generalizing to a broader population. Finally, too little is known about how factors like timing of treatment (cyclic vs. continuous) and addition of progestagens influence the effectiveness of hormone therapy, so addressing these issues will be critical in future clinical studies.

3 Nonhuman Primates

Nonhuman primates, such as rhesus monkeys (*Macacamulatta*) and cynomolgus monkeys (*Macaca fascicularis*), are the most common model systems for the study of reproductive aging. Similar to humans, female macaques exhibit a 28 day menstrual cycle and experience ovarian hormone fluctuations comparable to human women (Gilardi et al. 1997; Goodman et al. 1977; Knobil and Neill 1988). Further, menopause in macaque females is very similar to that of women

(Gilardi et al. 1997), although the post-menopausal life of these monkeys is much shorter, given that the onset of menopause occurs after age 25 in species with lifespans of around 30 (Tigges et al. 1988). As in humans, the transition to reproductive senescence involves multiple parallel processes, including increasing irregularity of the menstrual cycle, depletion of the follicular reserves, decreased levels of circulating estrogen, and elevated levels of FSH, LH, and gonadotrophin releasing hormone (GnRH) (Downs and Urbanski 2006; Gore et al. 2004). Given these multiple commonalities, macaque females are an excellent model system to examine the effects of reproductive senescence and hormone therapy on cognitive and neural function. However, monkey research is limited by the high cost and low availability of animals, particularly postmenopausal females (Bellino and Wise 2003). Thus, much of the work examining effects of reproductive aging on the brain has been conducted in younger ovariectomized monkeys. Nevertheless, studies in young and aging macaque monkeys provide a valuable glimpse into how reproductive senescence may affect human women, and serve to bridge the gap between rodent models and human clinical studies.

E_2 can profoundly affect the hippocampus and prefrontal cortex of both young and aging female macaques. Cyclic E_2 treatment increased the number of CA1 dendritic spines by over 1 billion in both young (6–8 yrs) and aged (19–23 yrs) ovariectomized rhesus monkeys (Hao et al. 2003), suggesting a similar responsiveness to E_2 in the young and aged primate brain. In 7–15 year-old ovariectomized rhesus monkeys, chronic E_2 treatment increased the expression in CA1 of pre- and post-synaptic proteins such as syntaxin, synaptophysin, and spinophilin (Choi et al. 2003). Progesterone blocked these changes, suggesting that progesterone interferes with the positive effects of E_2 on hippocampal synapses. Interestingly, progesterone alone increased the expression of synaptophysin only (Choi et al. 2003), indicating that each hormone may be beneficial to synaptic spine morphology alone, but detrimental in combination. However, chronic treatment with E_2 alone or E_2 plus progesterone in adult (7–14.5 yrs) ovariectomized rhesus monkeys tended to increase neurogenesis in the dentate gyrus, showing that the combination of both hormones may be beneficial for hippocampal neurogenesis (Kordower et al. 2010). Together, these findings support the notion that E_2 can positively regulate hippocampal morphology in both young and aging female monkeys.

An increasing body of work has shown similar effects of E_2 on the prefrontal cortex. Long-term cyclic treatment of young and aged ovariectomized rhesus monkeys significantly increases relative to vehicle both apical and basal dendritic spine density, number and morphology within layer III pyramidal cells of area 46 in the prefrontal cortex (Hao et al. 2007; Hao et al. 2006; Tang et al. 2004). This effect has been demonstrated in behaviorally characterized aged monkeys who exhibited improved performance in a prefrontal cortex-dependent spatial-delayed response task (Rapp et al. 2003a), suggesting that E_2-induced increases in prefrontal spine density may lead to enhanced prefrontal-dependent memory in aged females. This conclusion is supported by a recent paper demonstrating a positive correlation in these same monkeys between ERα expression in the postsynaptic densities of prefrontal excitatory synapses and memory in the delayed response task (Wang et al. 2008).

Other studies in aged rhesus females have also revealed positive effects of ovarian hormones on cognitive processes mediated by the prefrontal cortex. The only study to examine gonadally intact females found that peri-/post-menopausal monkeys (20–27 years) performed significantly worse on the spatial-delayed response task than age-matched premenopausal monkeys and young monkeys (Roberts et al. 1997). Interestingly, different investigators using a battery of spatial and non-spatial memory tasks found that aged rhesus females (19–27 years) who had been ovariectomized for 12 years were better than age-matched intact females on spatial delayed response, but impaired on a delayed non-match to sample task (Lacreuse et al. 2000). The effects of E_2 treatment on naturally menopausal monkeys have not yet been tested, but several studies have examined the effects of E_2 treatment on surgically menopausal monkeys and report beneficial effects of E_2 on tasks mediated by the prefrontal cortex. For example, in aged rhesus monkeys ovariectomized 7–13 years prior to treatment, E_2 improved working memory in the spatial-delayed response task (Rapp et al. 2003a) and spatial-delayed recognition span test (Lacreuse et al. 2002), in some cases to the level of young monkeys (Rapp et al. 2003a). The benefits of E_2 treatment on working memory tasks, even 13 years after ovariectomy, suggests that the aged prefrontal cortex remains sufficiently responsive to E_2 such that memory function can be enhanced even after periods of prolonged ovarian hormone deprivation (Lacreuse 2006). However, these studies found that the E_2-induced improvement did not extend to other processes mediated by the prefrontal cortex, as delayed non-matching-to-sample, object discrimination, and Wisconsin Card Sort tests showed no or moderate improvement after treatment (Lacreuse et al. 2004, 2002; Rapp et al. 2003a). Interestingly, studies by Voytko and colleagues of aged cynomolgous monkeys have shown that E_2 can enhance several of these prefrontal-dependent abilities including visuospatial attention, visual recognition memory, and executive function (Voytko 2000, 2002, 2008). The Voytko (2002) laboratory has also shown that E_2 interacts with the basal forebrain cholinergic system to affect attention, but not memory. The discrepancies among the macaque studies in the effects of E_2 on specific prefrontal-dependent tasks may be due to methodological issues or differences between rhesus and cynomolgous monkeys. Nevertheless, the monkey data collectively demonstrate that E_2 can reverse ovariectomy-induced cognitive dysfunction in aged females, and may benefit several cognitive domains including memory, attention, and executive functioning.

4 Rodents

Rats (*Rattus norvegicus*) and mice (*Mus musculus*) are the most common animal models used to study the effects of hormones on cognition because of their compact size, short life spans, and abundant supply. For studies of cognitive aging, rats and mice are typically considered "aged" at approximately 2 years, "middle-aged" at approximately 16–18 months, and "young" at approximately 3–4 months (Frick 2009). Rodents do not exhibit a true menstrual cycle, as they lack a luteal phase and uterine wall sloughing (Wise 2000). Instead, they undergo 4–5 day-long estrous cycles that feature surges of estradiol and progesterone just prior to ovulation (McCarthy and Becker 2002). Further, rodents experience significant changes in their regular reproductive cycle with aging. Although they do not experience complete follicle loss (Wise 2000) and maintain relatively normal gonadotrophin levels (Wise 2000), reproductive senescence in rodents is similar to menopause in several respects, including increases in FSH, LH, and estradiol levels, variability of cycle length prior to acyclicity, and ultimate cessation of hormone cycling (LeFevre and McClintock 1988; Nelson et al. 1995). In rats, reproductive decline begins at 9–12 months of age, with 70% of 12-month-olds exhibiting irregular cycles or acyclicity, and nearly 75% of females acyclic by 24 months (Markowska 1999). In mice, alterations begin at 13–14 months of age (Nelson et al. 1995), with 80% of 17-month-olds exhibiting irregular cycles or acyclicity, and all females acyclic by 25 months (Frick et al. 2000).

As in humans and nonhuman primates, memory decline has been associated with the loss of reproductive cycling in both rats and mice. This relationship has been particularly well described for spatial memory tested in the Morris water maze, which declines at an earlier age in females than in males. Significant deficits in females are observed by 12 months in rats and 17 months in mice, whereas such deficits are not apparent in male rats until 18 months and in male mice until 25 months (Frick et al. 2000; Markowska 1999). Moreover, the onset of this premature spatial memory decline in females coincides with the cessation of ovarian hormone cycling, as illustrated by the fact that the age at which spatial memory deficits first appear in both species is marked by a sharp decline in regular estrous cycling (Frick et al. 2000; Markowska 1999). Further, performance among 12–24-month-old rats in a daily probe trial was best in regularly cycling females, intermediate in irregularly cycling females, and worst in acyclic females (Markowska 1999), suggesting that the disruption of estrous cycling is detrimental to spatial memory throughout the aging process.

The age-related cognitive decline accompanying the loss of estrous cycling in rodents has most often been attributed to reduced estradiol levels in the hippocampus. In the hippocampus of young rodents, E_2 increases CA1 dendritic spine density (Woolley and McEwen, 1992, 1993), enhances long-term potentiation (Warren et al. 1995; Foy et al. 1999), increases neurogenesis (Tanapat et al. 1999), and rapidly activates cell signaling cascades including extracellular signal-regulated kinase/mitogen activated protein kinase (ERK/MAPK) and protein kinase A (PKA) (Fernandez et al. 2008; Lewis et al. 2008a). E_2 also enhances the function of hippocampal- and cortically-projecting cholinergic neurons (e.g., Gibbs and Aggarwal 1998; Wu et al. 1999; Gibbs 2000), which are involved in attention and cortical information processing, as well as some aspects of learning and memory (e.g., Bartus et al. 1985; Baxter and Chiba 1999; Berger-Sweeney et al. 2000). However, the effects of E_2 in the aging hippocampus are not identical to that of young females, perhaps due to reduced hippocampal expression of ERα and ERβ in aged females (Yamaguchi-Shima and Yuri 2007). For example, E_2 in aged females does not increase dendritic spine density in hippocampal CA1, but does increase density in the dentate gyrus (Adams et al. 2001; Miranda et al. 1999). Other studies have shown that the hippocampus of aging female rodents remains generally responsive to E_2, which can increase hippocampal levels of synaptophysin and nerve growth factor, augment dentate gyrus dendritic spine density, activate protein kinases, normalize intracellular calcium homeostasis, and phosphorylate NMDA receptors in aging females (Bi et al. 2003; Fernandez and Frick 2004; Foster 2005; Frick et al. 2002; Miranda et al. 1999). Of these changes, E_2-induced increases in hippocampal synaptophysin protein levels in aged female mice have been associated with improved spatial memory (Frick et al. 2002).

A growing literature has examined the effects of E_2, and to a lesser extent progesterone, on hippocampal-dependent memory in aging rats and mice. To date, E_2 treatments of varying dose, duration, route of administration, and timing relative to testing have improved memory in middle-aged or aged rodents tested in tasks of spatial reference memory, spatial working memory, and object recognition [reviewed in (Frick 2009)]. However, several key factors appear to contribute to treatment effectiveness in rodents. Age at treatment may be a key variable, as several studies that compared the effects of E_2 in rodents of multiple ages report memory improvements in middle-aged, but not aged, ovariectomized females (Gresack et al. 2007; Savonenko and Markowska 2003; Talboom et al. 2008). The duration of hormone loss prior to treatment also appears to be a particularly critical factor; long delays in treatment after ovariectomy in aging females impair basal forebrain cholinergic functioning (Gibbs 1998) and reduce E_2's ability to improve spatial memory (Daniel et al. 2006; Gibbs 2000; Markowska and Savonenko 2002). Collectively, these data support the "critical period hypothesis" of estrogen action originally proposed to explain why hormone therapy in women appears to work best when initiated near menopause (Maki 2006). The origins of this critical period may lie in biochemical alterations in the aged brain, as suggested by a recent study from our laboratory which found that the ability of E_2 to enhance object memory consolidation in young and middle-aged ovariectomized mice was associated with its ability to activate cell signaling cascades that are critical for long-term memory formation; in aged mice, E_2 had no effect on memory or cell signaling (Fan et al. 2010).

Beyond the critical period, the rodent literature has begun to shed light on other important issues, like whether cyclic or continuous hormone administration is most effective, whether treatment should include a progestagen, and whether certain populations might benefit more from treatment than others (Frick 2009). With regard to this last point, we recently showed that exposure to a cognitively and physically enriching environment can reduce the mnemonic benefits of E_2 in young and middle-aged female mice (Gresack and Frick 2004; Gresack et al. 2007). Such data are consistent with clinical data suggesting that estrogen therapy may be most effective in women with less education (Matthews et al. 1999). Data on the mnemonic effects of progesterone are inconsistent, with several studies reporting that it blocks the beneficial effects of E_2 on spatial memory in aging females (Bimonte-Nelson et al. 2006; Harburger et al. 2007), and others reporting no such interference (Gibbs 2000; Markham et al. 2002). When given alone, acute progesterone treatment can improve spatial and object recognition memory in aged female mice (Lewis et al. 2008b).

The relative ease and flexibility of the rodent model has also allowed for the development and testing of alternatives to traditional hormone therapy, such as selective estrogen receptor modulators (SERMs). SERMs are non-steroidal compounds that act as estrogen agonists in some tissues and antagonists in others. The most commonly tested SERMs include tamoxifen, raloxifene, phytoestrogens, and ICI 182,780. Although these SERMs exhibit neuroprotective properties in vitro, none have consistently improved memory in women or rodent models (Frick 2009; Zhao et al. 2005). Drugs selective for ERα or ERβ have been developed and are currently being tested in rodents; thus far, ERβ agonism appears to most consistently improve hippocampal memory in young rats and mice (Frick et al. 2010; Walf et al. 2006), but none of these agonist compounds have yet been tested in aging females. An alternative to further refining SERMs is to elucidate the molecular mechanisms underlying memory-enhancing effects of hormones and then develop drugs that target those mechanisms. Our laboratory has published a series of studies in this regard, demonstrating that E_2-induced alterations of cell signaling, epigenetic mechanisms, and gene expression are necessary for this hormone to enhance object memory consolidation in young and middle-aged ovariectomized mice (Fan et al. 2010; Fernandez et al. 2008; Frick 2009; Zhao et al. 2010). In particular, E_2-induced activation of ERK/MAPK signaling, histone acetylation, and DNA methylation are especially critical. As such, these mechanisms may be useful targets to which non-steroidal drugs can be designed that mimic the beneficial effects of E_2 on memory. Because such drugs would ideally modulate the downstream effectors of estrogen receptors, rather than the receptors themselves, this approach may ultimately lead to the development of hormone-based treatments that can safely and effectively reduce age-related memory decline in women.

5 Conclusions

The past 20 years has seen an explosion of research on the roles of hormones on cognition, and much progress has been made. As the studies discussed above illustrate, there is considerable evidence in humans, nonhuman primates, and rodents that the age-related loss of ovarian hormones, particularly estrogens, is detrimental for cognitive function. Although this work has yielded many important insights, much more needs to be done to gain a comprehensive understanding of how these hormones affect cognition in aging females. For example, research on naturally reproductively senescent women and animals is sorely lacking, thereby limiting conclusions about the impact of reproductive aging on cognition. Further, clinical trials such as the WHI have raised critical questions about how and when hormone therapy should be administered, and who should receive treatment. Future research on hormone therapy should be directed toward better assessing the efficacy of various treatments on specific cognitive functions in women, pinpointing the best age to begin and length of time to conduct treatment, understanding the role of progestagens in modulating cognitive functions, and identifying specific populations of women (e.g., less well educated) that might benefit the most from treatment. Many of these issues can be addressed relatively easily in rodent and primate models as a first step.

As the numbers of menopausal and postmenopausal women skyrocket in the coming years, addressing these issues will become of paramount importance. The health risks associated with commonly prescribed conjugated equine estrogens, such as increases in breast cancer, heart disease, and stroke (Rossouw et al. 2002), also warrant the accelerated development of alternative approaches to hormone replacement, including SERMs and other treatments that target the molecular mechanisms underlying hormonal modulation of cognitive function (Frick 2009; Frick et al. 2010). The groundbreaking empirical research of the past 20 years has laid the foundation for the next generation of promising breakthroughs in the science of hormones and cognition, which will hopefully lead to a better understanding of the impact of reproductive senescence on cognition and more effective hormone treatments for the prevention of age-related cognitive decline.

Acknowledgments This work was sponsored by the University of Wisconsin-Milwaukee and Yale University.

References

Abe T, Bereczki D, Takahashi Y, Tashiro M, Iwata R, Itoh M (2006) Medial frontal cortex perfusion abnormalities as evaluated by positron emission tomography in women with climacteric symptoms. Menopause 13:891–901

Adams MM, Shah RA, Janssen WG, Morrison JH (2001) Different modes of hippocampal plasticity in response to estrogen in young and aged female rats. Proc Natl Acad Sci, USA 98:8071–8076

Amore M, Di Donato P, Berti A, Palareti A, Chirico C, Papalini A, Zucchini S (2007) Sexual and psychological symptoms in the climacteric years. Maturitas 56:303–311

Barrett-Connor E, Kritz-Silverstein D (1993) Estrogen replacement therapy and cognitive function in older women. J Am Med Assoc 269:2637–2641

Barrett-Connor E, Laughlin GA (2009) Endogenous and exogenous estrogen, cognitive function, and dementia in postmenopausal women: Evidence from epidemiologic studies and clinical trials. Semin Reprod Med 27:275–282

Barrett-Connor E, Goodman-Gruen D, Patay B (1999) Endogenous sex hormones and cognitive function in older men. J Clin Endocrinol Metab 84:3681–3685

Bartus RT, Dean RT, Pontecorvo MJ, Flicker C (1985) The cholinergic hypothesis: A historical overview, current perspective, and future directions. Ann NYAcad Sci 444: 332–358

Baxter MG, Chiba AA (1999) Cognitive functions of the basal forebrain. Curr Opin Neurobiol 9: 178–183

Bellantoni MF, Blackman MR (1996) Menopause and its consequences. In: Schneider EL, Rowe JW (eds) Handbook of the biology of aging. Academic Press, New York, pp 415–430

Bellino FL, Wise PM (2003) Nonhuman primate models of menopause workshop. Biol Reprod 68:10–18

Berent-Spillson A, Persad C, Love T, Tkaczyk A, Wang H, Reame N, Frey K, Zubieta J, Smith Y (2010) Early menopausal hormone use influences brain regions used for visual working memory. Menopause 17:692–699

Berger-Sweeney J, Stearns NA, Frick KM, Beard B, Baxter MG (2000) Cholinergic basal forebrain is critical for social transmission of food preferences. Hippocampus 10: 729–738

Bi R, Foy MR, Thompson RF, Baudry M (2003) Effects of estrogen, age, and calpain on MAP kinase and NMDA receptors in female rat brain. Neurobiol Aging 24:977–983

Bimonte-Nelson HA, Francis KR, Umphlet CD, Granholm A-C (2006) Progesterone reverses the spatial memory enhancements initiated by tonic and cyclic oestrogen therapy in middle-aged ovariectomized female rats. Eur J Neurosci 24:229–242

Binder E, Schechtman K, Birge S, Williams D (2001) Effects of hormone replacement therapy on cognitive performance in elderly women. Maturitas 38:137–146

Boccardi M, Ghidoni R, Govoni S, Testa C, Benussi L, Bonetti M, Binetti G, Frisoni G (2006) Effects of hormone therapy on brain morphology of healthy postmenopausal women: A Voxel-based morphometry study. Menopause 13:584–591

Brodie AM, Schwarzel WC, Brodie HJ (1976) Studies on the mechanism of estrogen biosynthesis in the rat ovary–I. J Steroid Biochem 7:787–793

Choi J, Romeo R, Brake W, Bethea C, Rosenwaks Z, McEwen B (2003) Estradiol increases pre- and post-synaptic proteins in the CA1 region of the hippocampus in female rhesus macaques (Macaca mulatta). Endocrinology 144:4734–4738

Daniel JM, Hulst JL, Berbling JL (2006) Estradiol replacement enhances working memory in middle-aged rats when initiated immediately after ovariectomy but not after a long-term period of ovarian hormone deprivation. Endocrinology 147:607–614

den Heijer T, Geerlings MI, Hofman A, de Jong FH, Launer LJ, Pols HA, Breteler MM (2003) Higher estrogen levels are not associated with larger hippocampi and better memory performance. Arch Neurol 60:213–220

deToledo L, Stoub TR, Wang C (2007) Hippocampal atrophy and disconnection in incipient and mild Alzheimer's disease. Prog Brain Res 163:741–753

Downs JL, Urbanski HF (2006) Neuroendocrine changes in the aging reproductive axis of female rhesus macaques (Macaca mulatta). Biol Reprod 75:539–546

Drake EB, Henderson VW, Stanczyk FZ, McCleary CA, Brown WS, Smith CA, Rizzo AA, Murdock GA, Buckwalter JG (2000) Associations between circulating sex steroid hormones and cognition in normal elderly women. Neurology 54:599–603

Driscoll I, Sutherland RJ (2005) The aging hippocampus: Navigating between rat and human experiments. Rev Neurosci 16:87–121

Duff SJ, Hampson E (2000) A beneficial effect of estrogen on working memory in postmenopausal women taking hormone replacement therapy. Horm Behav 38:262–276

Duka T, Tasker R, McGowan JF (2000) The effects of 3 week estrogen hormone replacement on cognition in elderly healthy females. Psychopharmacology 149:129–139

Dumitriu D, Rapp PR, McEwen BS, Morrison JH (2010) Estrogen and the aging brain: An elixir for the weary cortical network. Ann N Y Acad Sci 1204:104–112

Eberling J, Wu C, Haan M, Mungas D, Buonocore M, Jagust W (2003) Preliminary evidence that estrogen protects against age-related hippocampal atrophy. Neurobiol Aging 24:725–732

Eichenbaum H (1997) Declarative memory: Insights from cognitive neurobiology. Annu Rev Psychol 48:547–572

Eichenbaum H (2002) The cognitive neuroscience of memory. Oxford University Press, New york, NY

Elsabagh S, Hartley DE, File SE (2007) Cognitive function in late versus early postmenopausal stage. Maturitas 56:84–93

Erickson K, Voss M, Prakash R, Chaddock L, Kramer A (2010) A cross-sectional study of hormone treatment and hippocampal volume in postmenopausal women: Evidence for a limited window of opportunity. Neuropsychology 24:68–76

Espeland MA, Rapp SR, Shumaker SA, Brunner R, Manson JE, Sherwin BB, Hsia J, Margolis KL, Hogan PE, Wallace R, Dailey M, Freeman R, Hays J (2004) Conjugated equine estrogens and global cognitive function in postmenopausal women: Women's Health Initiative Memory Study. J Am Med Assoc 291:2959–2968

Fan L, Orr PT, Zhao Z, Chambers CH, Lewis MC, Frick KM (2010) Estradiol-induced object memory consolidation in middle-aged female mice requires dorsal hippocampal extracellular signal-regulated kinase and phosphatidylinositol 3-kinase activation. J Neurosci 30:4390–4400

Farkash Y, Timberg R, Orly J (1986) Preparation of antiserum to rat cytochrome P-450 cholesterol side chain cleavage, and its use for ultrastructural localization of the immunoreactive enzyme by protein A-gold technique. Endocrinology 118:1353–1365

Fernandez SM, Frick KM (2004) Chronic oral estrogen affects memory and neurochemistry in middle-aged female mice. Behav Neurosci 118:1340–1351

Fernandez SM, Lewis MC, Pechenino AS, Harburger LL, Orr PT, Gresack JE, Schafe GE, Frick KM (2008) Estradiol-induced enhancement of object memory consolidation involves hippocampal ERK activation and membrane-bound estrogen receptors. J Neurosci 28: 8660–8667

Foster TC (2005) Interaction of rapid signal transduction cascades and gene expression in mediating estrogen effects on memory over the lifespan. Front Neuroendocrinol 26:51–64

Foy MR, Xu J, Xie X, Brinton RD, Thompson RF, Berger TW (1999) 17ß-estradiol enhances NMDA receptor-mediated EPSPs and long-term potentiation. J Neurophysiol 81:925–929

Frick KM (2009) Estrogens and age-related memory decline in rodents: What have we learned and where do we go from here? Horm Behav 55:2–23

Frick KM, Burlingame LA, Arters JA, Berger-Sweeney J (2000) Reference memory, anxiety, and estrous cyclicity in C57BL/6NIA mice are affected by age and sex. Neuroscience 95:293–307

Frick KM, Fernandez SM, Bulinski SC (2002) Estrogen replacement improves spatial reference memory and increases hippocampal synaptophysin in aged female mice. Neuroscience 115:547–558

Frick KM, Fernandez SM, Harburger LL (2010) A new approach to understanding the molecular mechanisms through which estrogens affect cognition. Biochimica et Biophysica Acta—Gen Subj 1800:1045–1055

Fuh JL, Wang SJ, Lee SJ, Lu SR, Juang KD (2006) A longitudinal study of cognition change during early menopausal transition in a rural community. Maturitas 53:447–453

Gibbs RB, Aggarwal P (1998) Estrogen and basal forebrain cholinergic neurons: Implications for brain aging and Alzheimer's disease-related cognitive decline. Horm Behav 34: 98–111

Gibbs RB (1998) Impairment of basal forebrain cholinergic neurons associated with aging and long-term loss of ovarian function. Exp Neurol 151:289–302

Gibbs RB (2000) Long-term treatment with estrogen and progesterone enhances acquisition of a spatial memory task by ovariectomized aged rats. Neurobiol Aging 21:107–116

Gilardi KV, Shideler SE, Valverde CR, Roberts JA, Lasley BL (1997) Characterization of the onset of menopause in the rhesus macaque. Biol Reprod 57:335–340

Gold EB, Sternfeld B, Kelsey JL, Brown C, Mouton C, Reame N, Salamone L, Stellato R (2000) Relation of demographic and lifestyle factors to symptoms in a multi-racial/ethnic population of women 40–55 years of age. Am J Epidemiol 152:463–473

Goldman-Rakic PS (1992) Working memory and the mind. Sci Am 2647:110–117

Goodman AL, Descalzi CD, Johnson DK, Hodgen GD (1977) Composite pattern of circulating LH, FSH, estradiol, and progesterone during the menstrual cycle in cynomolgus monkeys. Proc Soc Exp Biol Med 155:479–481

Gore AC, Windsor-Engnell BM, Terasawa E (2004) Menopausal increases in pulsatile gonadotropin-releasing hormone release in a nonhuman primate (Macaca mulatta). Endocrinology 145:4653–4659

Grady D, Yaffe K, Kristof M, Lin F, Richards C, Barrett-Connor E (2002) Effect of postmenopausal hormone therapy on cognitive function: The Heart and Estrogen/progestin Replacement Study. Am J Med 113:543–548
Gresack JE, Frick KM (2004) Environmental enrichment reduces the mnemonic and neural benefits of estrogen. Neuroscience 128:459–471
Gresack JE, Kerr KM, Frick KM (2007) Life-long environmental enrichment differentially affects the mnemonic response to estrogen in young, middle-aged, and aged female mice. Neurobiol Learn Mem 88:393–408
Grodstein F, Chen J, Pollen D, Albert M (2000) Postmenopausal hormone therapy and cognitive function in healthy older women. J Am Geriatr Soc 48:746–752
Halbreich U, Lumley LA, Palter S, Manning C, Gengo F, Joe SH (1995) Possible acceleration of age effects on cognition following menopause. J Psychiatr Res 29:153–163
Hao J, Janssen WG, Tang Y, Roberts JA, McKay H, Lasley B, Allen PB, Greengard P, Rapp PR, Kordower JH, Hof PR, Morrison JH (2003) Estrogen increases the number of spinophilin-immunoreactive spines in the hippocampus of young and aged female rhesus monkeys. J Comp Neurol 465:540–550
Hao J, Rapp PR, Leffler AE, Leffler SR, Janssen WG, Lou W, McKay H, Roberts JA, Wearne SL, Hof PR, Morrison JH (2006) Estrogen alters spine number and morphology in prefrontal cortex of aged female rhesus monkeys. J Neurosci 26:2571–2578
Hao J, Rapp PR, Janssen WG, Lou W, Lasley BL, Hof PR, Morrison JH (2007) Interactive effects of age and estrogen on cognition and pyramidal neurons in monkey prefrontal cortex. Proc Natl Acad Sci, USA 104:11465–11470
Harburger LL, Bennett JC, Frick KM (2007) Effects of estrogen and progesterone on spatial memory consolidation in aged females. Neurobiol Aging 28:602–610
Herlitz A, Thilers P, Habib R (2007) Endogenous estrogen is not associated with cognitive performance before, during, or after menopause. Menopause 14:425–431
Hogervorst E, Boshuisen M, Riedel WJ, Willekes C, Jolles J (1999) The effect of hormone replacement therapy on cognitive function in elderly women. Psychoneuroendocrinology 24:43–68
Hojo Y, Hattori TA, Enami T, Furukawa A, Suzuki K, Ishii HT, Mukai H, Morrison JH, Janssen WG, Kominami S, Harada N, Kimoto T, Kawato S (2004) Adult male rat hippocampus synthesizes estradiol from pregnenolone by cytochromes P45017alpha and P450 aromatase localized in neurons. Proc Natl Acad Sci, USA 101:865–870
Ishunina TA, Swaab DF (2007) Alterations in the human brain in menopause. Maturitas 57:20–22
Ishunina TA, Fischer DF, Swaab DF (2007) Estrogen receptor alpha and its splice variants in the hippocampus in aging and Alzheimer's disease. Neurobiol Aging 28:1670–1681
Janowsky JS, Chavez B, Orwoll E (2000) Sex steroids modify working memory. J Cogn Neurosci 12:407–414
Joffe H, Hall J, Gruber S, Sarmiento I (2006) Estrogen therapy selectively enhances prefrontal cognitive processes: A randomized, double-blind, placebo-controlled study with functional magnetic resonance imaging in perimenopausal and recently postmenopausal women. Menopause 13:411–422
Kampen DL, Sherwin BB (1994) Estrogen use and verbal memory in healthy postmenopausal women. Obstet Gynecol 83:979–983
Kandel ER, Schwartz JH, Jessell TM (2000) Principles of neural science, 4th ed. McGraw Hill, New York
Keating NL, Cleary PD, Rossi AS, Zaslavsky AM, Ayanian JZ (1999) Use of hormone replacement therapy by postmenopausal women in the United States. Ann Intern Med 130:545–553
Knobil E, Neill J (1988) The menstrual cycle and its neuroendocrine control. The physiology of reproduction. Raven Press, New York, pp 1971–1994
Koike S, Sakai M, Muramatsu M (1987) Molecular cloning and characterization of rat estrogen receptor cDNA. Nucleic Acids Res 15:2499–2513

Kordower JH, Chen EY, Morrison JH (2010) Long-term gonadal hormone treatment and endogenous neurogenesis in the dentate gyrus of the adult female monkey. Exp Neurol 224:252–257

Kretz O, Fester L, Wehrenberg U, Zhou L, Brauckmann S, Zhao S, Prange-Kiel J, Naumann T, Jarry H, Frotscher M, Rune GM (2004) Hippocampal synapses depend on hippocampal estrogen synthesis. J Neurosci 24:5913–5921

Lacreuse A (2006) Effects of ovarian hormones on cognitive function in nonhuman primates. Neuroscience 138:859–867

Lacreuse A, Herndon JG, Moss MB (2000) Cognitive function in aged ovariectomized female rhesus monkeys. Behav Neurosci 114:506–513

Lacreuse A, Wilson ME, Herndon JG (2002) Estradiol, but not raloxifene, improves aspects of spatial working memory in aged ovariectomized rhesus monkeys. Neurobiol Aging 23:589–600

Lacreuse A, Chhabra RK, Hall MJ, Herndon JG (2004) Executive function is less sensitive to estradiol than spatial memory: Performance on an analog of the card sorting test in ovariectomized aged rhesus monkeys. Behav Process 67:313–319

Laughlin GA, Kritz-Silverstein D, Barrett-Connor E (2010) Endogenous oestrogens predict 4 year decline in verbal fluency in postmenopausal women: The Rancho Bernardo Study. Clin Endocrinol 72:99–106

LeBlanc ES, Janowsky J, Chan BKS, Nelson HD (2001) Hormone replacement therapy and cognition. Systematic review and meta-analysis. J Am Med Assoc 285:1489–1499

LeBlanc E, Neiss M, Carello P, Samuels M (2007) Hot flashes and estrogen therapy do not influence cognition in early menopausal women. Menopause 14:191–202

Lebrun CE, van der Schouw YT, de Jong FH, Pols HA, Grobbee DE, Lamberts SW (2005) Endogenous oestrogens are related to cognition in healthy elderly women. Clin Endocrinol 63:50–55

LeFevre J, McClintock MK (1988) Reproductive senescence in female rats: A longitudinal study of individual differences in estrous cycles and behavior. Biol Reprod 38:780–789

Lewis MC, Kerr KM, Orr PT, Frick KM (2008a) Estradiol-induced enhancement of object memory consolidation involves NMDA receptors and protein kinase A in the dorsal hippocampus of female C57BL/6 mice. Behav Neurosci 122:716–721

Lewis MC, Orr PT, Frick KM (2008b) Differential effects of acute progesterone administration on spatial and object memory in middle-aged and aged female C57BL/6 mice. Horm Behav 54:455–462

Lord C, Buss C, Lupien S, Pruessner J (2008) Hippocampal volumes are larger in postmenopausal women using estrogen therapy compared to past users, never users and men: A possible window of opportunity effect. Neurobiol Aging 29:95–101

Maki P (2005) A systematic review of clinical trials of hormone therapy on cognitive function: effects of age at initiation and progestin use. Ann N Y Acad Sci 1052:182–197

Maki PM (2006) Hormone therapy and cognitive function: Is there a critical period for benefit? Neuroscience 138:1027–1030

Maki PM, Zonderman AB, Resnick SM (2001) Enhanced verbal memory in nondemented elderly women receiving hormone-replacement therapy. Am J Psychiatry 158:227–233

Markham JA, Pych JC, Juraska JM (2002) Ovarian hormone replacement to aged ovariectomized female rats benefits acquisition of the Morris water maze. Horm Behav 42:284–293

Markowska AL (1999) Sex dimorphisms in the rate of age-related decline in spatial memory: Relevance to alterations in the estrous cycle. J Neurosci 19:8122–8133

Markowska AL, Savonenko AV (2002) Effectiveness of estrogen replacement in restoration of cognitive function after long-term estrogen withdrawal in aging rats. J Neurosci 22:10985–10995

Matthews KA, Kuller LH, Wing RR, Meilahn EN, Plantinga P (1996) Prior to use of estrogen replacement therapy, are users healthier than nonusers? Am J Epidemiol 143:971–978

Matthews K, Cauley J, Yaffe K, Zmuda JM (1999) Estrogen replacement therapy and cognitive decline in older community women. J Am Geriatr Soc 47:518–523

McCarthy MM, Becker JB (2002) Neuroendocrinology of sexual behavior in the female. In: Becker JB, Breedlove SM, Crews D, McCarthy MM (eds) Behavioral endocrinology. MIT Press, Cambridge, MA, pp 117–151

Milner TA, McEwen BS, Hayashi S, Li CJ, Reagan LP, Alves SE (2001) Ultrastructural evidence that hippocampal alpha estrogen receptors are located at extranuclear sites. J Comp Neurol 429:355–371

Milner TA, Ayoola K, Drake CT, Herrick SP, Tabori NE, McEwen BS, Warrier S, Alves SE (2005) Ultrastructural localization of estrogen receptor beta immunoreactivity in the rat hippocampal formation. J Comp Neurol 491:81–95

Miranda P, Williams CL, Einstein G (1999) Granule cells in aging rats are sexually dimorphic in their response to estradiol. J Neurosci 19:3316–3325

Nelson JF, Karelus K, Bergman MD, Felicio LS (1995) Neuroendocrine involvement in aging: Evidence from studies of reproductive aging and caloric restriction. Neurobiol Aging 16:837–843

Osterlund MK, Keller E, Hurd YL (2000) The human forebrain has discrete estrogen receptor alpha messenger RNA expression: High levels in the amygdaloid complex. Neuroscience 95:333–342

Phillips SM, Sherwin BB (1992) Effects of estrogen on memory function in surgically menopausal women. Psychoneuroendocrinology 17:485–495

Rapp PR, Morrison JH, Roberts JA (2003a) Cyclic estrogen replacement improves cognitive function in aged ovariectomized rhesus monkeys. J Neurosci 23:5708–5714

Rapp SR, Espeland MA, Shumaker SA, Henderson VW, Brunner RL, Manson JE, Gass MLS, Stefanick ML, Lane DS, Hays J, Johnson KC, Coker LH, Dailey M, Bowen D (2003b) Effect of estrogen plus progestin on global cognitive function in postmenopausal women. The Women's Health Initiative Memory Study: A randomized controlled trial. J Am Med Assoc 289:2663–2672

Resnick SM, Maki PM, Rapp SR, Espeland MA, Brunner R, Coker LH, Granek IA, Hogan P, Ockene JK, Shumaker SA (2006) Effects of combination estrogen plus progestin hormone treatment on cognition and affect. J Clin Endocrinol Metab 91:1802–1810

Resnick S, Espeland M, An Y, Maki P, Coker L, Jackson R, Stefanick M, Wallace R, Rapp S (2009a) Effects of conjugated equine estrogens on cognition and affect in postmenopausal women with prior hysterectomy. J Clin Endocrinol Metab 94:4152–4161

Resnick SM, Espeland MA, Jaramillo SA, Hirsch C, Stefanick ML, Murray AM, Ockene J, Davatzikos C (2009b) Postmenopausal hormone therapy and regional brain volumes: The WHIMS-MRI Study. Neurology 72:135–142

Roberts JA, Gilardi KV, Lasley B, Rapp PR (1997) Reproductive senescence predicts cognitive decline in aged female monkeys. Neuroreport 8:2047–2051

Rossouw JE, Anderson GL, Prentice RL, LaCroix AZ, Kooperbert C, Stefanick ML, Jackson RD, Beresford SA, Howard BV, Johnson KC, Kotchen JM, Ockene J (2002) Risks and benefits of estrogen plus progestin in healthy postmenopausal women. J Am Med Assoc 288:321–333

Savonenko AV, Markowska AL (2003) The cognitive effects of ovariectomy and estrogen replacement are modulated by aging. Neuroscience 119:821–830

Sherwin BB (1988) Estrogen and/or androgen replacement therapy and cognitive functioning in surgically menopausal women. Psychoneuroendocrinology 13:345–357

Sherwin BB, Henry JF (2008) Brain aging modulates the neuroprotective effects of estrogen on selective aspects of cognition in women: A critical review. Front Neuroendocrinol 29:88–113

Shughrue PJ, Merchenthaler I (2000) Evidence for novel estrogen binding sites in the rat hippocampus. Neuroscience 99:605–612

Shughrue P, Scrimo P, Lane M, Askew R, Merchenthaler I (1997) The distribution of estrogen receptor-ß mRNA in forebrain regions of the estrogen receptor-α knockout mouse. Endocrinology 138:5649–5652

Shughrue PJ, Scrimo PJ, Merchenthaler I (2000) Estrogen binding and estrogen receptor characterization (ERα and ERß) in the cholinergic neurons of the rat basal forebrain. Neuroscience 96:41–49

Shumaker SA, Legault C, Kuller L, Rapp SR, Thal L, Lane DS, Fillit H, Stefanick ML, Hendrix SL, Lewis CE, Masaki K, Coker LH (2004) Conjugated equine estrogens and incidence of probable dementia and mild cognitive impairment in postmenopausal women: Women's Health Initiative Memory Study. J Am Med Assoc 291:2947–2958

Soules MR, Sherman S, Parrott E, Rebar R, Santoro N, Utian W, Woods N (2001) Stages of reproductive aging workshop (STRAW). J Womens Health Gend Based Med 10:843–848

Spencer JL, Waters EM, Romeo RD, Wood GE, Milner TA, McEwen BS (2008) Uncovering the mechanisms of estrogen effects on hippocampal function. Front Neuroendocrinol 29:219–237

Spreafico E, Bettini E, Pollio G, Maggi A (1992) Nucleotide sequence of estrogen receptor cDNA from Sprague-Dawley rat. Eur J Pharmacol 227:353–356

Squire LR (1992) Memory and the hippocampus: A synthesis from findings with rats, monkeys, and humans. Psychol Rev 99:195–231

Steffens D, Norton M, Plassman B, Tschanz J (1999) Enhanced cognitive performance with estrogen use in nondemented community-dwelling older women. J Am Geriatr Soc 47: 1171–1175

Talboom JS, Williams BJ, Baxley ER, West SG, Bimonte-Nelson HA (2008) Higher levels of estradiol replacement correlate with better spatial memory in surgically menopausal young and middle-aged rats. Neurobiol Learn Mem 90:155–163

Tanapat P, Hastings NB, Reeves, AJ, Gould E (1999) Estrogen stimulates a transientincrease in the number of new neurons in the dentate gyrus of the adult female rat. J Neurosci 19: 5792–5801

Tang Y, Janssen WG, Hao J, Roberts JA, McKay H, Lasley B, Allen PB, Greengard P, Rapp PR, Kordower JH, Hof PR, Morrison JH (2004) Estrogen replacement increases spinophilin-immunoreactive spine number in the prefrontal cortex of female rhesus monkeys. Cereb Cortex 14:215–223

Thilers PP, Macdonald SW, Nilsson LG, Herlitz A (2010) Accelerated postmenopausal cognitive decline is restricted to women with normal BMI: Longitudinal evidence from the Betula project. Psychoneuroendocrinology 35:516–524

Tigges J, Gordon TP, McClure HM, Hall EC, Peters A (1988) Survival rate and life span of rhesus monkeys at the Yerkes Regional Primate Research Center. Am J Primatol 15: 263–273

Tremblay GB, Tremblay A, Copeland NG, Gilbert DJ, Jenkins NA, Labrie F, Giguere V (1997) Cloning, chromosomal localization, and functional analysis of the murine estrogen receptor beta. Mol Endocrinol 11:353–365

United States Census Bureau (2008) National Population Projections. http://www.census.gov/population/www/projections/2008projections.html

United States National Center for Health Statistics (2009) Deaths: Final Data for 2006. National Vital Statistics Reports (NVSR) 57 (14)

Viscoli C, Brass L, Kernan W, Sarrel P (2005) Estrogen therapy and risk of cognitive decline: Results from the women's estrogen for stroke trial (WEST). Am J Obstet Gynecol 192: 387–393

Voytko ML (2000) The effects of long-term ovariectomy and estrogen replacement therapy on learning and memory in monkeys (Macaca fascicularis). Behav Neurosci 114:1078–1087

Voytko ML (2002) Estrogen and the cholinergic system modulate visuospatial attention in monkeys (Macaca fascicularis). Behav Neurosci 116:187–197

Voytko ML, Higgs CJ, Murray R (2008) Differential effects on visual and spatial recognition memory of a novel hormone therapy regimen of estrogen alone or combined with progesterone in older surgically menopausal monkeys. Neuroscience 154:1205–1217

Voytko ML, Tinkler GP, Browne C, Tobin JR (2009) Neuroprotective effects of estrogen therapy for cognitive and neurobiological profiles of monkey models of menopause. Am J Primatol 71:794–801

Walf AA, Rhodes ME, Frye CA (2006) Ovarian steroids enhance object recognition in naturally cycling and ovariectomized, hormone-primed rats. Neurobiol Learn Mem 86:35–46

Wang JM, Liu L, Brinton RD (2008) Estradiol-17beta-induced human neural progenitor cell proliferation is mediated by an estrogen receptor beta-phosphorylated extracellularly regulated kinase pathway. Endocrinology 149:208–218

Warren SG, Humphreys AG, Juraska JM, Greenough WT (1995) LTP varies across the estrous cycle: Enhanced synaptic plasticity in proestrus rats. Brain Res 703: 26–23

Wise PM (2000) New understanding of the complexity of the menopause and challenges for the future. In: Bellino FL (ed) Proceedings of the international symposium on the biology of menopause. Springer, Norwell, MA, pp 1–8

Wise PM, Dubal DB, Wilson ME, Rau SW, Liu Y (2001) Estrogens: Trophic and protective factors in the adult brain. Front Neuroendocrinol 22:33–66

Wolf OT, Kirschbaum C (2002) Endogenous estradiol and testosterone levels are associated with cognitive performance in older women and men. Horm Behav 41:259–266

Wolf OT, Kudielka BM, Hellhammer DH, Törber S, McEwen BS, Kirschbaum C (1999) Two weeks of transdermal estradiol treatment in postmenopausal elderly women and its effect on memory and mood: Verbal memory changes are associated with the treatment induced estradiol levels. Psychoneuroendocrinology 24:727–741

Woodruff-Pak DS (1997) Neuropsychology of Aging. Blackwell Publishers, Malden, MA

Woolley CS, McEwen BS (1992) Estradiol mediates fluctuation in hippocampal synapse density during the estrous cycle in the adult rat. J Neurosci 12:2549–2554

Woolley CS, McEwen BS (1993) Roles of estradiol and progesterone in regulation of hippocampal dendritic spine density during the estrous cycle in the rat. J Comp Neurol 336: 293–306

Wu X, Glinn MA, Ostrowski NL, Su Y, Ni B, Cole HW, Bryant HU, Paul SM (1999) Raloxifene and estradiol benzoate both fully restore hippocampal cholineacetyltransferase activity in ovariectomized rats. Brain Res 847:98–104

Yaffe K, Sawaya G, Lieberburg I, Grady D (1998) Estrogen therapy in postmenopausal women: Effects on cognitive function and dementia. J Am Med Assoc 279:688–695

Yaffe K, Lui LY, Grady D, Cauley J, Kramer J, Cummings SR (2000) Cognitive decline in women in relation to non-protein-bound oestradiol concentrations. Lancet 356:708–712

Yaffe K, Barnes D, Lindquist K, Cauley J, Simonsick EM, Penninx B, Satterfield S, Harris T, Cummings SR (2007) Endogenous sex hormone levels and risk of cognitive decline in an older biracial cohort. Neurobiol Aging 28:171–178

Yamaguchi-Shima N, Yuri K (2007) Age-related changes in the expression of ER-β mRNA in the female rat brain. Brain Res 1155:34–41

Zandi PP, Carlson MC, Plassman BL, Welsh-Bohmer KA, Mayer LS, Steffens DC, Breitner JCS (2002) Hormone replacement therapy and incidence of Alzheimer disease in older women. J Am Med Assoc 288:2123–2129

Zhao L, O'Neill K, Brinton RD (2005) Selective estrogen receptor modulators (SERMs) for the brain: Current status and remaining challenges for developing NeuroSERMs. Brain Res Rev 49:472–493

Zhao Z, Fan L, Frick KM (2010) Epigenetic alterations regulate the estradiol-induced enhancement of memory consolidation. Proc Natl Acad Sci, USA 107:5605–5610

Part IV
Medical and Psychiatric Factors in Aging

Neuropsychological Features of Mild Cognitive Impairment and Preclinical Alzheimer's Disease

David P. Salmon

Abstract Detectable cognitive decline occurs in patients with Alzheimer's disease (AD) well before the clinical diagnosis can be made with any certainty. Studies examining this preclinical period identify decline in episodic memory as the earliest manifestation of the disease (i.e., a condition of amnestic Mild Cognitive Impairment). The episodic memory impairment is characterized by deficits in a number of processes including delayed recall, the recollective aspect of recognition memory, associative memory necessary for "binding" representations of two or more stimuli, pattern separation necessary to distinguish between two similar memory representations, prospective memory required to remember a delayed intention to act at a certain time in the future, and autobiographical memory for specific episodes that occurred in one's past. A growing body of evidence suggests that cognitive changes in preclinical AD may be more global in nature. Deterioration of semantic knowledge is evident on demanding naming and category fluency tasks, and "executive" dysfunction is apparent on tasks that require concurrent mental manipulation of information (e.g., working memory) or cue-directed behavior (e.g., set-shifting). Asymmetric cognitive test performance may also be apparent prior to significant decline in cognitive ability. The pattern and progression of these neuropsychological changes fit well with the proposed distribution and spread of AD pathology and serve as important cognitive markers of early disease.

Keywords Alzheimer's disease · Cognition · Dementia · Mild cognitive impairment · Memory

---

D. P. Salmon (✉)
Department of Neurosciences (0948), University of California,
9500 Gilman Drive, La Jolla, CA 92093-0948, USA
e-mail: dsalmon@ucsd.edu

Curr Topics Behav Neurosci (2012) 10: 187–212
DOI: 10.1007/7854_2011_171

Published Online: 1 November 2011

Contents

1 Episodic Memory Decline in Preclinical AD and MCI ... 189
2 Semantic Memory Decline in Preclinical AD and MCI ... 198
3 Executive Dysfunction and Attention Deficits in Preclinical AD and MCI ... 199
4 Decline in Visual Cognition in Preclinical AD and MCI ... 201
5 Asymmetry in Cognitive Abilities in Preclinical AD and MCI ... 203
6 Summary and Conclusions ... 204
References ... 205

It is widely accepted that the neurodegenerative changes of Alzheimer's disease (AD) begin well before the clinical manifestations of the disease become apparent (Katzman 1994; Jack et al. 2010). In the usual case, the neuronal atrophy, synapse loss, and abnormal accumulation of diffuse and neuritic amyloid plaques and neurofibrillary tangles associated with AD begin primarily in medial temporal lobe limbic structures (e.g., entorhinal cortex, hippocampus) and then progress to the association cortices of the frontal, temporal, and parietal lobes (Braak and Braak 1991). As these pathologic changes gradually accumulate, a threshold for the initiation of the clinical symptoms of the disease is eventually reached. Once this threshold is crossed, cognitive deficits become evident and gradually worsen in parallel with continued neurodegeneration. When the cognitive deficits become global and severe enough to interfere with normal social and occupational functioning, established criteria for dementia and a clinical diagnosis of AD are met.

It is apparent from this sequence of events that subtle cognitive decline is likely to occur in a patient with AD well before the clinical diagnosis can be made with any certainty. Such decline has been noted by a number of investigators with terms such as "questionable dementia", age-associated memory impairment, and others, but is now largely known as Mild Cognitive Impairment (MCI). As originally proposed by Petersen et al. (1999), MCI is characterized primarily as an amnestic disorder that represents a cognitive gray area between normal aging and AD. Criteria include a memory complaint corroborated by an informant, evidence of objective memory impairment, preservation of general cognitive functioning (commonly defined as a Mini-Mental State Exam (MMSE) score at or above 24), and intact activities of daily living, but all in the absence of dementia (Petersen et al. 2001). Broader conceptualizations of MCI have now emerged that encompass cognitive domains other than memory (Petersen and Morris 2005). Clinical subtypes of MCI now include amnestic and non-amnestic forms that can involve single or multiple cognitive domains.

A considerable amount of neuropsychological research has identified cognitive changes that reliably distinguish MCI from normal aging, particularly when MCI is considered to be a preclinical stage of AD. Because not all patients with MCI go on to develop the dementia of AD (Petersen et al. 1999), it is important to identify those features that predict which patients will progress or "convert" to evident dementia. One of the major approaches to this research is longitudinal examination of cognitive performance in healthy elderly individuals who go on to develop AD in order to retrospectively identify the nature of their earliest cognitive decline. This allows investigators to determine which aspects of cognition are first affected and how cognition changes over time prior to the development of frank, clinically diagnosable dementia. A second approach is to look for subtle differences in cognition in healthy elderly individuals with or without a risk factor for the development of AD (e.g., positive vs. negative family history, Apolipoprotein E (ApoE) ε4+ genotype vs. ε4- genotype). This approach presumes that the group with the risk factor contains more "preclinical" cases than the group without the risk factor and will perform more poorly in those cognitive domains that are first affected by AD. Research using these methods avoids the inherent circularity of defining a group as MCI based on neuropsychological criteria and then examining differences in neuropsychological performance between that group and elderly without MCI.

As will be reviewed, the majority of studies examining cognitive changes in preclinical AD identify decline in episodic memory as the earliest manifestation of the condition, a finding that drove the initial development of criteria for amnestic MCI (Albert and Blacker 2006; Collie and Maruff 2000). However, a growing body of evidence suggests that cognitive changes in preclinical AD may be more global in nature and can involve language, attention and executive functions, and visuospatial abilities (Backman et al. 2004, 2005; Twamley et al. 2006). The global nature of decline was demonstrated in a recent study by Mickes et al. (2007) that examined the relative degree to which various cognitive functions are impaired and the speed with which they decline during the preclinical period. The results of detailed neuropsychological evaluations of eleven normal elderly individuals who went on to develop AD dementia were retrospectively examined. This included evaluations over the course of three years up to and including the first year of a non-normal diagnosis (i.e., MCI or dementia). The results showed that performance falls off rapidly in all areas of cognitive functioning prior to the time a diagnosis of dementia can be made, but abilities thought to be mediated by the medial and lateral temporal lobes (episodic and semantic memory, respectively) appear to be substantially more impaired than those thought to be dependent upon the frontal lobes (Fig. 1).

These findings map well onto the proposed distribution and spread of AD pathology early in the disease process (Braak and Braak 1991) and imaging evidence that multiple brain regions (e.g., medial temporal lobes, frontal lobes, anterior cingulate cortex) are impaired in preclinical AD (Albert et al. 2001; Andrews-Hanna et al. 2007; Small et al. 2003). Given the involvement of multiple cognitive domains in preclinical AD and MCI, each of the major domains will be discussed in turn.

1 Episodic Memory Decline in Preclinical AD and MCI

An impaired ability to learn and retain new information (i.e., an episodic memory deficit or anterograde amnesia) is usually the earliest and most prominent feature of AD (for review, see Salmon 2000). The episodic memory deficit in early AD is

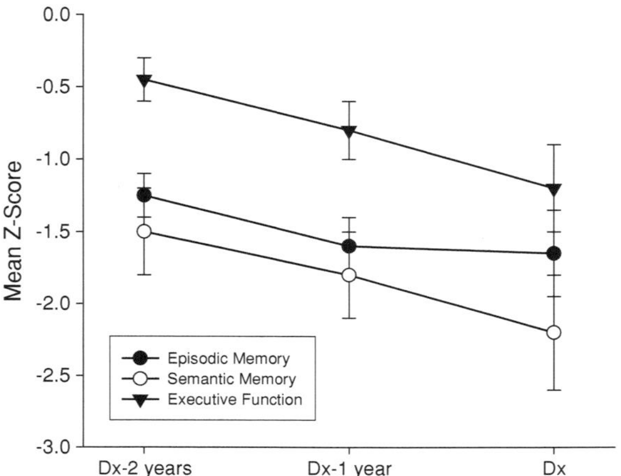

Fig. 1 The mean z-scores (relative to normal control subjects) on composite measures of episodic memory, semantic memory, and executive function achieved by eleven non-demented elderly individuals in the two-year period immediately preceding a diagnosis of AD (i.e., the preclinical AD period). Although episodic memory and semantic memory were more impaired than executive function, significant decline was observed in all three cognitive domains from two years prior to diagnosis (Dx-2 years), to one year prior to diagnosis (Dx-1 year), to the year of diagnosis (Dx) (Adapted from Mickes et al. 2007)

qualitatively and quantitatively similar to the anterograde amnesia exhibited by patients with circumscribed damage to medial temporal lobe structures (e.g., hippocampus, entorhinal cortex) or patients with alcoholic Korsakoff's syndrome (Delis et al. 1991). This is not surprising given that the earliest pathologic changes in AD usually occur in these same medial temporal lobe brain structures that are known to be critical for episodic memory function (for review, see Squire 1992). From a cognitive perspective, the episodic memory impairment in AD is thought to be largely due to ineffective consolidation of new information because patients with the disease exhibit abnormally rapid forgetting on tests of delayed recall (e.g., Welsh et al. 1991; Butters et al. 1988), and to-be-remembered information is not accessible after a delay even if retrieval demands are reduced by the use of recognition testing (e.g., Delis et al. 1991). Additional processing deficits can adversely influence memory performance in patients with AD. For example, they have difficulty utilizing semantic information to improve encoding in episodic memory tasks (e.g., Knopman and Ryberg 1989; Buschke et al. 1997), and they have an enhanced tendency to produce intrusion errors (i.e., producing previously learned information when attempting to recall new material) due to increased sensitivity to interference or diminished inhibitory processes (Fuld et al. 1982).

As mentioned, the concept of amnestic MCI arose from numerous observations that subtle decline in episodic memory often occurs prior to the emergence of the obvious cognitive and behavioral changes required for a clinical diagnosis of AD (for review, see Twamley et al. 2006). In one of the earliest studies to show this effect, Fuld et al. (1990) found that poor performance by non-demented elderly individuals on recall measures from the Fuld Object Memory Test or the Selective Reminding Test correctly predicted the subsequent development of AD within the next 5 years. An extensive follow-up to this study showed that delayed recall measures from the Selective Reminding Test and the Fuld Object Memory Test, in conjunction with Digit Symbol Substitution and verbal fluency performance, were moderately effective in identifying individuals who later developed AD (32/64 subjects; 50%) and provided excellent specificity for identifying individuals who remained free of dementia (238/253 subjects; 94%) over a subsequent 11-year period (Masur et al. 1994). Similar results have been observed in a number of large epidemiological studies which show that poor episodic memory performance (particularly poor delayed recall) at the initial neuropsychological evaluation of non-demented elderly individuals predicted those who subsequently developed dementia (e.g., Bäckman et al. 2001; Grober and Kawas 1997; Howieson et al. 1997; Jacobs et al. 1995; Kawas et al. 2003; Linn et al. 1995; Small et al. 2000; Tabert et al. 2006).

The importance of episodic memory measures in identifying preclinical AD has also been shown in studies that compared the neuropsychological test performances of non-demented elderly individuals with or without at least one ApoE ε4 allele. In a study by Bondi et al. (1999), for example, cognitively normal elderly ε4+ subjects performed significantly worse than comparable ε4- subjects on measures of delayed recall from the California Verbal Learning Test (CVLT), but not on tests of other cognitive abilities (Fig. 2). Cox proportional hazards analysis showed that ApoE ε4 status and measures of delayed recall were significant independent predictors of subsequent conversion to AD. Importantly, when subjects who developed AD were removed from analyses of baseline performance, there was no difference between the two ApoE groups on memory or other cognitive measures supporting the idea that poor recall is an early sensitive neuropsychological marker of AD and not simply a cognitive phenotype of the ε4 genotype.

A number of studies suggest that episodic memory performance may decline rapidly in the period immediately preceding the diagnosis of AD dementia (Chen et al. 2001; Lange et al. 2002). Lange et al. (2002), for example, compared the rate of decline in episodic memory during the preclinical phase of AD in individuals with or without at least one APOE ε4 allele. Non-demented normal control participants, non-demented older adults who subsequently developed dementia within one or two years, and patients with mild AD were examined with two commonly employed tests of episodic memory, the Logical Memory subtest of the Wechsler Memory Scale-Revised and the CVLT. Results revealed a precipitous decline in verbal memory abilities one to two years prior to the onset of the dementia syndrome (Fig. 3; also see Albert et al. 2007). Interestingly, there was little effect of ApoE genotype on the rate of preclinical memory decline. This suggests that ApoE genotype may influence the onset of AD, but once the disease is present it proceeds at a similar rate in those with or without the ε4 genotype.

These and similar results suggest that the rate of decline in episodic memory performance in patients with MCI may predict an imminent dementia diagnosis

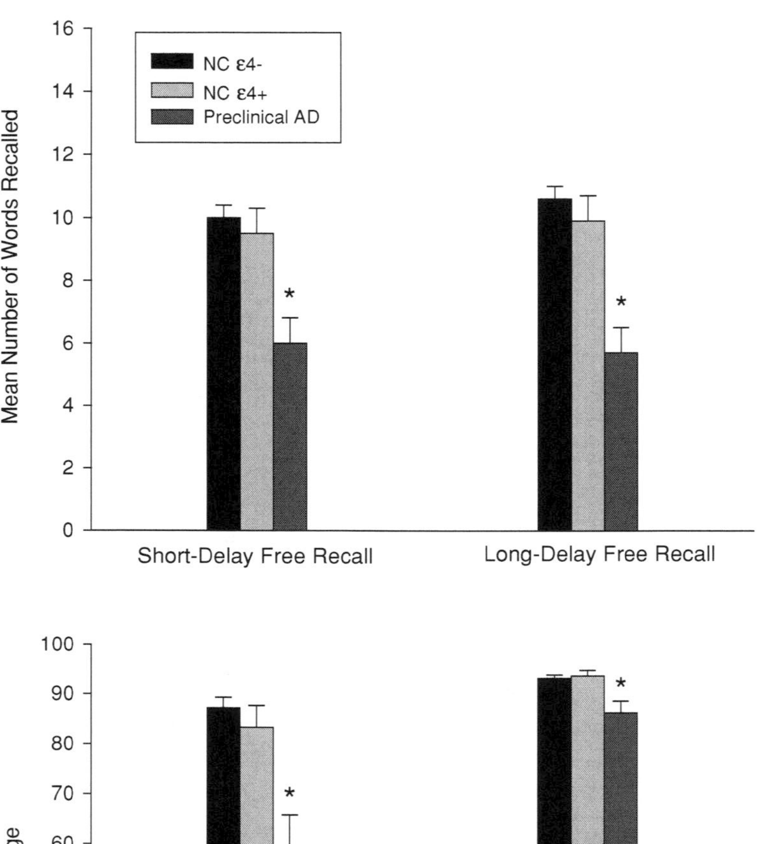

Fig. 2 The mean scores achieved on key measures from the California Verbal Learning Test (CVLT) by elderly normal control subjects with (NC ε4+) or without (NC ε4-) the apolipoprotein (ApoE) ε4 allele who remained cognitively stable and elderly normal control subjects who subsequently developed AD dementia (i.e., preclinical AD). Long Delay Retention refers to the percentage of words recalled on trial 5 of the learning condition that were recalled again after a 20-minute delay interval. The preclinical AD group performed significantly worse (* $p < .05$) than both NC groups on each of the CVLT measures. This demonstrates that poor episodic memory performance may be a sensitive marker for the early detection of AD. The lack of difference in the two stable control groups suggests that ApoE genotype is not associated with a phenotype of poor memory, but is a marker of AD risk (Adapted from Bondi et al. 1999)

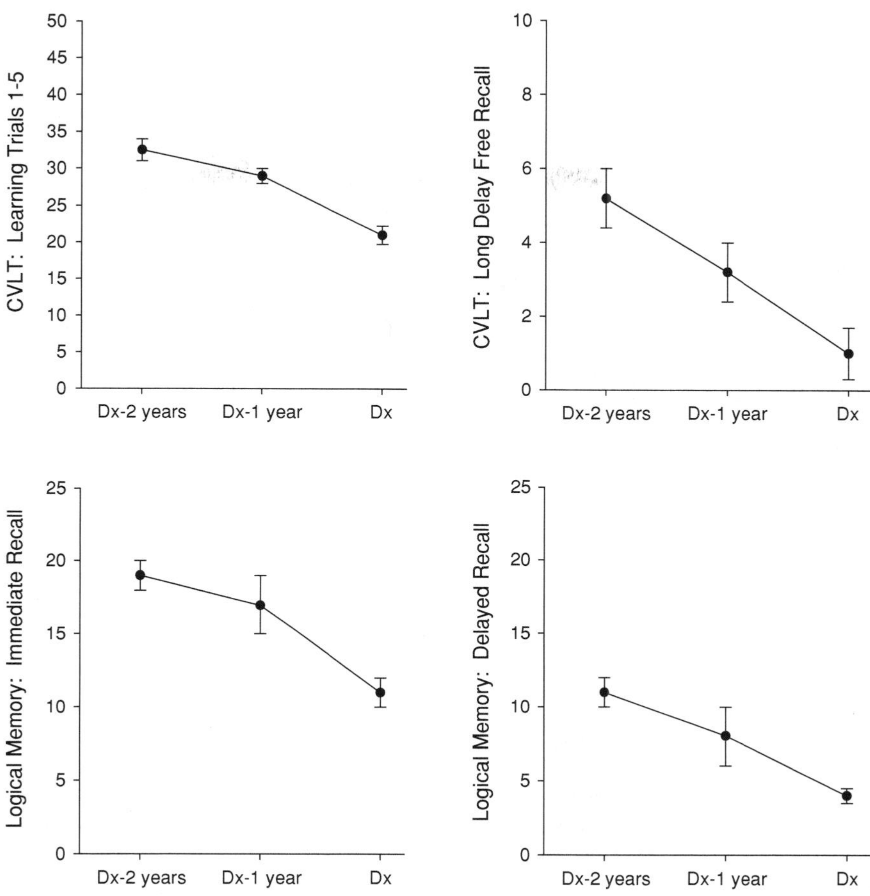

Fig. 3 The mean scores achieved on key measures from the California Verbal Learning Test (CVLT) and the WMS-R Logical Memory Test by non-demented elderly individuals in the two-year period immediately preceding a diagnosis of AD (i.e., the preclinical AD period). Each measure showed significant and relatively linear decline from two years prior to diagnosis (Dx-2 years), to one year prior to diagnosis (Dx-1 year), to the year of diagnosis (Dx). These results suggest that the rate of decline in episodic memory performance in non-demented elderly may predict an imminent dementia diagnosis more effectively than absolute level of memory performance (Adapted from Lange et al. 2002)

more effectively than absolute level of memory performance. This possibility was recently supported by a longitudinal study of MCI patients which showed that delayed recall measures, and particularly one-year decline in delayed recall, were more effective than neuroimaging (e.g., hippocampal volume) and cerebrospinal fluid (CSF) protein (e.g., Aβ 1-42 and tau) biomarkers in predicting conversion from MCI to AD dementia within the subsequent four years (Gomar et al. 2011; also see Landau et al. 2010).

Although episodic memory performance declines precipitously just before the diagnosis of dementia can be made, the years preceding this sharp decline appear to be characterized by poor but relatively stable memory performance (e.g., Bäckman et al. 2001; Small et al. 2000). This pattern of results across studies suggests that there may be a plateau in the course of decline in episodic memory rather than a monotonic or linear decrease in memory ability over many years. Such a model of mild but stable episodic memory decline followed by more abrupt decline in the years proximal to diagnosis was validated in a large-scale study by Smith et al. (2007) who found that a plateau was evident on tests of episodic memory, but not on tests of other cognitive domains. A number of functional neuroimaging studies (Bookheimer et al. 2000; Bondi et al. 2005; Han et al. 2007) suggest that the plateau period may result from compensatory brain responses to the development of AD pathology. That is, a wider network of activation in medial temporal lobe and other cortical regions is necessary to maintain the same level of memory performance in those who subsequently develop dementia compared to those who do not progress. Once these compensatory responses are overwhelmed, memory decline becomes evident again. This idea is consistent with the notion that an abrupt decline in memory in an elderly individual might better predict the imminent onset of dementia than poor but stable memory ability.

Although there is considerable evidence that decline in episodic memory is an early manifestation of preclinical AD, there is less evidence regarding the specific aspects of memory that are affected. Better understanding of the nature of the memory deficit in MCI could help tie the impairment to specific areas of brain pathology and improve the ability to determine whether or not the deficit is likely to be related to preclinical AD. Accordingly, recent studies have examined the memory processes (e.g., encoding, retrieval, binding of associations, pattern separation) and types of memory (e.g., prospective memory, remote memory) that might differ in normal elderly individuals and patients with amnestic MCI. Several of these studies have shown that the episodic memory deficit in amnestic MCI is characterized, in most cases, by abnormally rapid forgetting on tests of delayed recall (Libon et al. 2011; Perri et al. 2007; Manes et al. 2008) and comparable levels of impairment on tests of free recall and recognition (Libon et al. 2011). This pattern is usually attributed to ineffective encoding and consolidation of new information and is virtually identical to the pattern shown by patients with circumscribed amnesia arising from bilateral damage to medial temporal lobe (e.g., hippocampus, entorhinal cortex) or diencephalic structures (e.g., anterior and dorsomedial thalamic nuclei, mammillary bodies) (for review, see Salmon and Squire 2009). This similarity is not surprising given that the earliest pathological changes of AD that are thought to underlie amnestic MCI typically occur in these same medial temporal lobe structures (Braak and Braak 1991). Like patients with early AD, those with amnestic MCI also fail to benefit in a normal fashion from deep semantic encoding in episodic memory tasks (Hudon et al. 2011), and they have an enhanced tendency to produce prototypical intrusion errors during free recall (Libon et al. 2011).

Given that damage to medial temporal lobe structures is the primary neural substrate of episodic memory deficits in patients with amnestic MCI, it is not surprising that they have deficits in associative memory. Associative memory refers to the ability to remember relationships between two or more items or between an item and its context (e.g., when or where something was seen). This form of memory "binding" is thought to be critically dependent upon the hippocampus and is impaired in patients with circumscribed amnesia (Eichenbaum 1997). The associative memory deficit in amnestic MCI was recently shown in a study that examined the ability of patients and control subjects to remember six simple geometric forms (memory for items) and their location within a spatial array (associative memory), or to remember nine symbols and the digits with which they were paired (Troyer et al. 2008). The results showed that in both tasks amnestic MCI patients were impaired relative to controls and the impairment was greater for associative information than for item information. Similar deficits in associative memory in amnestic MCI patients have been shown using various paired-associate learning tasks that use word or object pairs (Pike et al. 2008; de Rover et al. 2011), and these deficits have been tied to hippocampus and entorhinal cortex abnormalities through structural and functional imaging (de Rover et al. 2011). In addition, a deficit in face-name paired-associate learning was recently observed in non-demented elderly individuals carrying AD pathology as indicated by PET imaging with Pittsburgh compound-B ([^{11}C]-PIB), an agent that binds to β amyloid (the main constituent of the plaque) in the brain (Rentz et al. 2011). These results suggest that a deficit in associative memory may be a particularly early marker of AD.

Just as the hippocampus is important for binding item and context information in memory, it also plays an important role in separating similar representations (or patterns) into distinct representations that can be reliably distinguished from one another (Marr 1971). Pattern separation and its neural substrates were recently examined in patients with amnestic MCI using high resolution functional MRI during a difficult recognition memory task in which subjects had to distinguish between previously shown stimuli ("old" items), "lure" stimuli very similar to old items (Yassa et al. 2010), and completely novel ("new") items. The results showed that patients with MCI were more likely than controls to report a lure item as previously seen, indicative of pattern completion rather than pattern separation, and had an overall reduction in separation bias (i.e., the difference between the probability of calling a lure item similar and the probability of calling a novel item similar). This occurred despite normal performance in discriminating novel items from previously seen items. This selective impairment of pattern separation in patients with amnestic MCI was accompanied by hyperactivity in the left dentate gyrus and CA3 field of the hippocampus. Furthermore, the degree of this hyperactivity was negatively correlated with separation bias scores. Because medial temporal lobe structures are often the first affected in AD, these results suggest that poor performance on memory tasks that place heavy demands on pattern separation may be a particularly sensitive marker of early disease.

Although many studies have shown that recognition memory is impaired in patients with amnestic MCI (for review, see Twamley et al. 2006), few studies have examined the qualitative nature of this deficit. Current theories of recognition memory postulate that the ability to recognize whether or not an event or stimulus previously occurred involves relatively independent processes of recollection and familiarity (e.g., Mandler 1980). Recollection refers to the conscious re-experience of a recent event, while familiarity is the feeling of having previously encountered an event with no associated contextual information. Westerberg et al. (2006) compared recognition memory under yes/no recognition and forced-choice recognition procedures with the assumption that forced-choice was more likely to be mediated by familiarity. They found that amnestic MCI patients were impaired in the yes/no condition, but normal in the forced choice condition, suggesting that they have impaired recollection with relatively preserved familiarity (also see Bennett et al. 2006). The same conclusion was reached in a study that used a process dissociation procedure to separate the effects of recollection and familiarity in amnestic MCI (Anderson et al. 2008). In this study, analysis of recognition for words presented either auditorally or visually was tested under two sets of instructions: (1) to identify repeated words regardless of presentation modality, or (2) to identify repeated words only if they were repeated in the original modality. By comparing the two conditions, the ability to recognize a word in its original context (recollection) or independent of its context (familiarity) could be contrasted. Patients with amnestic MCI showed impaired recollection and normal familiarity under these conditions (also see Serra et al. 2010). Finally, a study that used a Remember/Know paradigm in which subjects indicated if their recognition of an event was based (Remember/recollection) or not based (Know/familiarity) on remembering contextual information about the event showed that recollection was impaired and familiarity was spared in patients with amnestic MCI (Hudon et al. 2009; also see Serra et al. 2010).

The majority of studies of the contributions of recollection and familiarity to recognition memory in patients with amnestic MCI suggest that recollection is impaired and familiarity is spared (but see Algarabel et al. 2009). This conclusion is bolstered by studies that show normal feeling-of-knowing judgments (Anderson and Schmitter-Edgecombe 2010; but see Perrotin et al. 2007), susceptibility to false recognition lures (Hudon et al. 2006), and incidental recognition of shallowly encoded words (Mandzia et al. 2009) in patients with amnestic MCI. Each of these processes is thought to depend upon familiarity to some degree. There are, however, several studies that show impaired familiarity in patients with amnestic MCI using some of the same procedures described above (e.g., Wolk et al. 2008; Ally et al. 2009). These discrepant results could be related to different degrees of extra-hippocampal pathology related to more advanced disease and suggest the need for further study.

A number of recent studies have investigated the impact of amnestic MCI on the ability to remember a delayed intention to act at a certain time or when some external event occurs in the future (Troyer and Murphy 2007; Costa et al. 2010; Thompson et al. 2010; Schmitter-Edgecombe et al. 2009; Karantzoulis et al. 2009). This form of prospective memory is essential for carrying out activities critical for independent living such as remembering to pay bills on a certain date, remembering to take medication at a certain time of day or remembering to make a turn when a particular land-mark is spotted. Various components of prospective memory are thought to involve separate cognitive functions that could be affected by damage to distinct brain regions. For example, an episodic memory component is involved in remembering the specific act to be performed at the appropriate time, while an executive and attention-related component supports cognitive operations such as monitoring the passage of time, planning sequential activities, and switching from an on-going activity to the intended activity. Studies of prospective memory have shown that both time-based and event-based tasks are impaired in patients with amnestic MCI (Troyer and Murphy 2007; Costa et al. 2010; Thompson et al. 2010; Schmitter-Edgecombe et al. 2009; Karantzoulis et al. 2009). Several of these studies also indicate that time-based prospective memory is more impaired than event-based (Troyer and Murphy 2007; Karantzoulis et al. 2009). Because time-based prospective memory tasks place greater demands on executive function and attention (e.g., self-initiation, time monitoring) than event-based tasks, this pattern of results suggests that deficits in both hippocampus-dependent and frontal cortex-dependent functions contribute to the prospective memory deficit in patients with amnestic MCI. This possibility is supported by a study that showed that patients with MCI were more impaired on recall of the intention to act (an executive-based function) than on execution of the action (a memory-based function), and this discrepancy was greater in amnestic MCI patients with additional executive impairment than in those without additional impairment (Costa et al. 2010). A study that showed a correlation between prospective memory performance and other frontal cortex mediated functions (e.g., temporal order memory, source memory) provides additional support (Schmitter-Edgecombe et al. 2009). Taken together, these findings suggest that time-based prospective memory impairment in patients with amnestic MCI may reflect a more advanced neurodegenerative process that may help to predict imminent development of dementia.

Mildly demented patients with AD often exhibit a severe and temporally graded retrograde amnesia (i.e., a deficit in the ability to remember past events that were successfully remembered prior to a brain injury or the onset of a neurological disease) with memories from the distant past better retained than memories from the more recent past (e.g., Beatty et al. 1988). The temporal gradient is similar to the pattern of loss exhibited by patients with circumscribed amnesia and has been attributed to the interruption of a long-term consolidation process that is critically dependent upon the hippocampus (for review, see Salmon 2000). The remote memory deficit in patients with AD reflects both the loss of remote autobiographical memory that is episodic in nature (e.g., specific details of an event) and the loss of personal semantics (e.g., general information regarding an event that does not have spatiotemporal context).

A number of studies indicate that retrograde memory loss also occurs in patients with amnestic MCI. Leyhe and colleagues, for example, showed that patients with amnestic MCI were impaired on a test of memory for historic public events that occurred over the past 60 years (Leyhe et al. 2010), as well as on a test of remote autobiographical memory (Leyhe et al. 2009). A temporal gradient was observed for the autobiographical information with older remote memories better retained than newer remote memories. The remote autobiographical memory loss that occurs in patients with MCI is associated with functional reorganization of the neural network that underlies this aspect of memory. Poettrich et al. (2009) found the pattern of activation observed with functional MRI during recall of remote autobiographical events not only involved a left-lateralized network of frontal, temporal, and parietal areas and the medial frontal cortex in normal control subjects, but also included the symmetrical activation of these areas on the right and activation of the precuneus and supplementary motor areas in patients with MCI. This spread of activation may indicate a compensatory response to maintain remote memory performance in the patients with MCI.

While studies consistently show that remote autobiographical memory is impaired in amnestic MCI, there is less consistency in the evidence for a loss of personal semantics. Murphy et al. (2008) found that personal semantics were preserved in amnestic MCI, but Irish et al. (2010) found that they were impaired. The inconsistency in these findings could be related to degree of spread of AD pathology from the hippocampus that mediates remote autobiographical memory to temporal association cortices that mediate semantic memory. If this is the case, the presence of impairment in both remote autobiographical memory and personal semantics could indicate an increased likelihood of imminent progression to dementia.

2 Semantic Memory Decline in Preclinical AD and MCI

Semantic memory refers to our general fund of knowledge which consists of the meanings and representations of words, concepts, and over-learned facts that are not dependent upon contextual cues for their retrieval. This knowledge is usually assumed to be organized as a complex associative network in which concepts that have many attributes in common are more strongly associated than those that share fewer attributes. Semantic knowledge is thought to be stored in a distributed manner in neocortical association areas of the temporal and parietal lobes, and is not dependent upon the medial temporal lobe structures that are important for episodic memory.

Consistent with the widely distributed neocortical damage that occurs in AD, semantic memory is often impaired relatively early in the course of the disease (for reviews, see Salmon and Chan 1994). Semantic memory impairment is evident in AD patients' reduced ability to recall over-learned facts (e.g., the number of days in a year), and in their impairment on tests of confrontation naming and verbal fluency. The impairment is thought to reflect the loss of semantic knowledge, rather than inefficient retrieval because studies that have probed for knowledge of particular concepts across different modes of access and output (e.g., fluency, confrontation naming, sorting, word-to-picture matching, definition generation) show that patients with AD are significantly impaired across all tasks, and there is item-to-item correspondence so that when a particular stimulus item is missed (or correctly identified) in one task, it is likely to be missed (or correctly identified) in other tasks that access the same information in a different way (Chertkow and Bub 1990; Hodges et al. 1992).

A growing body of evidence suggests that subtle deficits in semantic memory occur in patients with amnestic MCI and may be an indication of imminent progression to dementia (e.g., Dudas et al. 2005; Joubert et al. 2010; Thompson et al. 2002). Although patients with amnestic MCI do not consistently show deficits in confrontation naming on clinical tests such as the Boston Naming Test (Balthazar et al. 2008, 2010), they are impaired on more semantically demanding naming tasks that require producing proper nouns such as the names of famous people or buildings (Ahmed et al. 2008; Adlam et al. 2006; Borg et al. 2010; Joubert et al. 2010; Seidenberg et al. 2009). Amnestic MCI patients are also often impaired when required to generate exemplars from a specific semantic category (e.g., "animals"), but are not impaired when required to rapidly generate words beginning with a particular letter (e.g., F, A, or S) (Adlam et al. 2006; Biundo et al. 2011; Murphy et al. 2006). This pattern of performance is also present in mildly demented patients with AD (e.g., Butters et al. 1987) and is often interpreted as a reflection of semantic knowledge loss because the semantic fluency task places greater demands than the letter fluency task on the use of semantic organization to efficiently generate words from a small and highly related set of exemplars.

Some studies do not find differential verbal fluency impairment in patients with amnestic MCI, but rather slightly impaired performance on both letter and semantic category fluency tasks (Brandt and Manning 2009; Nutter-Upham et al. 2008). In these studies, disproportionately impaired semantic category fluency is only observed in patients with multi-domain MCI who may be further along in the course of the disease. These studies suggest that greater semantic than letter fluency impairment in patients with MCI may mark those who are most likely to convert to AD.

An atypical form of AD sometimes occurs in which patients initially present with a form of primary progressive aphasia (PPA) characterized by hesitant, grammatically correct speech, and spared language comprehension (Gorno-Tempini et al. 2004; Mesulam et al. 2009). These language deficits occur in the context of preserved memory, executive functions, and visuospatial abilities, and could be considered a form of single domain MCI. This so-called "logopenic" PPA is usually associated with AD pathology disproportionately distributed in language-related cortical areas (Mesulam et al. 2008). As the AD pathology progresses, other neocortical areas become involved and additional cognitive domains are affected. Eventually, patients take on a more typical presentation of AD.

3 Executive Dysfunction and Attention Deficits in Preclinical AD and MCI

Deficits in "executive" functions responsible for concurrent mental manipulation of information, concept formation, problem solving, and cue-directed behavior occur early in the course of AD (see Perry and Hodges 1999). The ability to perform concurrent manipulation of information on tests that require set-shifting, self-monitoring, or sequencing appears to be particularly vulnerable

(Lefleche and Albert 1995). This may be related to a fundamental decrease in the ability to inhibit prepotent responses in early AD due to a decline in attentional control mechanisms. Patients with AD exhibit deficits in aspects of attention relatively early in the course of the disease, particularly on dual-processing tasks, tasks that require the disengagement and shifting of attention, and working memory tasks that are dependent upon the control of attentional resources (for reviews, see Parasuraman and Haxby 1993; Perry and Hodges 1999).

Similar deficits in executive functions have been noted in patients with preclinical AD or amnestic MCI. Brandt and colleagues (2009), for example, found that amnestic MCI patients performed worse than normal control subjects on composite executive function measures of planning/problem solving and working memory, although not on a composite measure of judgment. These executive function deficits were even more apparent in patients with amnestic multi-domain MCI, suggesting that those patients might be at highest risk for the imminent onset of dementia. As mentioned previously, Mickes et al. (2007) found that a composite executive function measure declined significantly several years prior to the diagnosis of dementia in patients with preclinical AD. Similarly, Albert et al. (2007) found that the combination of a composite executive function measure and a composite episodic memory measure accurately predicted the development of AD dementia in a group of non-demented elderly individuals (also see Chapman et al. 2011).

Evidence for deficits in inhibition or attentional control in patients with preclinical AD comes from a study that compared the performances of non-demented elderly individuals who subsequently developed AD dementia (i.e., preclinical AD) and those who did not (i.e., controls) on a computerized version of the Stroop interference task (Balota et al. 2010). In the usual version of this task, participants are timed for how quickly they can read words which are the names of colors (i.e., color words), name the color of ink patches (or a series of three x's), and finally name the color of the ink in which non-congruent color words are printed (i.e., say "red" when the word green is printed in red ink). The effects of response inhibition are indicated by slower response times when naming the color of the ink of non-congruent color words than when reading words which are the names of colors or naming the color of ink patches. This response slowing presumably occurs because activation of the non-congruent color word (reading the word is the prepotent response) interferes with the production of the correct name of the color and must be actively inhibited. In the version of the task used in this study, the production of errors on the non-congruent trials was used to indicate a decline in inhibition and attentional control. The results of this study showed that preclinical AD patients made more errors than control subjects on the non-congruent trials and had a larger Stroop effect (i.e., difference between congruent and non-congruent trial reaction times). A similar pattern of impairment was shown in a study that compared non-demented elderly with or without an ApoE $\varepsilon 4$ genotype on a new and difficult Inhibition/Switching condition of the Stroop test that required simultaneous response inhibition and cognitive set switching (i.e., in the non-congruent condition, name the color of the word, unless it appears in a box, then read the word; Wetter et al. 2005). The $\varepsilon 4+$ group committed more errors than the $\varepsilon 4$- group on the Inhibition/Switching condition, even though performance on other conditions was comparable in the two groups. In a related study (Fine et al. 2008), scores on these new Stroop task switching measures predicted which non-demented elderly individuals would decline over the next year on a measure of global mental status (i.e., the Mattis Dementia Rating Scale) and which would remain cognitively stable.

Deficits in inhibition and attentional control also occur in patients with amnestic MCI. In some cases, these patients are impaired relative to normal control subjects on non-congruent color naming conditions of the Stroop task (Belanger et al. 2010), on the Hayling task that requires subjects to complete a sentence with a non-dominant response (e.g., He hit the nail with a ____.; Belanger and Belleville 2009), on an expectancy violation task in which subjects must rapidly judge the relatedness of coordinate word pairs (e.g., apple-pear) that occasionally (i.e., unexpectedly) occur in the context of judging category word pairs (e.g., peach-fruit) (Davie et al. 2004), and on a digit-number identification cognitive set switching task (Sinai et al. 2010). In this latter task, subjects were cued on randomly interleaved trials to make a decision regarding either the number (i.e., odd or even) or letter (i.e., vowel or consonant) of a compound number-letter stimulus. Although the task could be completed successfully by all normal control subjects, only 16 of 27 patients with MCI could perform above 80% correct in a given block of trials. Furthermore, 5 of these 16 patients could not maintain the cue in working memory and required an additional cue at the time of response to achieve successful performance. Compared to normal control subjects, the MCI patients who were unable to perform the task had significant cortical atrophy in frontal, temporal, and superior parietal lobe regions on MRI, poorer performance on standard neuropsychological tests of executive function, and an increased risk of transition to AD dementia.

A deficit in working memory has been observed in patients with preclinical AD (Grober et al. 2008; Rapp and Reischies 2005) or amnestic MCI (e.g., Gagnon and Belleville 2011; Saunders and Summers 2011; Sinai et al. 2010; Darby et al. 2002). Working memory refers to a limited capacity memory system in which information that is the immediate focus of attention can be temporarily held in limited-capacity language-based or visual-based buffers while being manipulated through a primary central executive system (Baddeley 1986). The working memory deficit in MCI is consistent with that seen in early AD (Baddeley et al. 1991; Collette et al. 1999) and is usually mild and limited to disruption of the central executive consistent with a decline in attentional control. This interpretation is supported by evidence that patients with amnestic MCI are impaired on divided attention tasks and that this impairment grows during the transition to AD dementia (Belleville et al. 2007; Saunders and Summers 2011).

4 Decline in Visual Cognition in Preclinical AD and MCI

Deficits in visuospatial abilities and constructional praxis occur in patients with AD, but they usually emerge after the early stages of the disease (e.g., Locascio et al. 1995) and are not present in patients with preclinical AD or MCI. However, a

Table 1 The mean and standard deviation (in parentheses) of scores achieved by patients with preclinical Alzheimer's disease (AD) and normal control subjects on tests of verbal (i.e., Boston Naming Test) and visuospatial (i.e., Block Design Test) abilities, and on a measure of asymmetry in these abilities (i.e., the absolute value of the difference between verbal and visuospatial z-scores)

	Preclinical AD	Normal controls	
Boston naming test	26.4	27.6	n.s.
	(2.9)	(7.6)	
Block design test	37.8	41.5	n.s.
	(8.8)	(8.3)	
Asymmetry score	1.42	0.64	$p < 0.01$
	(1.11)	(0.55)	

Although test scores of the two groups were similar in both conditions, the asymmetry score was significantly higher in the preclinical AD subjects. An asymmetric profile was twice as likely in preclinical AD as in normal control subjects (Chi-square = 4.80; $p < 0.05$). Verbal greater than visuospatial and visuospatial greater than verbal asymmetric profiles were equally likely in the preclinical AD group (Adapted from Jacobson et al. 2002)

few studies have shown that relatively subtle visual processing deficits can occur in patients with amnestic MCI (e.g., Bonney et al. 2006; Bublak et al. 2011; Mapstone et al. 2003). For example, the threshold for the presentation time necessary for a visual stimulus to be perceived was found to be increased in patients with amnestic MCI relative to age-matched normal control subjects using both a letter-series identification task (Bublak et al. 2011) and a symbol-component identification task (i.e., inspection time paradigm; Bonney et al. 2006). The ability of amnestic MCI patients to detect visual motion may also be abnormal. Mapstone and colleagues (2003) found that MCI patients performed worse than normal elderly subjects on a task in which they had to detect the direction of motion of a number of small white dots that coherently moved radially toward or away from a central location while embedded in a large number of randomly moving dots. The threshold for the number of dots that had to move coherently within the noise to ensure accurate detection of direction was highest for patients with AD, intermediate for patients with MCI, and lowest for normal elderly. Motion detection threshold was correlated with performance on a test of visuospatial orientation (Money Road Map test), but not tests of episodic memory.

There are somewhat rare instances when AD initially presents with circumscribed posterior cortical atrophy with cognitive impairment dominated by higher-order visual dysfunction (Caine 2004; Mendez et al. 2002; Tang-Wai et al. 2004). This clinical syndrome of Posterior Cortical Atrophy (PCA) could be considered a form of single domain MCI because memory, language, and judgment and insight are relatively preserved until the late stages of disease. Patients with PCA usually have prominent visual agnosia (sometimes including prosopagnosia) and constructional apraxia, and exhibit many or all of the features of Balint's syndrome including optic ataxia, gaze apraxia, and simultanagnosia (i.e., can detect visual details of an object but cannot organize them into a meaningful whole). They may also exhibit components of Gerstmann's syndrome including acalculia, right-left disorientation, finger agnosia, and agraphia (Caine 2004; Mendez et al. 2002; Renner et al. 2004; Tang-Wai et al. 2004). A visual field defect, decreased visual attention, impaired color perception, or decreased contrast sensitivity may also occur (Della Sala et al. 1996).

PCA is usually associated with AD pathology, but may also occur in the presence of neuropathological changes of dementia with Lewy bodies or Creutzfeld-Jakob disease (Renner et al. 2004). Neuropathologic examination of the brains of patients with PCA reveals disproportionate atrophy and pathologic lesions in the occipital cortex and posterior parietal cortex relative to other cortical association areas (Hof et al. 1997; Renner et al. 2004). Posterior cortical hypometabolism in patients with PCA has also been shown with PET imaging, with particular involvement of the dorsal visual stream (Nestor et al. 2003). In the case of PCA due to AD, the neurofibrillary tangles and neuritic plaques in the posterior cortical regions are qualitatively identical to those in typical AD and have the same laminar distribution in the cortex (Hof et al. 1997). The disproportionately posterior cortical distribution of AD pathology in PCA has recently been demonstrated in living patients using PET imaging of β amyloid binding (Tenovuo et al. 2008). The cause of the focal presentation of PCA is unknown but is the focus of ongoing research efforts.

5 Asymmetry in Cognitive Abilities in Preclinical AD and MCI

A growing body of research suggests that subtle cognitive changes during the preclinical phase of AD can be detected as an asymmetric profile of performance across cognitive domains in non-demented older adults who are destined to develop dementia. These studies were motivated by earlier research that documented lateralized cognitive deficits (e.g., greater verbal than visuospatial deficits or vice versa) in subgroups of mildly demented AD patients (e.g., Haxby et al. 1985). Jacobson et al. (2002) reasoned that subtle asymmetric cognitive decline due to AD might be detectable during the preclinical period in a subset of individuals, even if their performance in all cognitive domains remained above normal because the individual would be essentially serving as their own control. To examine this possibility, Jacobson et al. (2002) compared 20 cognitively normal elderly adults who were in a preclinical phase of AD (i.e., they were diagnosed with AD approximately one year later) and 20 age- and education-matched normal control subjects on a number of cognitive tests and a derived score that reflected cognitive asymmetry: the absolute difference between verbal (Boston Naming Test) and visuospatial (Block Design Test) standardized scores. The results showed that the two groups performed similarly on individual tests of memory, language, and visuospatial ability. In contrast, cognitive asymmetry (in either direction) was significantly greater in the preclinical AD patients than in normal controls (Table 1). These results suggest that there is a subgroup of patients with preclinical AD who have asymmetric cognitive decline that may be obscured when cognitive scores are averaged over the entire group.

The longitudinal course of cognitive asymmetry in preclinical AD was examined in a recent case study of an elderly woman who was examined annually for six years prior to the development of clinically evident dementia (Jacobson et al. 2009). During the preclinical period, this woman exhibited intact learning and memory (relative to age and education appropriate normative data) and had no evidence of overall cognitive or functional decline. Despite this relatively high level of performance, a pattern of increasing cognitive asymmetry was apparent with stable performance on non-verbal/visuospatial tests relative to declining (but still normal) scores on verbally based tasks. This case supports the notion that an asymmetric profile of cognitive test performance can be a sensitive marker of early neuropsychological changes that may be apparent prior to significant decline in either memory or overall cognitive ability.

A series of studies showed that greater cognitive asymmetry is apparent in non-demented individuals with genetic risk of AD than in those without that risk. Jacobson et al. (2005a), for example, found a significantly larger discrepancy between WAIS-R Digit Span and WMS-R Visual Memory Span performance in ApoE $\varepsilon4+$ than in $\varepsilon4$- groups of non-demented elderly individuals. In a subsequent study, more non-demented $\varepsilon4+$ subjects than $\varepsilon4$- subjects had an asymmetric profile (i.e., a z-score difference >1 standard deviation) on the new switching conditions of the Verbal Fluency and Design Fluency tests of the Delis-Kaplan Executive Function System (Houston et al. 2005). Finally, groups of ApoE $\varepsilon4+$ and $\varepsilon4$- non-demented elderly were compared on a Global Local Memory Test (GLMT) in which participants were asked to recall and draw complex stimuli that contained both a large global component (thought to primarily require right hemisphere processing) and smaller, detailed local elements (thought to primarily require left hemisphere processing) (Jacobson et al. 2005b). The results showed that $\varepsilon4+$ and $\varepsilon4$- groups had comparable levels of immediate recall for both the global and the local features of the stimuli. However, more $\varepsilon4+$ subjects (64%) than $\varepsilon4$- subjects (23%) had a 1 standard deviation or more difference between immediate recall of the global and local features.

Taken together, these studies suggest that preclinical AD may be detectable as a discrepancy (or asymmetry) in an individual's own performance across cognitive domains, even when absolute levels of performance in all cognitive domains is normal. They further suggest that consideration of non-memory asymmetric changes, in addition to consideration of subtle declines in memory, might improve the ability to detect AD in its earliest stages.

6 Summary and Conclusions

The studies reviewed above clearly indicate that there is a preclinical phase of detectable cognitive decline that can precede the clinical diagnosis of AD by several years. The detection of incipient dementia is most effectively accomplished with sensitive measures of learning and memory, and may be enhanced if risk factors such as a positive family history or the presence of the ApoE ε4 allele are also considered. The ability of cognitive measures to detect AD in its earliest, preclinical stage continues to be an important topic of neuropsychological research, particularly since neuroprotective agents designed to impede the progression of the disease are being developed (Raffi and Aisen 2009). Identification of the initial cognitive changes would help to reliably detect AD in its earliest stages when disease modifying therapies would be most effective, and would provide a target for very early symptomatic treatment.

Advances in the neuropsychological detection of preclinical AD have occurred in parallel with advances in detecting reliable biological markers of the disease. It is now possible to use MRI to detect reductions in hippocampal volume and cortical thickness typically associated with AD, to use PET imaging with PIB or other similar agents to detect the deposition of β amyloid in the brain, and to use biochemical assays of CSF to detect abnormal levels of the β amyloid and tau proteins that constitute the plaques and tangles of AD (for review, see Jack et al. 2010). Each of these biomarkers is quite effective at predicting the development of AD dementia in non-demented elderly individuals. These discoveries have prompted a revision of the research diagnostic criteria for AD that have been widely used for nearly 30 years (McKhann et al. 1984). The revised criteria continue to define AD dementia largely as before (McKhann et al. 2011), but now also incorporate the presence of a biomarker as supporting evidence. The intermediate stage of MCI that precedes frank dementia is now considered early AD with varying degrees of confidence determined by the presence of AD biomarkers (Albert et al. 2011; also see Dubois et al. 2007). Finally, an even earlier stage of "preclinical AD" is identified which is characterized by the presence of biomarkers in asymptomatic individuals (Sperling et al. 2011). Currently, the recommended use of biomarkers to detect early AD is limited to research, so the early clinical diagnosis continues to heavily depend upon neuropsychological assessment to provide reliable symptom markers of the disease. The combined use of neuropsychological assessment and biomarkers for the disease may be particularly effective in reliably detecting AD in its earliest stages (e.g., Nordlund et al. 2008).

References

Adlam AL, Bozeat S, Arnold R, Watson P, Hodges JR (2006) Semantic knowledge in mild cognitive impairment and mild Alzheimer's disease. Cortex 42:675–684

Ahmed S, Arnold R, Thompson SA, Graham KS, Hodges JR (2008) Naming of objects, faces and buildings in mild cognitive impairment. Cortex 44:746–752

Albert MS, Blacker D (2006) Mild cognitive impairment and dementia. Ann Rev Clin Psychol 2:379–388

Albert MS, Blacker D, Moss MB, Tanzi R, McArdle JJ (2007) Longitudinal change in cognitive performance among individuals with mild cognitive impairment. Neuropsychology 21:158–169

Albert MS, Dekosky ST, Dickson D, Dubois B, Feldman HH, Fox NC, Gamst A, Holtzman DM, Jagust WJ, Petersen RC et al (2011) The diagnosis of mild cognitive impairment due to Alzheimer's disease: recommendations from the National Institute on Aging-Alzheimer's

Association workgroups on diagnostic guidelines for Alzheimer's disease. Alzheimers Dement 7:270–279
Albert MS, Moss MB, Tanzi R, Jones K (2001) Preclinical prediction of AD using neuropsychological tests. J Int Neuropsychol Soc 7:631–639
Algarabel S, Escudero J, Mazon JF, Pitarque A, Fuentes M, Peset V, Lacruz L (2009) Familiarity-based recognition in young, healthy elderly, mild cognitive impaired and Alzheimer's patients. Neuropsychologia 47:2056–2064
Ally BA, McKeever JD, Waring JD, Budson AE (2009) Preserved frontal memorial processing for pictures in patients with mild cognitive impairment. Neuropsychologia 47:2044–2055
Anderson ND, Ebert PL, Jennings JM, Grady CL, Cabezza R, Graham S (2008) Recollection- and familiarity-based memory in healthy aging and amnestic mild cognitive impairment. Neuropsychology 22:177–187
Anderson JW, Schmitter-Edgecombe M (2010) Mild cognitive impairment and feeling-of-knowing in episodic memory. J Clin Exp Neuropsychol 32:505–514
Andrews-Hanna JR, Snyder AZ, Vincent JL et al (2007) Disruption of large-scale brain systems in advanced aging. Neuron 56:924–935
Backman L, Jones S, Berger AK, Laukka EJ, Small BJ (2004) Multiple cognitive deficits during the transition to Alzheimer's disease. J Inter Med 256:195–204
Backman L, Jones S, Berger AK, Laukka EJ, Small BJ (2005) Cognitive impairment in preclinical Alzheimer's disease: A meta-analysis. Neuropsychology 19:520–531
Backman L, Small BJ, Fratiglioni L (2001) Stability of the preclinical episodic memory deficit in Alzheimer's disease. Brain 124:96–102
Baddeley AD (1986) Working memory. Claredon Press, Oxford
Baddeley AD, Bressi S, Della Sala S, Logie R, Spinnler H (1991) The decline of working memory in Alzheimer's disease: a longitudinal study. Brain 114:2521–2542
Balota DA, Tse C, Hutchison KA, Spieler DH, Duchek JM, Morris JC (2010) Predicting conversion to dementia of the Alzheimer's type in a healthy control sample: the power of errors in stroop color naming. Psychol Aging 25:208–218
Balthazar MLF, Cendes F, Damasceno BP (2008) Semantic error patterns on the Boston naming test in normal aging, amnestic mild cognitive impairment, and mild Alzheimer's disease: is there semantic disruption? Neuropsychology 22:703–709
Balthazar MLF, Yasuda CL, Ramos F, Pereira S, Bergo FPG, Cendes F, Damasceno BP (2010) Coordinated and circumlocutory semantic naming errors are related to anterolateral temporal lobes in mild AD, amnestic mild cognitive impairment, and normal aging. J Int Neuropsychol Soc 16:1099–1107
Beatty WW, Salmon DP, Butters N, Heindel WC, Granholm EL (1988) Retrograde amnesia in patients with Alzheimer's disease or Huntington's disease. Neurobiol Aging 9:181–186
Belanger S, Belleville S (2009) Semantic inhibition impairment in mild cognitive impairment: a distinctive feature of upcoming cognitive decline? Neuropsychology 23:592–606
Belanger S, Belleville S, Gauthier S (2010) Inhibition impairments in Alzhiemer's disease, mild cognitive impairment and healthy aging: effect of congruency proportion in a Stoop task. Neuropsychologia 48:581–590
Bellieville S, Chertkow H, Gauthier S (2007) Working memory and control of attention in persons with Alzheimer's disease and mild cognitive impairment. Neuropsychology 21:458–469
Bennett IJ, Golob EJ, Parker ES, Starr A (2006) Memory evaluation in mild cognitive impairment using recall and recognition tests. J Clin Exp Neuropsychol 28:1408–1422
Biundo R, Gardini S, Caffarra P, Concari L, Martorana D, Neri TM, Shanks MF, Venneri A (2011) Influence of APOE status on lexical-semantic skills in mild cognitive impairment. J Int Neuropsychol Soc 17:423–430
Bondi MW, Houston WS, Eyler LT, Brown GG (2005) FMRI evidence of compensatory mechanisms in older adults at genetic risk for Alzheimer's disease. Neurology 64:501–508
Bondi MW, Salmon DP, Galasko D, Thomas RG, Thal LJ (1999) Neuropsychological function and apolipoprotein E genotype in the preclinical detection of Alzheimer's disease. Psychol Aging 14:295–303

Bonney KR, Almeida OP, Flicker L, Davies S, Clarnette R, Anderson M, Lautenschlager NT (2006) Inspection time in non-demented older adults with mild cognitive impairment. Neuropsychologia 44:1452–1456

Bookheimer SY, Strojwas MH, Cohen MS, Saunders AM, Pericak-Vance MA, Mazziotta JC, Small GW (2000) Patterns of brain activation in people at risk for Alzheimer's disease. New Eng J Med 343:450–456

Borg C, Thomas-Atherion C, Bogey S, Davier K, Laurent B (2010) Visual imagery processing and knowledge of famous names in Alzheimer's disease and MCI. Aging Neuropsychol Cog 17:603–614

Braak H, Braak E (1991) Neuropathological staging of Alzheimer-related changes. Acta Neuropathol 82:239–259

Brandt J, Aretouli E, Neijstrom E, Samek J, Manning K, Albert MS, Bandeen-Roche K (2009) Selectivity of executive function deficits in mild cognitive impairment. Neuropsychology 23:607–618

Brandt J, Manning KJ (2009) Patterns of word-list generation in mild cognitive impairment and Alzheimer's disease. Clin Neuropsychologist 23:870–879

Bublak P, Redel P, Sorg C, Kurz A, Forstl H, Muller HJ, Schneider WX, Finke K (2011) Staged decline of visual processing capacity in mild cognitive impairment and Alzheimer's disease. Neurobiol Aging 32:1219–1230

Buschke H, Sliwinski MJ, Kuslansky G, Lipton RB (1997) Diagnosis of early dementia by the double memory test. Neurology 48:989–997

Butters N, Granholm E, Salmon DP, Grant I, Wolfe J (1987) Episodic and semantic memory: a comparison of amnesic and demented patients. J Clin Exp Neuropsychol 9:479–497

Butters N, Salmon DP, Cullum CM, Cairns P, Troster AI, Jacobs D, Moss M, Cermak LS (1988) Differentiation of amnesic and demented patients with the Wechsler memory scale - revised. Clin Neuropsychol 2:133–148

Caine D (2004) Posterior cortical atrophy: a review of the literature. Neurocase 10:382–385

Chapman RM, Mapstone M, McCrary JW, Gardner MN, Porteinsson A, Sandoval TC, Guillily MD, Degrush E, Reilly LA (2011) Predicting conversion from mild cognitive impairment to Alzheimer's disease using neuropsychological tests and multivariate methods. J Clin Exp Neuropsychol 33:187–199

Chen P, Ratcliff G, Belle SH, Cauley JA, DeKosky ST, Ganguli M (2001) Patterns of cognitive decline in presymptomatic Alzheimer disease: a prospective community study. Arch Gen Psychiatr 58:853–858

Chertkow H, Bub D (1990) Semantic memory loss in dementia of Alzheimer's type. Brain 113:397–417

Collette F, Van der Linden M, Bechet S, Salmon E (1999) Phonological loop and central executive functioning in Alzheimer's disease. Neuropsychologia 37:905–918

Collie A, Maruff P (2000) The neuropsychology of preclinical Alzheimer's disease and mild cognitive impairment. Neurosci Biobehav Rev 24:365–374

Costa A, Perri R, Serra L, Barban F, Gatto I, Zabberoni S, Caltagirone C, Carlesimo GA (2010) Prospective memory functioning in mild cognitive impairment. Neuropsychology 24:327–335

Darby D, Maruff P, Collie A, McStephen M (2002) Mild cognitive impairment can be detected by multiple assessments in a single day. Neurology 59:1042–1046

Davie JE, Azuma T, Goldinger SD, Connor DJ, Sabbagh MN, Silverberg NB (2004) Sensitivity to expectancy violations in healthy aging and mild cognitive impairment. Neuropsychology 18:269–275

Delis DC, Massman PJ, Butters N, Salmon DP, Cermak LS, Kramer JH (1991) Profiles of demented and amnesic patients on the California verbal learning test: implications for the assessment of memory disorders. Psychol Assess 3:19–26

Della Sala S, Spinnler H, Trivelli C (1996) Slowly progressive impairment of spatial exploration and visual perception. Neurocase 2:299–323

de Rover M, Pironti VA, McCabe JA, Acosta-Cabronero J, Arana FS, Morein-Zamir S, Hodges JR, Robbins TW, Fletcher PC, Nestor PJ, Sahakian BJ (2011) Hippocampal dysfunction in patients with mild cognitive impairment: a functional neuroimaging study of a visuospatial paired associates learning task. Neuropsychologia 49:2060–2070

Dubois B, Feldman H, Jacova C, Dekosky ST, Barberger-Gateau P, Cummings J, Delacourte A, Galasko D, Gauthier S, Jicha G, Meguro K, O'Brien J, Pasquier F, Robert P, Rossor M, Salloway S, Stern Y, Visser PJ, Scheltens P (2007) Research criteria for the diagnosis of Alzheimer's disease: revising the NINCDS-ADRDA criteria. Lancet Neurol 6:734–746

Dudas RB, Clague F, Thompson SA, Graham KS, Hodges JR (2005) Episodic and semantic memory in mild cognitive impairment. Neuropsychologia 43:1266–1276

Eichenbaum H (1997) How does the brain organize memories? Science 277:333–335

Fine EM, Delis DC, Wetter SR, Jacobson MW, Jak AJ, McDonald C, Braga J, Thal LJ, Salmon DP, Bondi MW (2008) Cognitive discrepancies versus APOE genotype as predictors of cognitive decline in normal-functioning elderly individuals: a longitudinal study. Amer J Geriatr Psychiatry 16:366–374

Fuld PA, Katzman R, Davies P, Terry RD (1982) Intrusions as a sign of Alzheimer dementia: chemical and pathological verification. Ann Neurol 11:155–159

Fuld PA, Masur DM, Blau AD, Crystal H, Aronson MK (1990) Object-memory evaluation for prospective detection of dementia in normal functioning elderly: predictive and normative data. J Clin Exp Neuropsychol 12:520–528

Gagnon LG, Belleville S (2011) Working memory in mild cognitive impairment and Alzheimer's disease: contribution of forgetting and predictive value of complex span tasks. Neuropsychology 25:226–236

Gomar JJ, Bobes-Bascaran MT, Conejero-Goldberg C, Davies P, Goldberg TE (2011) Utility of combinations of biomarkers, cognitive markers, and risk factors to predict conversion from mild cognitive impairment to Alzheimer disease in patients in the Alzheimer's disease neuroimaging initiative. Arch Gen Psychiatry 68:961–969

Gorno-Tempini ML, Dronkers NF, Rankin KP, Ogar JM, Phengrasamy L, Rosen HJ, Johnson JK, Weiner MW, Miller BL (2004) Cognition and anatomy in three variants of primary progressive aphasia. Ann Neurol 55:335–346

Grober E, Hall CB, Lipton RB, Zonderman AB, Resnick SM, Kawas C (2008) Memory impairment, executive dysfunction, and intellectual decline in preclinical Alzheimer's disease. J Int Neuropsychol Soc 14:266–278

Grober E, Kawas C (1997) Learning and retention in preclinical and early Alzheimer's disease. Psychol Aging 12:183–188

Han SD, Houston WS, Jak AJ, Eyler LT, Nagel BJ, Fleisher AS, Brown GG, Corey-Bloom J, Salmon DP, Thal LJ, Bondi MW (2007) Verbal paired-associate learning by APOE genotype in non-demented older adults: fMRI evidence of a right hemisphere compensatory response. Neurobiol Aging 28:238–247

Haxby JV, Duara R, Grady CL, Cutler NR, Rapoport SI (1985) Relations between neuropsychological and cerebral metabolic asymmetries in early Alzheimer's disease. J Cereb Blood Flow Metabol 5:193–200

Hodges JR, Salmon DP, Butters N (1992) Semantic memory impairment in Alzheimer's disease: failure of access or degraded knowledge? Neuropsychologia 30:301–314

Hof PR, Vogt BA, Bouras C, Morrison JH (1997) Atypical form of Alzheimer's disease with prominent posterior cortical atrophy: a review of lesion distribution and circuit disconnection in cortical visual pathways. Vis Res 37:3609–3625

Houston WS, Delis DC, Lansing A, Cobell KR, Bondi MW, Salmon DP, Jacobson MW (2005) Executive function asymmetry in elderly adults genetically at-risk for Alzheimer's disease: Verbal vs. design fluency. J Int Neuropsychol Soc 11:863–870

Howieson DB, Dame A, Camicioli R, Sexton G, Payami H, Kaye JA (1997) Cognitive markers preceding Alzheimer's dementia in the healthy oldest old. J Am Geriatr Soc 45:584–589

Hudon C, Belleville S, Gauthier S (2009) The assessment of recognition memory using the remember/know procedure in amnestic mild cognitive impairment and probable Alzheimer's disease. Brain Cogn 70:171–179

Hudon C, Belleville S, Souchay C, Gely-Nargeot M, Chertkow H, Gauthier S (2006) Memory for gist and detail information in Alzheimer's disease and mild cognitive impairment. Neuropsychology 20:566–577

Hudon C, Villeneuve S, Belleville S (2011) The effect of orientation at encoding on free-recall performance in amnestic mild cognitive impairment and probable Alzheimer's disease. J Clin Exp Neuropsychol 33:631–638

Irish M, Lawlor BA, O'Mara SM, Coen RF (2010) Exploring the recollective experience during autobiographical memory retrieval in amnestic mild cognitive impairment. J Int Neuropsychol Soc 16:546–555

Jack CR, Knopman DS, Jagust WJ, Shaw LM, Aisen PS, Weiner M, Petersen RC, Trojanowski JQ (2010) Hypothetical model of dynamic biomarkers of the Alzheimer's pathological cascade. Lancet Neurol 9:119–128

Jacobs DM, Sano M, Dooneief G, Marder K, Bell KL, Stern Y (1995) Neuropsychological detection and characterization of preclinical Alzheimer's disease. Neurology 45:957–962

Jacobson MW, Delis DC, Bondi MW, Salmon DP (2002) Do neuropsychological tests detect preclinical Alzheimer's disease?: individual-test versus cognitive discrepancy analyses. Neuropsychology 16:132–139

Jacobson MW, Delis DC, Bondi MW, Salmon DP (2005a) Asymmetry in auditory and spatial attention span in normal elderly genetically at risk for Alzheimer's disease. J Clin Exp Neuropsychol 27:240–253

Jacobson MW, Delis DC, Lansing A, Houston W, Olsen R, Wetter S, Bondi MW, Salmon DP (2005b) Asymmetries in global and local processing ability in elderly people with the apolipoprotein E e4 allele. Neuropsychology 19:822–829

Jacobson MW, Delis DC, Peavy GM, Wetter SR, Bigler ED, Abildskov TJ, Bondi MW, Salmon DP (2009) The emergence of cognitive discrepancies in preclinical Alzheimer's disease: a six year longitudinal study. Neurocase 15:278–293

Joubert S, Brambati SM, Ansado J, Barbeau EJ, Felician O, Didac M, Lacombe J, Goldstein R, Chayer C, Kergoat M (2010) The cognitive and neural expression of semantic memory impairment in mild cognitive impairment and early Alzheimer's disease. Neuropsychologia 48:978–988

Karantzoulis S, Troyer AK, Rich JB (2009) Prospective memory in amnestic mild cognitive impairment. J Int Neuropsychol Soc 15:407–415

Katzman R (1994) Apolipoprotein E and Alzheimer's disease. Cur Opin Neurobiol 4:703–707

Kawas CH, Corrada MM, Brookmeyer R, Morrison A, Resnick SM, Zonderman AB, Arenberg D (2003) Visual memory predicts Alzheimer's disease more than a decade before diagnosis. Neurology 60:1089–1093

Knopman DS, Ryberg S (1989) A verbal memory test with high predictive accuracy for dementia of the Alzheimer type. Arch Neurol 46:141–145

Landau SM, Harvey D, Madison CM, Reiman EM, Foster NL, Aisen PS, Petersen RC, Shaw LM, Trojanowski JQ, Jack CR, Weiner MW, Jagust WJ (2010) Comparing predictors of conversion and decline in mild cognitive impairment. Neurology 75:230–238

Lange KL, Bondi MW, Salmon DP, Galasko D, Delis DC, Thomas RG, Thal LJ (2002) Decline in verbal memory during preclinical Alzheimer's disease: examination of the effect of APOE genotype. J Int Neuropsychol Soc 8:943–955

Lefleche G, Albert MS (1995) Executive function deficits in mild Alzheimer's disease. Neuropsychology 9:313–320

Leyhe T, Muller S, Eschweiler GW, Saur R (2010) Deterioration of the memory for historic events in patients with mild cognitive impairment and early Alzheimer's disease. Neuropsychologia 48:4093–4101

Leyhe T, Muller S, Milian M, Eschweiler GW, Saur R (2009) Impairment of episodic and semantic autobiographical memory in patients with mild cognitive impairment and early Alzheimer's disease. Neuropsychologia 35:547–557

Libon DJ, Bondi MW, Price CC, Lamar M, Eppig J, Wambach DM, Nieves C, Delano-Wood L, Giovannetti T, Lippa C, Kabasakalian A, Cosentino S, Swenson R, Penney DL (2011) Verbal

serial list learning in mild cognitive impairment: a profile analysis of interference, forgetting, and errors. J Int Neuropsychol Soc 17:905–914
Linn RT, Wolf PA, Bachman DL, Knoefel JE, Cobb JL et al (1995) The 'preclinical phase' of probable Alzheimer's disease. Arch Neurol 52:485–490
Locascio JJ, Growdon JH, Corkin S (1995) Cognitive test performance in detecting, staging, and tracking Alzheimer's disease. Arch Neurol 52:1087–1099
Mandler G (1980) Recognizing: the judgment of previous occurrence. Psychol Rev 87:252–271
Mandzia JL, McAndrews MP, Grady CL, Graham SJ, Black SE (2009) Neural correlates of incidental memory in mild cognitive impairment: an fmri study. Neurobiol Aging 30:717–730
Manes F, Serrano C, Calcagno ML, Cardozo J, Hodges JR (2008) Accelerated forgetting in subjects with memory complaints. J Neurol 255:1067–1070
Mapstone M, Steffenella TM, Duffy CJ (2003) A visuospatial variant of mild cognitive impairment: getting lost between aging and AD. Neurology 60:802–808
Marr D (1971) Simple memory: a theory for archicortex. Phil Trans Royal Soc Series B Biol Sci 262:23–81
Masur DM, Sliwinski M, Lipton RB, Blau AD, Crystal HA (1994) Neuropsychological prediction of dementia and the absence of dementia in healthy elderly persons. Neurology 44:1427–1432
McKhann G et al (1984) Clinical diagnosis of Alzheimer's disease: report of the NINCDS-ADRDA Work Group under the auspices of department of health and human services task force on Alzheimer's disease. Neurology 34:939–944
McKhann GM, Knopman DS, Chertkow H, Hyman BT, Jack CR, Kawas CH, Klunk WE, Koroshetz WJ, Manly JJ, Mayeux R et al (2011) The diagnosis of dementia due to Alzheimer's disease: Recommendations from the National Institute on Aging-Alzheimer's Association workgroups on diagnostic guidelines for Alzheimer's disease. Alzheimers Dement 7:263–269
Mendez MF, Ghajarania M, Perryman KM (2002) Posterior cortical atrophy: clinical characteristics and differences compared to Alzheimer's disease. Dem Geriat Cogn Disord 14:33–40
Mesulam M, Wicklund A, Johnson N, Rogalski E, Leger GC, Rademaker A, Weintraub S, Bigio EH (2008) Alzheimer and frontotemporal pathology in subsets of primary progressive aphasia. Ann Neurol 63:709–719
Mesulam M, Wieneke C, Rogalski E, Cobia D, Thompson CK, Weintraub S (2009) Quantitative template for subtyping primary progressive aphasia. Arch Neurol 66:1545–1551
Mickes L, Wixted JT, Fennema-Notestine C, Galasko D, Bondi MW, Thal LJ, Salmon DP (2007) Progressive impairment on neuropsychological tasks in a longitudinal study of preclinical Alzheimer's disease. Neuropsychology 21:696–705
Murphy KJ, Rich JB, Troyer AK (2006) Verbal fluency patterns in amnestic mild cognitive impairment are characteristic of Alzheimer's type dementia. J Int Neuropsychol Soc 12: 570–574
Murphy KJ, Troyer AK, Levine B, Moscovitch M (2008) Episodic, but not semantic, autobiographical memory is reduced in amnestic mild cognitive impairment. Neuropsychologia 46:3116–3123
Nestor PJ, Caine D, Fryer TD, Clarke J, Hodges JR (2003) The topography of metabolic deficits in posterior cortical atrophy (the visual variant of Alzheimer's disease) with FDG-PET. J Neurol Neurosurg Psychiatry 74:1521–1529
Nordlund A, Rolstad S, Klang O, Lind K, Pedersen M, Blennow K, Edman A, Hansen S, Wallin A (2008) Episodic memory and speed/attention deficits are associated with Alzheimer-typical CSF abnormalities in MCI. J Int Neuropsychol Soc 14:582–590
Nutter-Upham KE, Saykin AJ, Rabin LA, Roth RM, Wishart HA, Pare N, Flashman LA (2008) Verbal fluency performance in amnestic MCI and older adults with cognitive complaints. Arch Clin Neuropsychol 23:229–241
Parasuraman R, Haxby JV (1993) Attention and brain function in Alzheimer's disease. Neuropsychology 7:242–272

Perri R, Serra L, Carlesimo GA, Caltagirone C (2007) Amnestic mild cognitive impairment: difference of memory profile in subjects who converted or did not convert to Alzheimer's disease. Neuropsychology 21:549–558

Perrotin A, Belleville S, Isingrini M (2007) Metamemory monitoring in mild cognitive impairment: Evidence of a less accurate episodic feeling-of-knowing. Neuropsychologia 45:2811–2826

Perry RJ, Hodges JR (1999) Attention and executive deficits in Alzheimer's disease: a critical review. Brain 122:383–404

Petersen RC, Morris JC (2005) Mild cognitive impairment as a clinical entity and treatment target. Arch Neurol 62:1160–1163

Petersen RC, Smith GE, Waring SC, Ivnik RJ, Tangalos EG, Kokmen E (1999) Mild cognitive impairment: clinical characterization and outcome. Arch Neurol 56:303–308

Petersen RC, Stevens JC, Ganguli M, Tangalos EG, Cummings JC, DeKosky ST (2001) Practice parameter: Early detection of dementia: Mild cognitive impairment (an evidence-based review). Neurology 56:1133–1142

Pike KE, Rowe CC, Moss SA, Savage G (2008) Memory profiling with paired associate learning in Alzheimer's disease, mild cognitive impairment, and healthy aging. Neuropsychology 22:718–728

Poettrich K, Weiss PH, Werner A, Lux S, Donix M, Gerber J, von Kummer R, Fink GR, Holthoff VA (2009) Altered neural network supporting declarative long-term memory in mild cognitive impairment. Neurobiol Aging 30:284–298

Raffi MS, Aisen PS (2009) Recent developments in Alzheimer's disease therapeutics. BMC Medicine 7(7):1741

Rapp MA, Reischies FM (2005) Attention and executive control predict Alzheimer's disease in late life: results from the Berlin aging study (BASE). Amer J Geriatr Psychiatry 13:134–141

Renner JA, Burns JM, Hou CE, McKeel DW, Storandt M, Morris JC (2004) Progressive posterior cortical dysfunction: a clinicopathologic series. Neurology 63:1175–1180

Rentz DM, Amariglio RE, Becker JA, Frey M, Olson LE, Frishe K, Carmasin J, Maye JE, Johnson KA, Sperling RA (2011) Face-name associative memory performance is related to amyloid burden in normal elderly. Neuropsychologia 49:2776–2783

Salmon DP (2000) Disorders of memory in Alzheimer's disease. In: Cermak LS (ed) Handbook of neuropsychology, 2nd edn (Vol 2): memory and its disorders. Elsevier, Amsterdam

Salmon DP, Chan AS (1994) Semantic memory deficits associated with Alzheimer's disease. In: Cermak LS (ed) Neuropsychological explorations of memory and cognition: essays in honor of Nelson Butters. Plenum Press, New York

Salmon DP, Squire LR (2009) The neuropsychology of memory dysfunction and its assessment. In: Grant I, Adams K (eds) Neuropsychological assessment of neuropsychiatric and neuromedical disorders, 3rd edn. Oxford University Press, New York

Saunders NLJ, Summers MJ (2011) Longitudinal deficits to attention, executive, and working memory in subtypes of mild cognitive impairment. Neuropsychology 25:237–248

Schmitter-Edgecombe M, Woo E, Greeley DR (2009) Characterizing multiple memory deficits and their relation to everyday functioning in individuals with mild cognitive impairment. Neuropsychology 23:168–177

Seidenberg M, Guidotti L, Nielson KA, Woodard JL, Durgerian S, Zhang Q, Gander A, Antuono P, Rao SM (2009) Semantic knowledge for famous names in mild cognitive impairment. J Int Neuropsychol Soc 15:9–18

Serra L, Bozzali M, Cercignani M, Perri R, Fadda L, Caltagirone C, Carlesimo GA (2010) Recollection and familiarity in amnestic mild cognitive impairment. Neuropsychology 24:316–326

Sinai M, Phillips NA, Chertkow H, Kabani NJ (2010) Task switching performance reveals heterogeneity amongst patients with mild cognitive impairment. Neuropsychology 24: 757–774

Small BJ, Fratiglioni L, Viitanen M, Winblad B, Bäckman L (2000) The course of cognitive impairment in preclinical Alzheimer disease: three- and 6-year follow-up of a population-based sample. Arch Neurol 57:839–844

Small BJ, Mobly JL, Laukka EJ, Jones S, Backman L (2003) Cognitive deficits in preclinical Alzheimer's disease. Acta Neurol Scand Suppl 179:29–33

Smith GE, Pankratz VS, Negash S et al (2007) A plateau in pre-Alzheimer memory decline: evidence for compensatory mechanisms? Neurology 69:133–139

Sperling RA, Aisen PS, Beckett LA, Bennett DA, Craft S, Fagan AM, Iwatsubo T, Jack CR, Kaye J, Montine TJ et al (2011) Toward defining the preclinical stages of Alzheimer's disease: recommendations from the National Institute on Aging-Alzheimer's Association workgroups on diagnostic guidelines for Alzheimer's disease. Alzheimers Dement 7:280–292

Squire LR (1992) Memory and the hippocampus: a synthesis from findings with rats, monkeys and humans. Psychol Rev 99:195–231

Tabert MH, Manly JJ, Liu X et al (2006) Neuropsychological prediction of conversion to alzheimer disease in patients with mild cognitive impairment. Arch Gen Psychiatry 63: 916–924

Tang-Wai DF, Graff-Radford NR, Boeve BF, Dickson DW, Parisi JE, Crook R, Caselli RJ, Knopman DS, Petersen RC (2004) Clinical, genetic, and neuropathologic characteristics of posterior cortical atrophy. Neurology 63:1168–1174

Tenovuo O, Kemppainen N, Aalto S, Nagren K, Rinne JO (2008) Posterior cortical atrophy: a rare form of dementia with in vivo evidence of amyloid-B accumulation. J Alz Dis 15: 351–355

Thompson C, Henry JD, Rendell PG, Withall A, Brodaty H (2010) Prospective memory function in mild cognitive impairment. J Int Neuropsychol Soc 16:318–325

Thompson SA, Graham KS, Patterson K, Sahakian BJ, Hodges JR (2002) Is knowledge of famous people disproportionately impaired in patients with early and questionable Alzheimer's disease? Neuropsychology 16:344–358

Troyer AK, Murphy KJ (2007) Memory for intentions in amnestic mild cognitive impairment: time- and event-based prospective memory. J Int Neuropsychol Soc 13:365–369

Troyer AK, Murphy KJ, Anderson ND, Hayman-Abello BA, Craik FIM, Moscovitch M (2008) Item and associative memory in amnestic mild cognitive impairment: performance on standardized memory tests. Neuropsychology 22:10–16

Twamley EW, Ropacki SAL, Bondi MW (2006) Neuropsychological and neuroimaging changes in preclinical Alzheimer's disease. J Int Neuropsychol Soc 12:707–735

Welsh K, Butters N, Hughes J, Mohs R, Heyman A (1991) Detection of abnormal memory decline in mild cases of Alzheimer's disease using CERAD neuropsychological measures. Arch Neurol 48:278–281

Westerberg CE, Paller KA, Weintraub S, Mesulam M, Holdstock JS, Mayes AR, Reber PJ (2006) When memory does not fail: familiarity-based recognition in mild cognitive impairment and Alzheimer's disease. Neuropsychology 20:193–205

Wetter SR, Delis DC, Houston WS, Jacobson MW, Lansing A, Cobell K, Salmon DP, Bondi MW (2005) Deficits in response inhibition and flexibility are associated with the APOE e4 allele in nondemented older adults. J Clin Exp Neuropsychol 27:943–952

Wolk DA, Signoff ED, Dekosky ST (2008) Recollection and familiarity in amnestic mild cognitive impairment: a global decline in recognition memory. Neuropsychologia 46: 1965–1978

Yassa MA, Stark SM, Bakker A, Albert MS, Gallagher M, Stark CEL (2010) High-resolution structural and functional MRI of hippocampal CA3 and dentate gyrus in patients with amnestic mild cognitive impairment. Neuroimage 51:1242–1252

Cerebrovascular Disease and Cognition in Older Adults

Gregory A. Seidel, Tania Giovannetti and David J. Libon

Abstract The well-established association between advanced age, cerebrovascular pathology, and cognitive decline is receiving greater attention as the population attains new levels of longevity. This chapter will provide an overview of vascular anatomy and age-related cerebrovascular disorders and diseases, including stroke and degenerative dementia. The cognitive and functional sequellae of these cerebrovascular disorders will also be described in detail. Throughout this review, we will emphasize topics that have been relatively underrepresented in the literature, including age-related diseases of the cerebral small vessels, nuanced characterization of cognitive impairment associated with insidious small-vessel vascular dementia, and the real-life functional consequences of cerebrovascular changes in older adults.

Keywords Cerebrovascular disease · Cognition · Older adults · Dementia · Small-vessel disease · Everyday action · Executive function · Episodic memory · Language

---

G. A. Seidel · T. Giovannetti
Department of Psychology, Temple University, Philadelphia, PA, USA
e-mail: gregory.seidel@temple.edu

T. Giovannetti
e-mail: tgio@temple.edu

D. J. Libon (✉)
Department of Neurology, Drexel University College of Medicine,
Philadelphia, PA, USA
e-mail: David.Libon@drexelmed.edu

Curr Topics Behav Neurosci (2012) 10: 213–241
DOI: 10.1007/7854_2011_140

Published Online: 20 July 2011

Contents

1 Introduction 214
2 Vascular Anatomy 214
2.1 Large-Vessel Anatomy 215
2.2 Small-Vessel Anatomy 215
3 Vascular Pathologies 217
3.1 Atherosclerosis and Vessel Disease Processes 217
3.2 Stroke 219
3.3 White Matter Disease 220
3.4 Derailment of the Blood–Brain Barrier 221
3.5 Lacunar Infarction 222
3.6 Cerebral Amyloid Angiopathy 222
3.7 Heart Disease 223
3.8 Genetic Factors 223
4 Cerebrovascular Disease in Older Adults 223
4.1 Acute Stroke and Multi-Infarct Dementia 224
5 Dementia Associated with Small-Vessel Disease 226
5.1 Frontal Systems/Executive Control Impairment 226
5.2 Episodic Memory 229
5.3 Semantic Knowledge and Language 231
5.4 Summary of the Cognitive Profile of smVaD 232
6 The Impact of Cerebrovascular Cognitive Deficits on Everyday Life 233
7 Conclusion 236
References 236

1 Introduction

Age is the most common and strongest risk factor for the large majority of cerebrovascular disorders and diseases, and the literature is replete with studies showing a clear association between age-related cerebrovascular changes/diseases and cognitive decline. These facts will be difficult to ignore as the U.S. population ages and our society is faced with the increasing health care demands associated with cerebrovascular diseases and cerebrovascular cognitive disorders. Greater understanding of cerebrovascular diseases and their cognitive and functional consequences is imperative to address these looming health care challenges.

2 Vascular Anatomy

An overview of vascular anatomy is a necessary first step in understanding the behavioral changes associated with age-related cerebrovascular phenomena. The complexity of vascular dysfunction and associated cognitive decline is realized even after brief consideration of the broad range of anatomical substrates and pathological processes involved. The historical focus has been on the large vessels that serve cortical gray matter and the effect of large-vessel stroke on higher-level cognitive functions. This has been to the exclusion of other important vascular and vascular-related substrates, including white matter tracts, subcortical gray matter structures, and the small-vessel cerebral vasculature. Extra-cerebral anatomy, most vitally the heart and peripheral vessels, must also be considered.

2.1 Large-Vessel Anatomy

In humans, the brain receives up to 20% of cardiac output (Girouard and Iadecola 2006), which delivers glucose and oxygen and eliminates heat and metabolic by-products. An intricate network of arteries, arterioles, veins, venules, and capillaries serves this purpose. The large arteries of this network—the anterior, middle, and posterior cerebral arteries—and the cortical regions served by each, have been well studied and represent an important component of neuropsychological knowledge and history. The left and right anterior and middle cerebral arteries can be traced back to the internal carotid, with its sister artery, the exterior carotid serving the face, and ultimately to the brachiocephalic artery and aortic arch. The left and right posterior cerebral arteries can be traced back to the basilar artery which runs along the ventral surface of the brain stem, where it is formed by the joining of the vertebral arteries, ultimately originating from the subclavian artery which connects to the aortic arch. Several arterial branches to the brainstem and cerebellum arise from the vertebrobasilar system. The three major cerebral arteries are interconnected in the Circle of Willis at the base of the brain, formed by the anterior and posterior communicating arteries. A fully-formed Circle of Willis is found in only about one-third of the population, but to the extent that it is present, it is thought to provide collateral circulation (Blumenfeld 2010; Rigs and Rupp 1963). Watershed regions are present where the coverage zones of the major cerebral arteries meet and, due to this distal location, are especially vulnerable to reduced perfusion.

2.2 Small-Vessel Anatomy

The acute nature of large artery stroke is likely responsible for the historical predominance for much of the 20th century of this aspect of the cerebrovasculature in the study of cognition. The epidemiological impact of dysfunction related to the small vessels, however, is arguably greater than acute stroke and certainly deserving of at least equal attention. Arterial small vessels have two origins: superficially from the subarachnoid circulation and deeper at the base of the brain (Pantoni 2010). As shown in Fig. 1, long penetrating small vessels arising from the leptomeningeal layers extend downward into the brain, perpendicular to its surface, passing through 3–5 mm of cortex into the white matter. Smaller distributing vessels branch perpendicularly from these penetrating arteries and perfuse most of the subcortical white matter (Pantoni and Garcia 1997). A separate set of short,

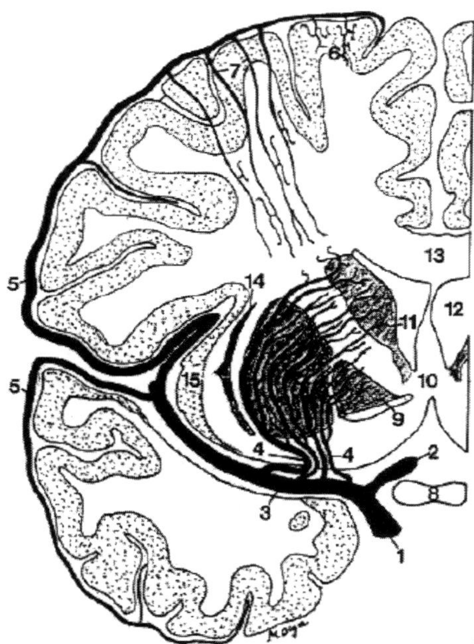

Fig. 1 Drawing of a coronal section of the right hemisphere showing arterial small vessels originating superficially from the subarachnoid circulation and deeper at the base of the brain. 1: Internal carotid artery; 2: anterior cerebral artery; 3: middle cerebral artery; 4: lenticulostriate small vessels; 5: leptomeningeal branches; 6: cortical penetrating small vessels; 7: long penetrating small vessels; 8: optic chiasm; 9: globus pallidus; 10: septal area; 11: head of the caudate nucleus; 12: frontal horn of the lateral ventricle; 13: corpus callosum; 14: claustrum; 15: insular cortex. Figure taken from Marinkovic et al. 2001. Reproduced with permission from the publisher

small arteries extend from the subarchnoid layer, but supply only cortex and the 3–4 mm strip of cerebral white matter just beneath (U-fibers). In this way, the cortex and white matter U-fibers receive collateral circulation, which could help explain why these regions of white matter are relatively spared from small-vessel disease compared to other brain areas (Pantoni and Garcia 1997).

Other small vessels arise from arteries at the base of the brain (see Fig. 1). The lenticulostriate vessels branch-off from the middle cerebral arteries proximal to their source at the internal carotid artery and penetrate upward to supply large portions of the basal ganglia and internal capsule (Blumenfeld 2010). Some investigators believe that the lenticulostriate system may, in fact, be veins, implying that the periventricular white matter serves as a distal irrigation field (Pantoni and Garcia 1997). Pantoni (2010) and de Reuck (1971) appear to side with an arterial lenticulostriate system, which creates a watershed region in the area where the small penetrating arteries arising from subarachnoid circulation and the lenticulostriate arteries projecting from basal areas meet, making this region including the periventricular white matter particularly vulnerable to the effects of small-vessel disease.

Other small vessels originating from the anterior cerebral arteries, internal carotid artery, and posterior cerebral arteries also supply deep structures (Blumenfeld 2010). Although there may be some variability, penetrating branches originate from the anterior cerebral arteries (e.g., recurrent artery of Huebner) to supply portions of the head of the caudate, anterior putamen, globus pallidus, and internal capsule. The anterior choroidal artery arises from the internal carotid artery to supply portions of the globus pallidus, putamen, thalamus (sometimes involving lateral geniculate nucleus), and posterior internal capsule. Small penetrating arteries arising from the proximal posterior cerebral arteries (close to the basilar artery) include the thalamoperforator, thalamogeniculate, and posterior choroidal arteries, which supply a large portion of the thalamus and posterior portions of internal capsule.

3 Vascular Pathologies

The incidence of the vascular pathologies described below increases dramatically with age. These pathologies are uniformly associated with reduced cerebral perfusion but are otherwise quite diverse. They may impact the large and/or small vessels, may be associated with acute or chronic disease processes, and may be ischemic or hemorrhagic in nature.

3.1 Atherosclerosis and Vessel Disease Processes

Atherosclerosis, the most common vascular disease (Fung and Poppas 2009), is the underlying cause of many cardiovascular events and complications and may affect vessels throughout the body (Miller et al. 2009). The blood vessels are composed of three tissue layers: tunica intima (inner layer which includes the endothelial cells), tunica media (smooth muscle), and tunica adventitia (outer layer of connective tissue). The center cavity of the vessel is called the lumen. Larger vessels (usually those greater than 0.5 mm luminal diameter) are served by their own vasculature, known as the vasa vasorum, comprised of small arteries entering the vascular wall either from the abluminal surface (vasa vasorum externa) or from the luminal surface (vasa vasorum interna), arborizing to the outer media (Ritman and Lerman 2007). Atherosclerosis is associated with a complex series of molecular events leading to cellular accumulation (plaque) that narrows or occludes the lumen, as well as functional changes within the vessel layers that disrupt the natural functioning of the vessel.

3.1.1 Atherosclerotic Plaque Formation

Multiple elements combine to form an atherosclerotic plaque in a process beginning with injury to the endothelium and lipid accumulation in the tunica intima (Fung and Poppas 2009). Cytokine release, activation of macrophages, uptake of oxidized lipoproteins and foam cell formation, and activation of platelets and injured endothelial cells leads to release of factors stimulating smooth muscle cells of the tunica media to migrate, proliferate, and produce extracellular matrix and connective tissue that results in plaque formation (Fung and Poppas 2009). The components of plaques, present in varying proportions, include connective tissue extracellular matrix, forms of cholesterol and phospholipids, cells such as monocyte-derived macrophages, T-lymphocytes, smooth-muscle cells, and thrombotic material containing platelets and fibrin (Fuster et al. 2005). While mainly affecting the intima, changes in the media and adventitia also occur, including growth of vasa vasorum (neovascularization; Fuster et al. 2005). Atherosclerotic plaques narrow the lumen and can lead to chronic hypoperfusion, but also to acute occlusion due to plaque rupture, an event leading to thrombus formation, particularly at high-risk sites (Fung and Poppas 2009). The thrombus can also travel through the vessel and lodge in a narrow section of lumen. The close linkage between atherosclerotic plaque formation, plaque rupture and erosion, and thrombosis has led to the integrating term atherothrombosis (Fuster et al. 2005). Atherosclerosis can be located in coronary, cerebral, or peripheral arteries and therefore can impact perfusion and create risk for thrombosis in multiple, but selective, locations throughout the body.

3.1.2 Endothelial Dysfunction and Autoregulatory Changes

The luminal narrowing and thrombosis are just one aspect of the impact of atherosclerosis. Healthy vasculature undergoes continuous structural changes and adaptations that enable the organism to respond to changing requirements for blood supply. Atherosclerosis disrupts these important vessel functions.

Autoregulation of vessel tone is one way the vasculature compensates for the imbalance in oxygen supply/demand that occurs in an organ suffering ischemia. Autoregulation is mediated by 3 main factors: sympathetic neural control, local metabolites, and endothelial (tunica intima) factors (balance between vaso-relaxing and vaso-constricting factors; Fung and Poppas 2009). Atherosclerosis may disrupt normal endothelial autoregulation of vascular resistance and lead to inappropriate vasoconstriction and reduced perfusion (Fung and Poppas 2009).

Other authors have suggested that endothelial dysfunction is not simply a consequence of atherosclerosis, but contributes to the pathogenesis of atherosclerosis and its later complications. Miller et al. (2009) cite clear evidence that changes in endothelial function *precede* development of atherosclerotic lesions in non-human primates. Given the tight linkage between atherosclerosis and endothelial dysfunction, it is not surprising that every known risk factor for atherosclerosis impairs endothelial function (Miller et al. 2009). Also of note, endothelial cells die after approximately 30 years and replacement cells do not function as effectively (Miller et al. 2009). The progressive decline in endothelial function as men and women approach the age of 40 or 50 (Miller et al. 2009) is a risk factor for the development of atherosclerosis, and likely, dementia. Aging is strongly associated with both endothelial dysfunction and atherosclerosis (Miller et al. 2009).

Once the vasculature is appreciated as not simply a passive transport system, but as a dynamic and active participant in our response to constantly changing metabolic demands, the far-reaching and complex impact of atherosclerosis can be more fully realized. In the formation of plaques and development of endothelial autoregulatory dysfunction, atherosclerosis can affect both large and small vessels and can lead to acute and chronic changes in perfusion that can be ischemic and hemorrhagic in nature.

3.2 Stroke

Stroke refers to an acute cerebrovascular event, typically involving the large cerebral arteries. Ischemic stroke, the most common stroke subtype, is the consequence of decreased blood flow to a portion of the brain; hemorrhagic stroke, the second stroke subtype, is the result of bleeding into the brain (Sharma et al. 2005). Ischemic stroke may be caused by a thrombus (stationary blood clot or plaque) or embolus (a thrombus or portion of thrombus that travels through the vasculature). Thrombus in the cerebral arteries is the most common cause of acute stroke. Arterial emboli can take many forms, including air bubbles, bone fragments, atherosclerotic plaque (as described above), and others. Causes of thrombus and embolus are numerous and can include hypercoagulability caused by genetic deficiencies or autoimmune disorders, endothelial cell injury caused by trauma to the vessel wall, infection, or turbulent blood flow due to heart conditions such as atrial fibrillation and heart failure, as well as cardiac surgery (Hatzinikolaou-Kotsakou et al. 2005).

Cardioembolism, in which blood clots arise from a damaged heart, is most commonly caused by atrial fibrillation (AF; Duffis and Fisher 2009), the most common persistent type of cardiac arrhythmia (Wolf et al. 1998). AF prevalence increases with age, doubling with each decade in adults older than 50 years. Whereas AF was long thought of as inconsequential, recently AF (excluding the form associated with rheumatic heart disease) has been identified as an important cause of death and a powerful independent risk factor for stroke, increasing stroke risk by a factor of 5 (Wolf et al. 1998).

Hemorrhagic stroke occurs when a cerebral blood vessel weakens and bursts open, causing blood to leak into surrounding brain tissue. The flow of blood which follows vessel rupture damages brain cells. Defects in the cerebral vessels present in some individuals make this more likely. Age increases the risk for hemorrahagic stroke (Ariesen et al. 2003), particularly among individuals treated with anticoagulants (Hylek and Singer 1994).

The ischemic and hemorrhagic stroke events discussed above have predominated in the study of relations between cerebrovascular disease and cognition. This is perhaps understandable, given the salient acute symptoms associated with stroke and that traditionally stroke is defined by clinical symptomatology (e.g., NIH Stroke Scale; Brott et al. 1989) rather than by an underlying disease process (Fisher 2010). A broadening of the conceptualization of stroke to include information gained from neuroimaging and neuropathology has been proposed, which would lead to an integration of clinical, subclinical, ischemic, and hemorrhagic elements, including subclinical white matter disease and microbleeds (Fisher 2010). Similar efforts have been made with transient ischemic attack (TIA), which is defined according to recent consensus guidelines as a brief episode of neurological dysfunction resulting from focal cerebral ischemia not associated with permanent cerebral infarction (Easton et al. 2009). Often thought of as benign events whose effects disappear completely, TIAs are more correctly viewed as next to stroke on a spectrum of serious ischemic brain conditions, with 10–15% of TIA patients going on to have a stroke within 3 months, half of these within 48 hours (Easton et al. 2009).

3.3 White Matter Disease

Ischemic and hemorrhagic effects of atherosclerosis and other pathological processes are also found in the small vessels. Arteriosclerotic small-vessel disease is mainly characterized by loss of smooth muscle cells, deposits of fibro-hyaline material, narrowing of the vessel lumen, and thickening of the vessel wall (Pantoni 2010). Small-vessel disease leads to chronic hypoperfusion and degeneration of white matter (Pantoni 2010). These white matter changes, which manifest as hyperintensities on T2-weighted MRI, have been termed "subcortical hyperintensities," "white matter lesions," and "unidentified bright objects" (Roman 1987). Hachinski et al. (1987) used the term leukoaraiosis (literally meaning "rarefied white matter") as a descriptive term for these signal changes. White matter changes are found in more than 95% of older adults over the age of 65 (Longstreth et al. 1996). Some authors have proposed a threshold of small-vessel lesion burden at which cognitive changes become symptomatic and impact daily function (Libon et al. 2008; Price et al. 2005; Roman et al. 1993).

Multiple processes have been implicated in ischemic forms of small-vessel disease, including inflammation and a breakdown in the blood–brain barrier (see Libon et al. 2004 for a review). Inflammation, which can occur in response to cell damage, is also being investigated in relation to white matter disease (Fisher 2010). Inflammation occurs when tissues are injured by ischemia, bacteria, trauma, toxins, heat, or any other cause. The damaged cells release chemicals including histamine, bradykinin, and prostaglandins, which cause blood vessels to leak fluid into surrounding tissues, causing swelling and demyelination. In the case of infection, inflammation may serve to isolate the bacteria or a foreign toxin from further contact with body tissues. The fluids that signal inflammatory processes may also have neurotoxic effects. Various inflammatory markers have been identified that may be related to cerebrovascular disease and be useful in future investigation (Cohen 2009).

3.4 Derailment of the Blood–Brain Barrier

Aging, independent of diseases of the cerebrovasculature, disrupts the structure and function of the blood–brain barrier (BBB; Mooradian 1988). Enhanced permeability of the BBB has been suggested as an important mechanism underlying cerebral white matter disease in older adults (Fisher 2010; Pantoni 2010). Research regarding the BBB can be traced to Lewandowsky (1900; cited in Zlokovic 2008) who commented on the absence of central nervous system pharmacological actions after the intravenous injection of certain bile acids. Lewandowsky (1900) described the BBB as a mechanical membrane whose function was to keep blood from the brain. Goldman (1909, 1913) conducted several experiments and observed that IV injection of certain substances distributed themselves widely throughout the body, but not the brain or spinal cord. Modern research regarding the BBB begins with the work of Davson (1976) who emphasized mechanical relations between spinal fluid produced in the ventricles and subarachnoid space.

Today the BBB is understood as a very complex endothelial structure that constitutes part of the neurovascular system. The BBB, along with cells that comprise and support the vasculature (e.g., pericytes, astrocytes, and microglia), separate components of the circulating blood from the brain. The BBB can be viewed as a number of different *tight junctions* between adjacent endothelial cells. Each of these tight junctions is associated with unique molecular attributes and transport systems. When functioning properly, these tight junctions/transport systems allow and disallow molecules into the brain. In the case of damage or dysfunction, the BBB may lead to neuronal disruption by allowing toxins to reach neurons or by preventing the clearance of toxins from the brain.

Investigators have posited a role for BBB damage and dysfunction in cognitive aging and dementia (Zlokovic 2004, 2008). Regional differences in BBB structure and permeability (Phares et al. 2006) may account for different dementia syndromes. For example, substantial BBB derailment in the posterior temporal lobe could result in semantic dementia, whereas BBB derailment in the medial temporal cortex might lead to Alzheimer's disease. Another consideration is how previously disallowed elements from plasma actually interact with neurons following BBB disruption. It could be that previously disallowed 'element A' from plasma has only a minimal effect on neuronal functioning, and this could be the anatomical basis for one of the mild cognitive impairment (MCI) syndromes. More severe cognitive disorders, such as dementia, may be the consequence of previously disallowed 'element A' in combination with previously disallowed 'element B'. Finally, it is possible that BBB dysfunction may result in particularly devastating combinations of previously disallowed elements and inefficient clearance of neuronal toxins. At present these ideas are conjecture requiring prospective research.

3.5 Lacunar Infarction

Research regarding lacunes has a long history reaching well back into the 19th century (see Libon et al. 2004 for a review). Lacunes, defined as hypointense foci on T1-weighted MRI, are typically seen in areas such as the basal ganglia, internal capsule, thalamus, and pons (Pantoni 2010). Consensus on size of lacunar infarcts has not been reached. Diameters vary by investigator, ranging from a minimum of 3 mm to a generally accepted maximum of 15 mm (Pantoni 2010). Fisher's lacunar hypothesis, though it remains unproven, explains lacunar infarcts as acute, complete occlusion of small vessels (Pantoni 2010). Authors have proposed that microatheroma, tiny foci of plaque material, are the most common mechanism of an arterial stenosis eventually leading to symptomatic lacunes (Marti-Vilalta et al. 2004). These atherosclerotic plaques have been found in arteries as small as 100 μm in diameter (Marti-Vilalta et al. 2004). Small, asymptomatic lacunes are thought to be most associated with lipohyalinosis and fibrinoid necrosis (Marti-Vilalta et al. 2004).

Although white matter disease and lacunar infarction have been viewed as distinct, recent evidence suggests this may be incorrect. Small deep infarcts and cerebral white matter disease may be often indistinguishable by brain imaging, with infarcts incorporated into white matter disease (Fisher 2010).

3.6 Cerebral Amyloid Angiopathy

The effects of small-vessel disease are not exclusively ischemic in nature and increased attention is being paid to hemorrhagic cerebral micro-events. If vessel wall damage reaches the point of rupture, microbleeds, major hematoma, or intracranial hemorrhage can result, with differences in wall thickness thought to determine the size of the rupture (Pantoni 2010). Cerebral amyloid angiopathy (CAA) is a collective term for a group of diseases with diverse aetiology and common pathology (Vasilevko et al. 2010). This pathology is characterized by congophilic deposition of amyloid in the walls of small and medium sized cerebral blood vessels and sometimes in the microvasculature, mostly in the leptomeningeal space, cortex, and, less often, in the capillaries and veins (Pantoni 2010). The amyloid can be formed by different peptides such as Aβ, cystatin C, gelsolin, prion protein, Abri, and ADan. CAA can lead to weakening of vessel walls leading to micro or macro hemorrhage and sometimes show luminal occlusion (Pantoni 2010). CAA, together with hypertension, is the most common cause of intracerebral hemorrhage in older adults (Vasilevko et al. 2010). While CAA is highly associated with microbleeds, it is also related to ischemic changes such as white matter lesions and lacunes (Pantoni 2010).

CAA can be sporadic or associated with rare genetic diseases. CAA appears with high frequency in the general older adult population, in as much as 50% of individuals in their 90s (Pantoni 2010). CAA is also a pathological hallmark of Alzheimer's disease, and this has been offered as an explanation for the increased risk of intracerebral hemorrhage in Alzheimer's disease (AD) and the high prevalence of

AD-associated microbleeds (Thoonsen et al. 2010). CAA would appear to be part of a series of findings that have contributed to a progressive blurring of the distinction between dementia of the Alzheimer type and vascular dementia. Indeed, the concept of 'mixed dementia' has been driven by the association between vascular risk factors and increased risk for AD, the identification of concomitant cerebral infarctions in many AD patients, and studies showing white matter lesions to be common in AD (Thoonsen et al. 2010). Greenberg et al. (2008) have recently proposed clinical criteria for the diagnosis of CAA (the so-called 'Boston criteria').

The relation between white matter disease and microbleeds remains a question. Fisher (2010) suggests that there are common mechanisms for cerebral white matter disease and cerebral microbleeds, at least in the presence of CAA. According to this hypothesis, the current consensus view is that cortical microbleeds represent CAA, whereas subcortical microbleeds stem from chronic hypertension.

3.7 Heart Disease

Although it would appear obvious that the heart and cerebral vasculature are interdependent systems, they are often treated as separate. Atherosclerosis affects any of the organs of the body which are deprived of constant circulation—including the heart. The heart is also the site of some of the mechanisms implemented by the body to compensate for circulatory dysregulation in distal organs. Coronary artery disease, myocardial infarction, heart failure, and cardiosurgical intervention impact the efficiency of the heart and thereby influence cerebral perfusion, as well as formation of emboli that travel to the brain (see Irani 2009 and Jefferson 2010 for reviews).

3.8 Genetic Factors

Finally, genetic small-vessel diseases, although not discussed in detail here, are an important consideration, as they could facilitate study in pathological processes that is highly applicable to acquired forms of small-vessel diseases that affect older adults (Pantoni 2010). Cerebral autosomal dominant arteriopathy with subcortical ischemic strokes and leukoencephalopathy (CADASIL) and Fabry's disease are most prominent in a growing list of examples.

4 Cerebrovascular Disease in Older Adults

Any vascular pathology in the brain falls under the rubric of cerebrovascular disease (CVD), encompassing dysfunction related to both acute events, such as ischemic or hemorrhagic stroke, and to insidious pathological processes, such as atherosclerotic changes and small-vessel disease. CVD-related cognitive impairment may vary according to course (stable vs. progressive) and severity (focal deficit vs. mild cognitive impairment vs. dementia). Acute stroke is typically associated with a relatively stable course and focal deficits. Vascular cognitive impairment (VCI) is a term used to refer to CVD-associated cognitive decline and behavioral change not reaching criteria for a dementia diagnosis. Once dementia diagnostic criteria are met, the term vascular dementia (VaD) is applied. Given the heterogeneity and complexity of CVD-related neuropathology discussed in the above sections, it is hardly surprising that the clinical neuropsychological manifestation of CVD is not represented by a single cognitive profile.

4.1 Acute Stroke and Multi-Infarct Dementia

A single strategic infarct (stroke) can result in the abrupt onset of symptoms reflecting one of the classic stroke syndromes. In this scenario, cognitive impairment is temporally linked to the infarct. Acute large-vessel infarcts have traditionally been viewed as not progressive; cognitive impairments often improve or stabilize in the days to months following the infarct. Infarct location within the vasculature and the cerebral territory affected are commonly associated with specific deficits. Lateralization of motor and sensory deficits is a common feature of the classic syndromes, with motor and sensory deficits observed on the side of the body contralateral to lesion location.

4.1.1 Middle Cerebral Artery Stroke

Large artery infarcts and ischemic events are more common in the middle cerebral artery (MCA) than in the anterior or posterior cerebral arteries (ACA; PCA), partly due to the relatively large area supplied by the MCA. Infarcts can be classified by the branch of the MCA in which the occlusion occurs: superior division, inferior division, or deep territory. Specific sensorimotor deficits are associated with the MCA division affected, with contralateral face and arm weakness associated with superior division occlusion, contralateral paralysis (arm and leg) with the deep division, and contralateral cortical-type sensory loss (hemianesthesia or homonymous hemianopia) with the inferior division. In individuals with left hemisphere language dominance, left MCA occlusion is classically associated with aphasias, specifically of the non-fluent, or Broca's, type (superior division) and fluent, or Wernicke's, type (inferior division). Right MCA occlusion is classically associated with contralateral hemineglect, particularly when the right MCA inferior division is affected. Proximal MCA occlusion, known as MCA stem infarct, can lead to a combination of these symptoms, and to global aphasia in the case of left hemisphere damage and profound left hemineglect in the case of right hemisphere damage. Infarcts affecting large parts of the MCA territory often result in a gaze preference toward the side of the lesion, particularly in the acute period shortly after onset (Blumenfeld 2010).

4.1.2 Anterior Cerebral Artery Stroke

ACA infarcts are typically associated with contralateral weakness and sensory loss affecting the leg more than the arm or face. Hemiplegia can occur with larger areas of damage. Manifestations of frontal lobe dysfunction can also be seen, including impaired judgment, flat affect, and apraxia. Transcortical motor aphasia can be observed in the case of left hemisphere damage or contralateral neglect with right hemisphere damage.

4.1.3 Posterior Cerebral Artery Stroke

PCA infarcts typically lead to a contralateral homonymous hemianopia. Involvement of the small penetrating vessels supplying the thalamus or posterior internal capsule can result in contralateral sensory loss, paralysis, or thalamic aphasia in the case of the language-dominant hemisphere. Alexia without agraphia can result from infarct of the left occipital cortex and the splenium of the corpus callosum. Midbrain dysfunction can also occur, such as third-nerve palsy, ataxia, decorticate posturing, and impaired consciousness. Anterograde amnesia is sometimes observed following PCA stroke, as the PCA supplies large portions of the hippocampus (Benson et al. 1974; Szabo et al. 2009).

When blood supply to two adjacent cerebral arteries is reduced or occluded, the border zone, or watershed zone, of their arterial territories is particularly susceptible to ischemia and infarction. ACA-MCA watershed infarcts, often due to sudden occlusion of the internal carotid artery, can lead to proximal arm and leg weakness ("man in the barrel" syndrome) because of involvement of regions of the motor strip serving trunk and proximal limbs.

4.1.4 Multi-Infarct Dementia

Multiple, serial infarcts may cause a progressive decline in cognition and functioning, with each new infarct associated with decline and subsequent stabilization. Dementia due to multiple strokes is referred to as multi-infarct dementia (MID). Individuals with MID typically present with neuroimaging evidence of focal infarcts and focal findings obtained from the neurological examination. Hachinski's Ischemic Scale, which notes the abrupt onset of symptoms, stepwise progression, vascular risk factors, and other signs, has been the traditional means by which MID has been diagnosed (Hachinski et al. 1987).

5 Dementia Associated with Small-Vessel Disease

The stepwise progression of MID can be contrasted with the slow, insidious decline typical of progressive small-vessel damage and AD (see Table 1). In fact, an insidious onset with progressive decline appears to be the most common expression of VaD (Cohen 2009), suggesting a prominent role of progressive small-vessel damage. Time course is useful for distinguishing multi-infarct dementia from AD, but it cannot be used to differentiate between AD and dementia due to small-vessel disease. In line with this, our clinical investigations of patients in an outpatient memory disorder clinic have shown that MRI scans with significant white matter alterations are rarely associated with an abrupt onset or step-wise decline in cognitive or neurological functioning. However, we have demonstrated that dementia patients with marked white matter pathology on MRI (hereafter small-vessel vascular dementia [smVaD]) may be distinguished from those with AD on neuropsychological measures of executive control, episodic memory, semantic knowledge and language, and everyday action. These differences are detailed below and summarized in Table 1.

5.1 Frontal Systems/Executive Control Impairment

Work from our laboratory suggests that *mental set* may be a construct that is useful in understanding the executive control impairment associated with smVaD. Mental set is the ability to appreciate and understand the nature and parameters of a task and to respond accordingly over time until the task is completed. In three recent studies, we have elucidated mental set difficulty in patients with smVaD.

5.1.1 Perseveration

Lamar et al. (1997) studied deficits in establishing and maintaining mental set by looking at the perseverative behavior produced by patients with AD and smVaD on the graphical sequence test, an assessment procedure inspired by the work of Luria (1980) in which participants are required to either draw the shapes or write the names of simple geometric figures, numerals, and other overlearned objects. The overall volume of errors was significantly greater among the smVaD patients, but the types of perseverative errors made by patients with smVaD were also quite distinctive. For example, patients with smVaD frequently persisted in producing responses even when there was no command to do so (i.e., *hyperkinetic/interminable perseverations*). The perseverative errors made by patients diagnosed with AD were different. For example, when asked to *write* the phrase "three squares and two circles" after drawing a series of simple figures in a prior item, AD patients *drew* three squares and two circles, reflecting perseveration of a prior

Table 1 Summary of neuropsychological patterns observed between dementia associated with small-vessel disease and Alzheimer's disease

	Dementia associated with small-vessel disease (smVaD)	Alzheimer's disease (AD)
Time course	Gradual, insidious decline	Gradual, insidious decline
MRI findings	White matter hyperintensities, prominent in the periventricular and deep region	Cortical atrophy with reduced hippocampal volumes
Executive control	Pervasive impairment affecting all response modalities	Impairment restricted to lexical/semantic operations
Perseveration	Highly perseverative, hyperkinetic/interminable motor perseveration	Mildly perseverative, conceptual perseveration
Concept formation	Fail to establish mental set and the abstract attitude to form concepts	Form vague, degraded concepts
Working memory/mental search	Difficulty maintaining mental set over time, reduced storage and rehearsal, and poor mental manipulation	Reduced storage and rehearsal
Episodic memory	Impaired retrieval	Impaired encoding and rapid forgetting
Cued free recall	Some benefit from cues	Minimal to no benefit from cues, high rates of high-frequency intrusion errors in response to category cues
Delayed recognition	Improvement with increased structure of the recognition format, false positives suggest interference effects and source memory failures	Little to no benefit from recognition test format, endorse unrelated foils
Semantic knowledge	Mild impairment secondary to executive control deficits	Moderate impairment due to semantic degradation
Naming	Mildly impaired, coordinate errors (acorn—peanut)	Moderately impaired, superordinate errors (acorn—nut)
Verbal fluency	Letter fluency more impaired than category fluency, words on category fluency highly associated	Category fluency more impaired than letter fluency, reduced association between consecutive words on category fluency
Everyday action	High rates of commission errors, interference from distractor objects	High rates of omission errors

category (i.e., figures); or when asked to draw a target figure such as a circle, AD patients often produced a different previously drawn figure such as a square or a triangle, reflecting perseveration of a prior item/element (i.e., *semantic/element perseverations*).

Among patients with smVaD, graphical sequence test performance correlated with performance on tests of motor functions, suggesting impaired regulation of motor behavior contributed to perseverative errors on the test. By contrast, among patients with AD, difficulty in performance was correlated with tests of naming and output on the 'animal' word generation task. Thus, problems in language and the lexical selection of semantically-related information may underlie difficulty in AD patients.

5.1.2 Concept Formation/Abstract Attitude

Giovannetti et al. (2001) examined the errors made by a heterogeneous group of dementia patients on the WAIS-R similarities subtest, and found 0-point responses provided by smVaD patients tended to be 'out-of-set errors' reflecting difficulty in establishing mental set and the formation of the necessary *abstract attitude* (Goldstein and Scheerer 1941) as required by the task parameters (e.g., *dog-lion*—"one barks and the other growls"). In contrast, the 0-point responses of AD patients, although vague, tended to be 'in-set', indicating an implied superordinate structure (e.g., *dog-lion*—"they're alive"). Factor analysis indicated that 'out-of-set errors' were associated with other measures of frontal system/executive control impairment while 'in-set' errors were related to impairment in language and lexical selection.

5.1.3 Working Memory/Mental Search

The Boston revision of the mental control subtest from the Wechsler memory scale (WMS) evaluates working memory and the capacity to establish and maintain a complex mental set through a series of tasks, such as asking patients to state letters that rhyme with the word 'key', to identify capital, printed letters with curved lines, and to recite the months of the year backward. Previous research has shown that patients with smVaD and patients with Parkinson's disease dementia perform less accurately on these tasks as compared to patients with AD (Libon et al. 1997). Further analysis of mental control performance has revealed that smVaD and AD patients perform differently over time, indicating differential impairment in maintaining mental set (Lamar et al. 2002). This performance pattern was also observed in letter fluency (Lamar et al. 2002). For example, patients with smVaD generated a larger proportion of words during the initial 15 s of the letter fluency task. After the first 15 s, however, the output of smVaD patients dropped precipitously below that of the AD patients. The proportion of responses generated by the AD patients within each 15 s quadrant was no different than the distribution of output generated by healthy control participants.

Lamar et al. (2007) assessed working memory using a new backwards digit span test. This test consisted of 3-, 4-, and 5-span trials. Short-term storage and rehearsal was assessed by tallying the total number of digits reported regardless of recall order (ANY-ORDER; e.g., 47981 recalled '18943', score = 4). Mental manipulation in the form of disengagement and temporal re-ordering was assessed by the total number of digits recalled in correct position (SERIAL-ORDER; e.g., 47981 recalled '18943', score = 3). Rather than using clinical diagnosis as the grouping variable, Lamar et al. (2007) used a visual rating scale of white matter hyperintensities on MRI to divide participants into groups with minimal-mild versus moderate-severe white matter disease. No between-group difference for ANY-ORDER recall was found, suggesting groups were equated in terms of short-term storage and rehearsal abilities. By contrast, participants with moderate-severe white matter disease scored lower for SERIAL-ORDER, suggesting differential impairment for mental manipulation and temporal re-ordering. Step-wise regression analyses showed ANY-ORDER performance variance was explained solely by dementia severity (MMSE). SERIAL-ORDER performance variance was explained by dementia severity (MMSE) and a composite score reflecting executive functioning. In a follow-up study, Lamar et al. (2008) found an association between increased left hemisphere white matter alterations around the posterior ventricular horn and frontal centrum semiovale and reduced SERIAL-ORDER recall.

This body of research suggests that inhibition of perseverations, concept formation/forming an abstract attitude, and establishing and sustaining a complex mental set are complex processes that may be differentially impaired in people with AD versus smVaD. The executive deficits associated with smVaD are extensive, pervasive, and context-independent whereas the executive deficits associated with AD are more restricted and context-dependent (i.e., specific to lexical/semantic operations). In other words, executive function deficits in dementia are hierarchically arranged in the sense that some deficits are primary and related to more rudimentary motor/cognitive functions (as in smVaD). Other executive deficits may be characterized as secondary to disorders of other domains of cognition, such as language or semantic knowledge (as in AD). These findings are consistent with the theoretical constructs put forth by Luria (1980) and Goldberg (1986) and empirically supported by factor analysis of executive function measures (Lamar et al. 2004). In terms of biological substrates, we have proposed that in smVaD ischemic damage to subcortical-cortical white matter projections and the basal ganglia disrupts the modulation operations of the prefrontal cortex. Consequently, the prefrontal cortex cannot effectively maintain or shift mental set to meet the needs that may be required over time or when task demands are changed or become complex.

5.2 Episodic Memory

Despite differential impairment on executive control tasks, past research suggests smVaD patients show relative preservation on tests of episodic memory (Bernard et al. 1992; Libon et al. 1996a, 1998; Tierney et al. 2001; Lafosse et al. 1997).

On list learning tasks, such as the California verbal learning test (CVLT; Delis et al. 1987; Libon et al. 1996b), AD patients display poor retention, rapid forgetting, little to no benefit from cued recall or recognition test conditions, and the production of many intrusion errors. In contrast, patients with smVaD show significantly higher scores on all measures of delayed free and cued recall episodic memory and significant improvement on recognition trials. These analyses are similar to prior research showing that AD patients are more impaired on measures of episodic memory relative to patients with Parkinson's disease (PD) and Huntington's disease (HD; Delis et al. 1991; Kramer et al. 1988; Massman et al. 1990).

Davis et al. (2002) examined the distribution of false positive responses produced by AD and smVaD patients on the delayed recognition task of the CVLT-9. Within-group analyses of the distribution of false positives indicated that patients with smVaD endorsed more interference (list B) foils than semantic or unrelated foils. By contrast, patients with AD endorsed more semantic and unrelated foils. The number of interference (list B) foils endorsed was positively correlated with perseverative errors on the graphical sequence test (Lamar et al. 1997). These findings suggest that poor performance on delayed recognition testing may be influenced by deficits in episodic memory, semantic knowledge, and executive control. Analysis of false positive errors may be necessary to identify the source of failure on delayed recognition testing. The list B foils endorsed by smVaD patients may reflect source memory failures or interference effects from executive function impairment. The semantic and unrelated foils endorsed by AD patients may reflect primary deficits in episodic memory and semantic knowledge.

Price et al. (2009) constructed a new 9-word verbal serial list learning test, the Philadelphia (repeatable) Verbal Learning Test (P[r]VLT). This test was modeled after the California verbal learning test (CVLT). Price et al. (2009) tested the hypothesis that patients with mild MRI white matter alterations would present with evidence of an amnesic syndrome while patients with severe MRI white matter alterations would present with serial list learning deficits consistent with dysexecutive impairment. Finally, patients with moderate MRI white matter alterations were expected to demonstrate a mixed amnesic/dysexecutive profile. Indeed, patients with only mild MRI white matter alterations presented with a flat learning curve on immediate free recall test trials, rapid forgetting with poor recall on delayed free recall/recognition test trials, and copious cued recall intrusion errors, a profile often associated with AD. By contrast, the severe MRI white matter group demonstrated some learning on immediate free recall test trials, exhibited less forgetting as assessed with delayed recognition versus delayed free recall test trials, and produced far fewer cued recall intrusion errors, a profile often associated with subcortical dementia. Patients in the moderate MRI white matter group presented with characteristics seen in both the mild and severe MRI white matter groups. Overall, this research suggests that MRI white matter disease can be associated with specific patterns of impairment on verbal serial list learning tests.

In sum, work from our laboratory (Libon et al. 1996a; 1998) and others (Bernard et al. 1992; Tierney et al. 2001; Lafosse et al. 1997) shows patients with smVaD demonstrate relative sparing of episodic memory performance relative to patients with AD. We also have shown that problems on measures of episodic memory among patients with smVaD may be largely explained by executive functioning deficits (Davis et al. 2002). However, the literature on episodic memory deficits in smVaD is not consistent, as some authors have not observed differences between smVaD and AD on measures of episodic memory (Reed et al. 2007). Some authors have suggested that frank episodic memory impairment may come later in the course of smVaD, while retrieval errors associated with executive dysfunction may appear early in the course (Cohen 2009).

5.3 Semantic Knowledge and Language

Deficits of language in patients with smVaD have not been as extensively studied as those of executive control or memory. While a variety of language-related tasks have been administered, the procedures most commonly reported include tests of naming and letter and/or category word fluency. Two general findings emerge from this literature. First, patients with smVaD sometimes produce better scores on tests of naming (Cannata et al. 2002; Kontiola et al. 1990). Second, output on tests of letter fluency produced by patients with smVaD is reduced as compared to AD patients (Carew et al. 1997; Lafosse et al. 1997). This finding is often interpreted within the context of greater executive control deficits associated with smVaD.

5.3.1 Visual Confrontation Naming

The observation that patients with smVaD make fewer semantically-related intrusion errors than patients with AD on tests of episodic memory may suggest that, in general, semantic knowledge may be less disrupted in subcortical VaD as compared to AD. Only a few studies have examined this issue. For example, Lukatela et al. (1998) found that patients with VaD made fewer errors on the Boston naming test (BNT) as compared to patients with AD. Also, there were distinct differences regarding BNT errors. Patients with AD tended to make superordinate errors, (acorn—nut), i.e., errors that tend to place the response within a broader semantic class than the stimulus. Patients with smVaD made more coordinate errors, (acorn—peanut), i.e., errors that tend to place the response within the same semantic class as the stimulus. Lukatela et al. (1998) interpreted their findings as evidence for relative preservation of semantic knowledge in smVaD as compared to AD.

Laine et al. (1997) also administered the BNT to patients with AD and smVaD. Unlike the data reported by Lukatela et al. (1998), there was no between-group difference on the BNT. Laine et al. (1997) administered a multiple choice task measuring word meaning. On this task, patients were asked to choose specific semantic features related to BNT target items. Compared to patients with smVaD, patients with AD tended to make more errors regarding semantic, as opposed to superordinate, features of target items. In this sense, these findings are similar to those reported by Lukatela et al. (1998).

5.3.2 Semantic Fluency

Carew et al. (1997) designed a paradigm to measure semantic organization on the ‘animal’ word list generation task. On this task, patients were asked to generate as many different animal names as they could in 1 min. All responses were coded into the following six attribute categories: size (big, small), geographic location (foreign, North America), diet (herbivore, carnivore, omnivore), zoological class (insect, mammal, bird, etc.), habitat (farm, Africa/jungle, widespread, etc.), and biological order/related groupings (feline, canine, bovine, etc.). An association index (AI) was calculated by totaling the number of shared attributes and then dividing by the number of total responses. The AI is believed to provide a measure of the semantic organization between successive responses independent of the number of words produced. Carew et al. (1997) found that the total number of responses made by patients with AD and smVaD did not differ. With respect to the AI, healthy control participants and smVaD patients did not differ; however, both groups obtained higher scores on this measure as compared to patients with AD. Carew et al. (1997) interpreted their data as consistent with the idea that semantic knowledge is relatively intact in smVaD as compared to AD.

5.4 Summary of the Cognitive Profile of smVaD

On the basis of differential impairment on tests of executive control as compared to other domains of cognitive functioning often observed in patients with white matter damage on MRI (i.e., smVaD), we believe that this kind of vascular dementia may be conceptualized as a subcortical dementia syndrome, similar to Huntington’s disease and Parkinson’s disease dementia. Furthermore, considering findings in the areas of executive functioning, episodic memory, and language/semantic knowledge, we conclude that the executive control deficits seen in smVaD tend to be ubiquitous or *pervasive*. The executive control deficits in AD are quantitatively and qualitatively different. AD patients show less severe executive deficits and the executive deficits in AD appear to be restricted to the response selection of lexical/semantic information. The pervasiveness of executive dysfunction in smVaD, or the means by which executive control deficits intrude into virtually all other aspects of cognition among such patient, is consistent with theoretical ideas suggested by Luria (1980). In this sense the executive control deficits in smVaD are context non-specific, whereas the executive control deficits in AD are, by contrast, rather context specific.

The executive deficits in smVaD may be further characterized as a deficit in regulating behavior over time. As noted by Lamar et al. (2002), as patients with smVaD attempt to work through various tasks, such as tasks of mental control, they tend to accumulate more and more errors. On tests of letter fluency, patients with smVaD tend to produce their output during the initial portion of the test. Again, these behaviors are different as compared to AD patients. The disruption of frontal lobe-basal ganglia-thalamic pathways may be the etiology of both the pervasiveness and poor regulation of executive control deficits in smVaD (Alexander et al. 1986; Sultzer et al. 1995).

6 The Impact of Cerebrovascular Cognitive Deficits on Everyday Life

Cerebrovascular disorders that impair cognition also negatively impact everyday functioning for older adults. In the case of vascular dementia, significant deficits in everyday functioning are a diagnostic criterion (American Psychiatric Association 2000), and more mild everyday difficulties are now clearly recognized in individuals with mild cognitive impairment (Giovannetti et al. 2008a, b). Difficulties with everyday life tasks are associated with a wide range of negative outcomes, including institutionalization and caregiver burden (Hope et al. 1998; Knopman et al. 1988; Noale et al. 2003; Severson et al. 1994). Thus, increased understanding of everyday functional deficits is imperative to improve outcomes associated with cerebrovascular disorders in older adults.

Neuropsychological research on daily functioning must begin with a clear definition of the activities under study. Some have used the term "everyday action" to denote behavior in the service of everyday tasks that involves sequencing multiple steps and using objects to achieve nested goals (Giovannetti et al. 2002). Activities of daily living (ADL) is a related term often used in the literature to denote everyday activities that are necessary for independent living, often subdivided into basic activities of daily living (BADL), such as bathing, dressing, and toileting, and instrumental activities of daily living (IADL), which are more complex activities such as cooking, housekeeping, and money management.

The neuropsychological literature on everyday functioning following acute, large-vessel stroke explores the relation between traditional cognitive test data and functional abilities assessed using clinician ratings of general functional abilities or the need for assistance (e.g., pass/fail or Likert scale). This important literature has consistently demonstrated an association between cognition and functional abilities (Fong et al. 2001; Galski et al. 1993; Man et al. 2006; Marcotte et al. 2010; Mysiw et al. 1989; Ozdemir et al. 2001; Zinn et al. 2004). However, these studies offer a limited view of functional abilities, as everyday functioning is not evaluated from a neuropsychological perspective. Commonly used functional measures do not address the *reasons* for functional deficits; therefore, they fail to elucidate *how* deficits in particular cognitive domains, such as executive control and episodic memory, lead to functional impairment (see Wilson 1993).

At least three studies are an exception to the mainstream literature on everyday action following stroke. These studies sought to elucidate the neurocognitive mechanisms underlying everyday action following stroke using detailed, performance-based methods that yield scores reflecting functional processes, such as the accomplishment of task steps as well as a broad range of error types. The results of these studies showed that patients with right versus left hemisphere stroke exhibited similar performance patterns on everyday tasks (Buxbaum et al. 1998; Hartmann et al. 2005; Schwartz et al. 1999). One conclusion from these studies is that specific cognitive deficits may not translate to differential patterns of everyday action impairment. However, detailed studies of everyday action performance in vascular dementia (described below) suggest that comparisons of right versus left hemisphere stroke may mask meaningful differences between individuals with anterior versus posterior strokes or between stroke patients with executive control deficits versus those with declarative memory deficits.

Before we describe the detailed studies of everyday action performance in people with vascular dementia, we will review the literature on caregiver ratings of everyday functioning and neuroimaging variables related to CVD (Boyle et al. 2003, 2004). In a cross sectional study considering subcortical hyperintensities and cortical volume in 29 participants, Boyle et al. (2003) reported a significant association between hyperintensities and IADL ratings (Lawton and Brody 1969) but not between cortical volume and IADL (Boyle et al. 2003). Subcortical hyperintensities were not related to the dementia rating scale (DRS) total or other subscale scores. In contrast, cortical volume correlated significantly with the DRS total and the memory subscale. After controlling for MMSE, executive dysfunction (DRS-I/P subscale) explained 28% of variance in IADL, and subcortical hyperintensities explained an additional 14% above and beyond executive dysfunction. A longitudinal study in a vascular dementia sample by Boyle et al. (2004) looked at the ability of subcortical hyperintensity volume to predict future IADL (1 year later) in 28 participants. They reported a trend towards significance in the relation between baseline subcortical hyperintensities and IADL at 1 year.

While Boyle et al. (2004) have suggested a prominent role for executive dysfunction in the performance of daily activities by patients with smVaD, the relation between everyday action and specific domains of cognitive performance is not yet conclusively determined. In their review of the literature, Farias et al. (2003) concluded that the evidence for relations between everyday action and specific cognitive domains is mixed, with studies finding an association with memory and visuospatial abilities, as well as executive functions. Farias et al. (2003) attributed this heterogeneity to methodological issues, including variability in functional measures, cognitive tests, and populations studied. An alternative explanation is that everyday action is a multidimensional construct that involves several neurocognitive processes (Hartmann et al. 2005).

As mentioned above, recent studies of performance-based deficits in smVaD and other dementia subtypes have led to a model that explicates the roles of different cognitive deficits in everyday action impairment (Giovannetti et al. 2008a, b; 2006; Kessler et al. 2007). This model of everyday action has emerged from research using performance-based error analysis methods, which provide opportunities for detailed coding and categorization of errors, facilitating a more sophisticated measure of everyday functioning than traditional caregiver questionnaires. Recent evidence has suggested that dissociable cognitive processes underlie everyday action deficits, with errors of omission (i.e., failure to perform a task step) distinct from those of commission (i.e., inaccurate execution of a step; Giovannetti et al. 2008a, b). Neuropsychological correlates of omission and commission errors have also been identified, indicating omissions are associated with poor performance on tests of declarative memory (i.e., episodic & semantic) and commissions with poor performance on tests of executive control. An additional class of errors, action-additions or off-task errors, share variance with both omissions and commissions and may be multi-determined.

A study by Giovannetti et al. (2006) examined differences between individuals diagnosed with smVaD versus AD on the naturalistic action test (NAT; Schwartz et al. 2003), a performance-based measure that includes three tasks of increasing complexity. The percentage of task steps accomplished, number of errors, and performance times were recorded for each task. While the groups did not differ in dementia severity or overall impairment on the NAT, the smVaD group committed more errors. The smVaD group also accomplished significantly fewer steps when salient distractor objects were present. Correlations between NAT variables and neuropsychological tests suggested that the executive control deficits associated with smVaD may have contributed to their specific pattern of everyday action difficulties, namely distractor interference and inefficient, error-prone performance on complex tasks. In AD, everyday action may be negatively influenced by episodic memory failures. These results demonstrated that individuals with different dementia diagnoses, associated with distinct neuropathology and cognitive profiles, exhibit differential deficits on complex tasks of everyday functioning.

A recent study reported by Seidel et al. (2011) investigated brain-behavior relations in the performance of everyday tasks in a sample of dementia participants including individuals with AD and/or smVaD. Relations were analyzed between volumes of cortical gray matter, hippocampus, caudate nucleus, and intact white matter in surface, deep, and periventricular regions, and two measures of functional performance: informant ratings on a modified-version IADL questionnaire and a direct assessment of everyday action (naturalistic action test, NAT). NAT coding allowed for quantification of discrete error types including omissions (failure to complete a task step), on-task commissions (inaccurate step execution), and off-task commissions (performance of steps inconsistent with task instructions). Both caregiver IADL ratings and on-task commission errors on the NAT showed a significant association with white matter integrity, specifically in the deep region. In contrast, off-task commission errors appeared to be multi-determined, showing a relation to both intact deep white matter and gray matter structures (cortex and hippocampus). Omission errors appeared most related to hippocampal volumes. Findings highlight an important role for deep white matter in functional deficits and suggest unique neurological substrates for discrete everyday action error types in dementia.

In sum, the study of everyday action, as an important and relatively understudied manifestation of cognitive deficits, may contribute to understanding vascular dementia and vascular cognitive impairment. A prominent role for executive dysfunction in the performance of daily activities by patients with vascular dementia has gained support from relations between ratings of their performance of daily tasks (IADL) and volumes of subcortical hyperintensities. Authors that view everyday action as multi-dimensional have used detailed analyses of performance-based methods rather than caregiver ratings to examine everyday action. These authors have found that smVaD patients may be particularly prone to distractor interference and inefficient, error-prone action on complex tasks, a pattern of performance that may reflect compromised executive functioning. A developing smVaD action profile is also supported by the association of inaccurate task execution (on-task commissions) with poor white matter integrity, specifically in deep regions.

7 Conclusion

We have attempted to describe the complexity of vascular anatomy and related pathologies, and the heterogeneity of their manifestation in cognitive function and everyday life. In our review, we have highlighted several topics that have been relatively underrepresented in the literature to date, including anatomy of the cerebral small vessels and the pathological impact of small-vessel disease, nuanced characterization of the cognitive profile associated with small-vessel disease, and the impact of cerebrovascular disease on everyday functioning (see Table 1). We hope that increased attention to the role of cerebrovascular disease in aging will lead to early detection, improved diagnosis, and effective prevention and treatment for the cognitive and functional deficits associated with these diseases. These clinical advances are essential to effectively address the looming health care needs of our aging population.

Acknowledgments Portions of chapter were presented at the 39th annual meeting of the International Neuropsychological Society, Boston, MA.

References

Alexander GE, DeLong MR, Strick PL (1986) Parallel organization of functionally segregated circuits linking basal ganglia and cortex. Annu Rev Neurosci 9:357–381

American Psychiatric Association (2000) Diagnostic and statistical manual of mental disorders: DSM-IV-TR, 4th edn. American Psychiatric Association, Washington, DC

Ariesen MJ, Claus SP, Rinkel GJE, Algra A (2003) Risk factors for intracerebral hemorrhage in the general population: a systematic review. Stroke 34:2060–2065

Benson DF, Marsden CD, Meadows JC (1974) The amnesic syndrome of posterior cerebral artery occlusion. Acta Neurol Scand 50(2):133–145

Bernard BA, Wilson RS, Gilley DW, Bennett DA, Fox JH (1992) Memory failure in Binswanger's disease and Alzheimer's disease. The Clin Neuropsychol 6:230–240

Blumenfeld H (2010) Neuroanatomy through clinical cases, 2nd edn. Sinauer Associates, Sunderland, MA

Boyle PA, Paul R, Moser D, Zawacki T, Gordon N, Cohen R (2003) Cognitive and neurologic predictors of functional impairment in vascular dementia. Am J Geriatr Psychiatry 11(1): 103–106

Boyle PA, Paul RH, Moser DJ, Cohen RA (2004) Executive impairments predict functional declines in vascular dementia. Clin Neuropsychol 18(1):75–82

Brott T, Adams HP Jr, Olinger CP, Marler JR, Barsan WG, Biller J et al (1989) Measurements of acute cerebral infarction: a clinical examination scale. Stroke 20(7):864–870

Buxbaum L, Schwartz M, Montgomery M (1998) Ideational apraxia and naturalistic action. Cogn Neuropsychol 15:617–643

Cannata AP, Alberoni M, Franceschi M, Mariani C (2002) Frontal impairment in subcortical ischemic vascular dementia in comparison to Alzheimer's disease. Dementia and Geriatr Cogn Disord 13:101–111

Carew TG, Lamar M, Cloud BS, Grossman M, Libon DJ (1997) Impairment in category fluency in ischaemic vascular dementia. Neuropsychology 11:400–412

Cohen RA (2009) Neuropsychology of cardiovascular disease. In: Cohen RA, Gunstad J (eds) Neuropsychology and cardiovascular disease. Oxford University Press, New York, pp 3–18

Davis KL, Price C, Kaplan E, Libon DJ (2002) Error analysis of the nine-word California verbal learning test (CVLT-9) among older adults with and without dementia. Clin Neuropsychol 16:81–89

Davson H (1976) The blood–brain barrier. J Physiol 255:1–28

de Reuck J (1971) The human periventricular arterial blood supply and anatomy of cerebral infarctions. Eur Neurol 5:321–334

Delis DC, Kramer JH, Kaplan E, Ober BA (1987) The California verbal learning test. Psychology Corporation, New York

Delis DC, Massman PJ, Butters N, Salmon DP, Cermak LS, Kramer JH (1991) Profiles of demented and amnesic patients on the California verbal learning test: implications for the assessment for the assessment of memory disorders. Psychol Assess: J Consult Clin Psychol 3:19–26

Duffis EJ, Fisher M (2009) Cardioembolic stroke. In: Cohen RA, Gunstad J (eds) Neuropsychology and cardiovascular disease. New York, Oxford University Press, pp 221–232

Easton JD, Saver JL, Albers GW, Alberts MJ, Chaturvedi S, Feldmann E et al (2009) Definition and evaluation of transient ischemic attack: a scientific statement for healthcare professionals from the American heart association/American stroke association stroke council; council on cardiovascular surgery and anesthesia; council on cardiovascular radiology and intervention; council on cardiovascular nursing; and the interdisciplinary council on peripheral vascular disease. Stroke 40(6):2276–2293

Farias ST, Harrell E, Neumann C, Houtz A (2003) The relationship between neuropsychological performance and daily functioning in individuals with Alzheimer's disease: ecological validity of neuropsychological tests. Arch Clin Neuropsychol 18(6):655–672

Fisher M (2010) The challenge of mixed cerebrovascular disease. Ann N Y Acad Sci 1207:18–22

Fong KNK, Chan CCH, Au DKS (2001) Relationship of motor and cognitive abilities to functional performance in stroke rehabilitation. Brain Injury 15(5):443–453

Fung WK, Poppas A (2009) Overview of cardiovascular physiology and pathology. In: Cohen RA, Gunstad J (eds) Neuropsychology and cardiovascular disease. Oxford University Press, New York, pp 66–80

Fuster V, Moreno PR, Fayad ZA, Corti R, Badimon JJ (2005) Atherothrombosis and high-risk plaque: part I: evolving concepts. J Am Coll Cardiol 46(6):937–954

Galski T, Bruno RL, Zorowitz R, Walker J (1993) Predicting length of stay, functional outcome, and aftercare in the rehabilitation of stroke patients—the dominant role of higher-order cognition. Stroke 24(12):1794–1800

Giovannetti T, Lamar M, Cloud BS, Swenson R, Fein D, Kaplan E et al (2001) Different underlying mechanisms for deficits in concept formation in dementia. Arch Clin Neuropsychol 16(6):547–560

Giovannetti T, Libon DJ, Buxbaum LJ, Schwartz MF (2002) Naturalistic action impairments in dementia. Neuropsychologia 40(8):1220–1232

Giovannetti T, Schmidt KS, Gallo JL, Sestito N, Libon DJ (2006) Everyday action in dementia: evidence for differential deficits in Alzheimer's disease versus subcortical vascular dementia. J Int Neuropsychol Soc 12(1):45–53

Giovannetti T, Bettcher BM, Brennan L, Libon DJ, Kessler RK, Duey K (2008a) Coffee with jelly or unbuttered toast: commissions and omissions are dissociable aspects of everyday action impairment in Alzheimer's disease. Neuropsychology 22(2):235–245

Giovannetti T, Bettcher BM, Brennan L, Libon DJ, Kessler RK, Duey K (2008b) Characterization of everyday functioning in mild cognitive impairment: a direct assessment approach. Dementia and Geriatric Cogn Disord 25:359–365

Girouard H, Iadecola C (2006) Neurovascular coupling in the normal brain and in hypertension, stroke, and Alzheimer disease. J Appl Physiol 100(1):328–335

Goldberg E (1986) Varieties of perseveration: a comparison of two taxonomies. J Clin Exp Neuropsychol 8(6):710–726

Goldmann EE (1909) Die äussere und innere Sekretion des gesunden und kranken organismus im Lichte der 'vitalen Farbung'. Beiträge zur klinischen Chirurgie 64:192–265

Goldmann EE (1913) Vitalfarbung am Zentralnervensystem. Abhandlungen der Preussischen Akademie der Wissenschaften Phys-Math 1:1–60

Goldstein K, Scheerer M (1941) Abstract and concrete behavior. Psychol Monogr 53:329–401

Greenberg SM, Grabowski T, Gurol ME, Skehan ME, Nandigam RN, Becker JA et al (2008) Detection of isolated cerebrovascular beta-amyloid with Pittsburgh compound B. Ann Neurol 64(5):587–591

Hachinski VC, Potter P, Merskey H (1987) Leuko-araiosis. Arch Neurol 44(1):21–23

Hartmann K, Goldenberg G, Daumuller M, Hermsdorfer J (2005) It takes the whole brain to make a cup of coffee: the neuropsychology of naturalistic actions involving technical devices. Neuropsychologia 43(4):625–637

Hatzinikolaou-Kotsakou E, Kartasis Z, Tziakas D, Stakos D, Hotidis A, Chalikias G et al (2005) Clotting state after cardioversion of atrial fibrillation: a haemostasis index could detect the relationship with the arrhythmia duration. Thrombosis J 3(1):2

Hope T, Keene J, Gedling K, Fairburn CG, Jacoby R (1998) Predictors of institutionalization for people with dementia living at home with a carer. Int J Geriatr Psychiatry 13(10):682–690

Hylek EM, Singer DE (1994) Risk factors for intracranial hemorrhage in outpatients taking warfarin. Ann Intern Med 120:897–902

Irani F (2009) Cardiac output and ejection fraction: impact on brain structure and function. In: Cohen RA, Gunstad J (eds) Neuropsychology and cardiovascular disease. Oxford University Press, New York, pp 233–239

Jefferson A (2010) Cardiac output as a potential risk factor for abnormal brain aging. J Alzheimer's Dis 20:813–821

Kessler RK, Giovannetti T, MacMullen LR (2007) Everyday action in schizophrenia: performance patterns and underlying cognitive mechanisms. Neuropsychology 21(4):439–447

Knopman DS, Kitto J, Deinard S, Heiring J (1988) Longitudinal study of death and institutionalization in patients with primary degenerative dementia. J Am Geriatr Soc 36(2): 108–112

Kontiola P, Laaksonen R, Sulkava R, Erkinjuntti T (1990) Pattern of language impairment is different in Alzheimer's disease and multi-infarct dementia. Brain and Language 38:364–383

Kramer JH, Delis DC, Blusewicz MJ, Brandt J, Ober BA, Strauss M (1988) Verbal memory errors in Alzheimer's and Huntington's dementias. Dev Neuropsychol 4:1–15

Lafosse JM, Reed BR, Mungas D, Sterling SB, Wahbeh H, Jagust WJ (1997) Fluency and memory differences between ischemic vascular dementia and Alzheimer's disease. Neuropsychology 11:514–522

Laine M, Vuorinen E, Rinne JO (1997) Picture naming deficits in vascular dementia and Alzheimer's disease. J Clin Exp Neuropsychol 19:126–140

Lamar M, Podell K, Carew TG, Cloud BS, Resh R, Kennedy C et al (1997) Perseverative behavior in Alzheimer's disease and subcortical ischemic vascular dementia. Neuropsychology 11(4):523–534

Lamar M, Price CC, Davis KL, Kaplan E, Libon DJ (2002) Capacity to maintain mental set in dementia. Neuropsychologia 40(4):435–445

Lamar M, Swenson R, Kaplan E, Libon DJ (2004) Characterizing alterations in executive functioning across distinct subtypes of cortical and subcortical dementia. Clin Neuropsycholo 18(1):22–31

Lamar M, Price CC, Libon DJ, Penney DL, Kaplan E, Grossman M, Heilman KM (2007) Alterations in working memory as a function of leukoaraiosis in dementia. Neuropsychologia 45:245–254

Lamar M, Catani M, Price CC, Heilman KM, Libon DJ (2008) The impact of region specific leukoaraiosis on working memory deficits in dementia. Neuropsychologia 46:2597–2601

Lawton MP, Brody EM (1969) Assessment of older people: self-maintaining and instrumental activities of daily living. Gerontologist 9(3):179–186

Lewandowsky M (1900) Zur Lehre der Cerebrospinalflussigkeit. Z Klin Med 40:480–494

Libon DJ, Malamut BL, Swenson R, Cloud BS (1996a) Further analyses of clock drawings among demented and non-demented subjects. Arch Clin Neuropsychol 11:193–211

Libon DJ, Mattson RE, Glosser G, Kaplan E, Malamut M, Sands LP, Swenson R, Cloud BS (1996b) A nine word dementia version of the California verbal learning test. Clin Neuropsychol 10:237–244

Libon DJ, Bogdanoff B, Bonavita J, Skalina S, Cloud BS, Resh R et al (1997) Dementia associated with periventricular and deep white matter alterations: a subtype of subcortical dementia. Arch Clin Neuropsychol 12(3):239–250

Libon DJ, Bogdanoff B, Cloud BS, Skalina S, Carew TG, Gitlin HL, Bonavita J (1998) Motor learning and quantitative measures of the hippocampus and subcortical white alterations in Alzheimer's disease and ischaemic vascular dementia. J Clin Exp Neuropsychol 20:30–41

Libon DJ, Price CC, Davis Garrett K, Giovannetti T (2004) From Binswanger's disease to leuokoaraiosis: what we have learned about subcortical vascular dementia. Clin Neuropsychol 18(1):83–100

Libon DJ, Price CC, Giovannetti T, Swenson R, Bettcher BM, Heilman KM et al (2008) Linking MRI hyperintensities with patterns of neuropsychological impairment: evidence for a threshold effect. Stroke 39(3):806–813

Longstreth WT Jr, Manolio TA, Arnold A, Burke GL, Bryan N, Jungreis CA et al (1996) Clinical correlates of white matter findings on cranial magnetic resonance imaging of 3301 elderly people. The cardiovascular health study. Stroke 27(8):1274–1282

Lukatela K, Malloy P, Jenkins M, Cohen R (1998) The naming deficit in early Alzheimer's and vascular dementia. Neuropsychology 12:565–572

Luria AR (1980) Higher cortical functions in man, 2nd edn. Basic Books Inc, New York, NY

Man DWK, Tam SF, Hui-Chan C (2006) Prediction of functional rehabilitation outcomes in clients with stroke. Brain Injury 20(2):205–211

Marcotte TD, Scott JC, Kamat R, Heaton RK (2010) Neuropsychology and the prediction of everyday functioning. In: Marcotte TD, Grant I (eds) Neuropsychology of everyday function. Guilford Press, New York, pp 5–38

Marinković S, Gibo H, Milisavljević M, Ćetković M (2001) Anatomic and clinical correlations of the lenticulostriate arteries. Clin Anat 14(3):190–195

Marti-Vilalta JL, Arboix A, Mohr JP (2004) Lacunes. In: Mohr JP, Choi DW, Grotta JC, Weir B, Wolf PA (eds) Stroke: pathophysiology, diagnosis, and management, 4th edn. Elsevier Health Sciences, Philadelphia, pp 275–300

Massman PJ, Delis DC, Butters N, Levin B, Salmon DP (1990) Are all subcortical dementias alike?: verbal learning and memory in Parkinson's and Huntington's disease patient's. J Clin Exp Neuropsychol 12:729–744

Miller IN, Haynes WG, Moser DJ (2009) Systemic vascular function and cognitive function. In: Cohen RA, Gunstad J (eds) Neuropsychology and cardiovascular disease. Oxford University Press, New York, pp 240–263

Mooradian AD (1988) Effect of aging on the blood–brain barrier. Neurobiol Aging 9:31–39

Mysiw WJ, Beegan JG, Gatens PF (1989) Prospective cognitive assessment of stroke patients before inpatient rehabilitation—the relationship of the neuro-behavioral cognitive status examination to functional improvement. Am J Phys Med Rehabil 68(4):168–171

Noale M, Maggi S, Minicuci N, Marzari C, Destro C, Farchi G et al (2003) Dementia and disability: impact on mortality. The Italian longitudinal study on aging. Dementia and Geriatr Cogn Disord 16(1):7–14

Ozdemir F, Birtane M, Tabatabaei R, Ekukle G, Kokino S (2001) Cognitive evaluation and functional outcome after stroke. Am J Phys Med Rehabil 80:410–415

Pantoni L (2010) Cerebral small vessel disease: from pathogenesis and clinical characteristics to therapeutic challenges. Lancet Neurol 9(7):689–701

Pantoni L, Garcia JH (1997) Cognitive impairment and cellular/vascular changes in the cerebral white matter. Ann N Y Acad Sci 826:92–102

Phares TW, Kean RB, Mikheeva T, Hooper DC (2006) Regional differences in blood-brain barrier permeability changes and inflammation in the apathogenic clearance of virus from the central nervous system. J Immunol 176(12):7666–7675

Price CC, Jefferson AL, Merino JG, Heilman KM, Libon DJ (2005) Subcortical vascular dementia: integrating neuropsychological and neuroradiologic data. Neurology 65(3):376–382

Price CC, Garrett KD, Jefferson AL, Cosentino S, Tanner J, Penney DL, Swenson R, Giovannetti T, Bettcher BM, Libon DJ (2009) The role of leukoaraiosis severity on learning and memory in dementia: performance differences on a 9-word list learning test. Clin Neuropsychol 23:1–18

Reed BR, Mungas DM, Kramer JH, Ellis W, Vinters HV, Zarow C et al (2007) Profiles of neuropsychological impairment in autopsy-defined Alzheimer's disease and cerebrovascular disease. Brain 130(Pt 3):731–739

Rigs HE, Rupp C (1963) Variation in form of Circle of Willis. The relation of the variations to collateral circulation: anatomic analysis. Archives of Neurology 8:8–14

Ritman EL, Lerman A (2007) The dynamic vasa vasorum. Cardiovasc Res 75(4):649–658

Roman GC (1987) Senile dementia of the Binswanger type. A vascular form of dementia in the elderly. JAMA 258(13):1782–1788

Roman GC, Tatemichi TK, Erkinjuntti T, Cummings JL, Masdeu JC, Garcia JH et al (1993) Vascular dementia: diagnostic criteria for research studies. Report of the NINDS-AIREN international workshop. Neurology 43(2):250–260

Schwartz MF, Buxbaum LJ, Montgomery MW, Fitzpatrick-DeSalme E, Hart T, Ferraro M et al (1999) Naturalistic action production following right hemisphere stroke. Neuropsychologia 37:51–66

Schwartz MF, Buxbaum LJ, Ferraro M, Veramonti T, Segal M (2003) The naturalistic action test. Thames Valley Test Company, Bury St. Edmunds, United Kingdom

Seidel GA, Giovannetti T, Price CC, Towler S, Tanner J, Mitchell S, Eppig J, Pennisi A, Libon DJ (2011) Structural neuroimaging correlates of everyday action in dementia. In: DJ Libon, The differential contribution of white and gray matter to the phenotypic expression of dementia. Symposium conducted at the 39th annual meeting of the international neuropsychological society, Boston, MA

Severson MA, Smith GE, Tangalos EG, Petersen RC, Kokmen E, Ivnik RJ et al (1994) Patterns and predictors of institutionalization in community-based dementia patients. J Am Geriatr Soc 42(2):181–185

Sharma M, Clark H, Armour T et al (2005) Acute Stroke: evaluation and treatment. Evidence reports/technology assessments, No. 127. Agency for Healthcare Research and Quality (US), Rockville, MD

Sultzer D, Mahler M, Cummings J, Van Gorp W, Hinkin C, Brown C (1995) Cortical abnormalities associated with subcortical lesions in vascular dementia. Arch Neurol 52: 773–780

Szabo K, Forster A, Jager T, Kern R, Griebe M, Hennerici MG et al (2009) Hippocampal lesion patterns in acute posterior cerebral artery stroke: clinical and MRI findings. Stroke 40(6): 2042–2045

Thoonsen H, Richard E, Bentham P, Gray R, van Geloven N, De Haan RJ et al (2010) Aspirin in Alzheimer's disease: increased risk of intracerebral hemorrhage: cause for concern? Stroke 41(11):2690–2692

Tierney MC, Black SE, Szalai JP, Snow G, Fisher RH, Nadon G, Chui HC (2001) Recognition memory and verbal fluency differentiate probable Alzheimer's disease from subcortical ischemic vascular dementia. Arch Neurol 58:1654–1659

Vasilevko V, Passos GF, Quiring D, Head E, Kim RC, Fisher M et al (2010) Aging and cerebrovascular dysfunction: contribution of hypertension, cerebral amyloid angiopathy, and immunotherapy. Ann N Y Acad Sci 1207:58–70

Wilson B (1993) Ecological validity of neuropsychological assessment: do neuropsychological indexes predict performance in everyday activities. Appl Prev Psychol 2:209–215

Wolf PA, Mitchell JB, Baker CS, Kannel WB, D'Agostino RB (1998) Impact of atrial fibrillation on mortality, stroke, and medical costs. Arch Intern Med 158(3):229–234

Zinn S, Dudley TK, Bosworth HB, Hoenig HM, Duncan PW, Horner RD (2004) The effect of poststroke cognitive impairment on rehabilitation process and functional outcome. Arch Phys Med Rehabil 85(7):1084–1090

Zlokovic BV (2004) Clearing amyloid through the blood–brain barrier. J Neurochem 89:807–811

Zlokovic BV (2008) The blood–brain barrier in health and chronic neurodegenerative disorders. Neuron 57(2):178–201

Psychiatric Disorders in Ageing

C. Wijeratne, S. Reutens, B. Draper and P. Sachdev

Abstract The ageing of the population brings particular challenges to psychiatric practice. Although the clinical presentation of common psychiatric disorders such as mood and psychotic disorders is largely similar to those in younger adults, late life presentations tend to be more complex as co-morbidity with dementia and physical illness is common. Suicide tends to increase with age in most countries. In this chapter we argue that the aetiology of disorders may be best understood within a stress vulnerability model in which neurobiological and psychosocial factors interplay. We further present that management strategies need to be comprehensive, incorporating physical, social, pharmacological, and psychological treatments appropriate to each case. We close with a call for the use of specialised multi-disciplinary services to improve the overall quality of care.

Keywords Elderly · Psychiatric disorders · Aetiology · Management

Contents

1 Introduction 244
2 Clinical Presentations in Late Life 245
3 Vulnerability and Resilience in Late Life 246
4 Epidemiology 248
5 Depressive Disorders 249
5.1 Aetiology 250
5.2 Prognosis 251
5.3 Suicide 252
6 Bipolar Disorder 253
7 Psychotic Disorders 254

C. Wijeratne · S. Reutens · B. Draper · P. Sachdev (✉)
School of Psychiatry, University of New South Wales, Sydney, Australia
e-mail: p.sachdev@unsw.edu.au

Curr Topics Behav Neurosci (2012) 10: 243–269
DOI: 10.1007/7854_2011_124

Published Online: 18 March 2011

7.1 Schizophrenia 254
7.2 Very Late Onset Schizophrenia (VLOS) 255
7.3 Delusional Disorder 256
8 Anxiety Disorders 256
9 Substance Use Disorders 257
10 Somatoform Disorders 258
11 Personality Disorders 259
12 Management 260
12.1 Psychiatric Services in Late Life 261
13 Conclusion 262
References 262

1 Introduction

The ageing of the population is a world-wide trend, with technologically advanced countries leading the way. In the United States, the proportion of the population aged 65 years and over is projected to increase from 12.4% in 2000 to 19.6% in 2030 (Kinsella and Velkoff 2001).

Ageing is associated with increased physical disability and an exponential increase in neurocognitive disorders. The picture in relation to psychiatric disorders is more complex, with both an increase and decrease in prevalence rates having been reported in different disorders and across studies. A number of epidemiological studies have reported a decline in 12 month and lifetime prevalence rates with age, in particular after 75 years (Kessler et al. 2005; Trollor et al. 2007). Such studies however do not capture those elderly with dementia or physical illness who are at increased risk of developing a psychiatric disorder. This chapter provides an overview of psychiatric disorders in elderly populations and notes that late life psychiatric disorders are associated with high levels of physical and functional disability, with significant implications for healthcare funding and service provision.

Classificatory systems, such as the Diagnostic and Statistical Manual (DSM) (American Psychiatric Association (APA) 1995), lists four major categories of psychiatric disorders affecting the elderly—(i) cognitive disorders, such as dementia which will not be the subject of specific discussion in this chapter; (ii) psychotic disorders, such as schizophrenia, characterised by delusions and hallucinations; (iii) mood disorders, such as depressive and bipolar disorders, characterised by disturbances of affect and vegetative changes; and (iv) anxiety disorders which are characterised by specific fears and autonomic arousal.

There are a number of limitations in this approach. Psychiatric symptoms, like physical measures such as blood pressure, are best understood as dimensional phenomena whereas psychiatric disorders are criteria-based categories. The threshold used to determine whether a disorder is present, is often arbitrary, a particular difficulty in older people who are more likely to report sub-syndromal depressive symptoms that are associated with significant disability (Chopra et al. 2005).

The DSM psychiatric disorders do not provide any explanatory mechanisms for psychiatric *disorders*, in contrast to other medical specialties which group *diseases* according to pathological processes such as infection or malignancy (McHugh 2005). The potential result is that patients presenting with differing aetiologies and symptoms may receive an identical diagnosis, despite varying treatment and prognostic implications.

The heterogeneity contained within one seemingly discrete disorder is particularly pertinent in older people who may continue to suffer an 'early onset' (EO) disorder that appeared in earlier adulthood, or develop a namesake but 'late onset' (LO) disorder in the context of ageing. Yet neither the DSM, nor the International Classification of Disorders (ICD), lists any diagnostic sub-type specific to senescence, although both highlight disorders of childhood and adolescence (APA 1995; World Health Organisation 1992).

2 Clinical Presentations in Late Life

There are a number of important factors that confound the way psychiatric disorders present in older people, resulting in significant differences from clinical presentation in younger adults.

The first is the effect of ageing on symptom patterns in patients with an EO disorder. The literature is limited with most studies being cross-sectional and comparing different cohorts, so it is difficult to determine, for instance, whether presentation in late life of someone with chronic episodic depression is similar to that in young adulthood. In schizophrenia, which tends to be associated with more persistent symptoms, most patients are stable with a modest improvement in symptomatology (Jeste et al. 2003).

The second is the differing symptom profiles in LO compared to EO disorders. LO schizophrenia has been associated with greater persecutory delusions, less formal thought disorder, fewer negative symptoms (Brodaty et al. 1999; Howard et al. 1993), and less severe baseline cognitive deficits than EO schizophrenia (EOS; Rajii et al. 2009). Agoraphobia appearing in late life tends to be preceded by a physical insult rather than panic disorder as is the case in EO cases (Lindesay 1991). The phenomenology of LO depression, however, by and large is identical to EO cases (Brodaty et al. 2001). It should be noted that the literature has used varying age cut-offs to designate LO disorders, ranging from 50 to 60 years (Brodaty et al. 2001; Sachdev et al. 1999).

Thirdly, physical illness complicates the presentation, diagnosis, and management of psychiatric disorders in late life (Jeste et al. 2005). Symptoms of chronic physical disease may confound the diagnosis of anxiety and depressive disorders; for instance, it may be difficult to determine whether vegetative symptoms like insomnia and anergia are due to depression or chronic pain.

The fourth factor is the intimate association between cognitive impairment and psychiatric disorders in older people (Table 1). Whilst this is not unique to late life, the effects of cognitive impairment are accentuated with age and its relationship with psychiatric disorders is complex. Cognitive impairment may be an integral part of the index disorder. For instance, impaired memory and executive function are objectively observed in the acute phase of depression (O'Brien et al. 2004). In other cases, depression may be an early manifestation of dementia that is not apparent clinically or undetected by cognitive screening instruments. Hippocampal atrophy may be one possible pathology shared by LO depression and dementia (Lloyd et al. 2004).

The index psychiatric disorder may also increase the risk of subsequently developing a cognitive disorder. Longitudinal studies have shown an almost fivefold greater risk of dementia in late life depression sufferers (Alexopoulos et al. 1993), and that nearly half of patients with LO schizophrenia go on to develop dementia (Brodaty et al. 2003a, b).

Finally, psychiatric disorders may complicate the presentation of an elder with an established diagnosis of dementia. In particular, psychotic symptoms in older people are most commonly due to dementia (Henderson et al. 1998). The distinction is made on the basis of variables such as the severity of baseline cognitive impairment, temporal variation in the development of the respective symptoms and specific phenomenological differences.

3 Vulnerability and Resilience in Late Life

The aetiology of psychiatric disorders of late life can be considered broadly as related to the effects of (i) neurobiological vulnerability, including damage to salient neural networks; (ii) life events or psychological trauma; and (iii) temperamental or personality vulnerability. These factors are by no means mutually exclusive; older people with personality vulnerability may be more sensitive to the

Table 1 Relationship between cognition and psychiatric disorders in ageing

Cognitive impairment integral part of psychiatric disorders
Depressive disorder
Early onset schizophrenia
Anxiety disorders
Psychiatric disorder may be early manifestation of dementia
Depressive disorder
Anxiety disorders
Some disorders may increase risk of dementia
Depressive disorder
Very late onset schizophrenia
Alcohol abuse
Psychiatric disorders more common in dementia
Visual hallucinations and delusions, especially in Alzheimer's and Lewy body dementia
Depressive disorder, especially in vascular dementia
Anxiety disorders

effects of life events. Equally, there are certain factors that make older people resilient to the onset of psychiatric disorders.

Vascular disease has been perhaps the most extensively studied aetiological factor in psychiatric disorders of late life. The concept of 'vascular depression' arose from the observation of associations between stroke, vascular risk factors like hypertension and diabetes, and depression, typically without a positive family history of depression and appearing after the age of 50 (Alexopoulos et al. 1997). In addition, vascular depression has been associated with diffuse cerebrovascular changes demonstrated by T2-weighted magnetic resonance imaging (MRI), on which an increase in signal hyperintensities has been equated with vascular damage (Kales et al. 2005). Similar MRI findings have been reported in LO schizophrenia (Sachdev et al. 1999), and in late life mania (de Asis et al. 2006).

In vascular depression, there is evidence for both lesion location, in particular in the frontal lobes (Firbank et al. 2004) and basal ganglia (Simpson et al. 1998), and lesion severity (Taylor et al. 2003) being pivotal. Other studies have correlated specific neuroanatomical changes with genetic vulnerability. Variations in the genotype of the serotonin transporter-linked promoter region (HTLPR) have been associated with depletion of hippocampal (Taylor et al. 2005) and caudate nucleus (Hickie et al. 2007) volumes.

The importance of vascular aetiological factors should not be overstated. Some studies have reported vascular risk factors to be significant only in episodes of depression not preceded by a life event (Oldehinkel et al. 2003), or of equal importance to subjective social support (Steffans et al. 2005). This highlights the necessity of using a stress vulnerability model in the development of late life psychiatric disorders, in which biological and psychosocial factors are associated with eventual clinical presentation.

With regard to psychosocial factors, older studies showed an association between depression and severe life events, usually bereavement or life threatening illness (Murphy 1982). The role of social network is less clear; for instance, older bipolar patients' perception of inadequate social support was disproportionate to the actual size of their social network and the number of interactions (Beyer et al. 2003).

The latter may reflect the importance of the capacity for intimacy which could be protective against the onset, or exacerbation, of psychiatric disorders. The absence of a confidante, also linked to depression, was greater in those who had previously failed to establish any intimate relationship than those who were bereaved (Murphy 1982). Another protective factor that has been postulated in the light of some epidemiological findings that rates of anxiety and depressive disorders decline with age is the concept of psychological immunisation, that is repeated exposure to stress may increase individual resilience (reduce emotional response) to life events (Henderson 1994). A distinct process, increased emotional control, whereby coping skills improve with age, may also be protective (Jorm 2000).

Finally, it is essential to consider the role of physical illness in aetiology. Apart from cerebrovascular disease, a number of physical diseases such as Parkinson's disease (PD) and thyroid disease increase neurobiological vulnerability. At a psychological level, chronic physical illness may be associated with reactions such as grief and demoralisation (Wilhelm et al. 2004), whilst associated physical disability may lead to social withdrawal.

4 Epidemiology

Epidemiological studies from around the world demonstrate that about 12.7–16% of community dwellings with older adults report symptoms of sufficient severity to meet criteria for a psychiatric disorder (Sandanger et al. 1999; Préville et al. 2008, Regier et al. 1988; Trollor et al. 2007). This can be compared to the 12 month prevalence rates of any psychiatric disorder in adults over 18 years, of 26.2% in the US National Comorbidity Survey Replication (Kessler et al. 2005), and 20% in the Australian National Mental Health and Well-being Survey (Andrews et al. 2001).

Any decline with age may be due to cohort effects such as a ‘healthy survivor effect’, or methodological reasons (Jeste et al. 2005). The latter include the exclusion of residents from aged care facilities, the use of age inappropriate diagnostic criteria, misattribution of symptoms to physical illness, and under-reporting due to stigma and poor recall. Such studies also do not capture those elderly with dementia or physical illness who are at increased risk of developing a psychiatric disorder. It is important to go beyond these bare prevalence rates to obtain a broader view of psychiatric morbidity in older people.

The first caveat is the high rate of psychiatric disorder in residential aged care facilities which are not included in standard epidemiological surveys. In a study of 454 consecutive admissions to institutions, 67.4% suffered dementia and 10% depressive disorder (Rovner et al. 1990). Of the patients with dementia, 40% had additional psychiatric symptoms such as delusions or depression. A review of studies conducted in long-term care facilities found the median rate of depression was 10% and of dementia 58% (Seitz et al. 2010).

Two, there may be differences in prevalence within the broad cohort of elderly, in particular an increase in morbidity in the ‘old old’ (those 85 years and over) compared with the ‘young old’ (65–84 years). It has been noted that the exclusion of institutionalised elderly in epidemiological surveys may have a significant effect on rates in the old old in particular (Jorm 2000).

Studies comparing the rates of depression within elderly cohorts have produced conflicting results. In an US study, the old old were found to have higher rates of depressive symptoms, although this was accounted for by factors such as gender, physical disability, socioeconomic status, and cognitive impairment (Blazer 2000). A Swedish study found an increase in the prevalence of Major Depressive Disorder in those aged 85 (13%), compared with those aged 70 (5.6%) (Pálsson et al. 2001). However other studies have found that those aged 75 years and over were less likely to experience a depressive disorder compared with people aged

65–69 years (Trollor et al. 2007), or that the prevalence remained stable with increasing age (Forsell and Winblad 1999).

The rate of psychotic symptoms may increase in the old old, with a Swedish study reporting rates of psychotic symptoms of 10% in non-demented individuals, aged 85 years and over (Ostling and Skoog 2002). The same group reported that 8% of 70 year olds developed psychotic symptoms over a 20 year period, and the cumulative rate was 19.8% in those who survived to 85 years (Ostling et al. 2002).

Three, sub-syndromal (also known as sub-threshold) disorders may be more common in older people, reflecting the unsuitability of diagnostic criteria in this age group. In particular, although rates of Major Depressive Disorder (diagnosed by the presence of five or more symptoms over a 2 week period) tend to be lower in the elderly, the reverse occurs with rates of Minor Depression (two to four symptoms) or sub-syndromal depression (Chopra et al. 2005). A survey of respondents from 68 countries (mean age 43 years) calculated an overall prevalence of 13.2% for all depressive disorders and 2.85% for sub-syndromal depression (Ayuso-Mateos et al. 2010). In the elderly, rates for sub-syndromal depression have ranged from 4% in the ECA (Blazer et al. 1987) to 12.9% in the Netherlands (Beekman et al. 1995).

Four, co-morbidity, the presence of two or more disorders concurrently, is common in older people. In a Canadian study, 13.6% with depressive disorders had co-morbid anxiety, whilst 15.9% with an anxiety disorder fulfilled criteria for depression (Préville et al. 2008). Another study found that 23% of depressed elderly had a concurrent anxiety disorder, predominantly panic disorder (9.3%), social anxiety (6.6%) or a specific phobia (8.8%) (Lenze et al. 2000). Even higher rates of co-morbidity were found in the Longitudinal Aging Study Amsterdam—47.5% of people with major depressive disorder had a co-morbid anxiety disorder, while 26.1% of those with an anxiety disorder met criteria for major depressive disorder (Beekman et al. 2000).

Finally, late life psychiatric disorders may be associated with disproportionately higher levels of physical and functional disability, including significant implications for healthcare funding. Women with hip fractures and persistent depression were only 1/3 as likely as those with no or transient depression to walk independently again (Mossey et al. 1990). Older schizophrenia patients incur the highest per capita expense for Medicare and Medicaid amongst all disorders (Bartels et al. 2003).

5 Depressive Disorders

Depression in late life is a significant public health issue (Chapman and Perry 2008). It is often untreated, in part due to the reluctance of many older people to seek help because of poor knowledge about depression, stoicism, shame, stigma, and lack of a supportive social network to facilitate their pathway to care (Cole and Yaffe 1996; Lawrence et al. 2006; Penter and Other-Gee 2005). There is also poor detection of depression in primary care, particularly in the presence of co-morbid physical disease (Shah and Harris 1997).

The criteria used to diagnose depression do not vary with age, but the way that they are applied requires an understanding of the nuances of symptom interpretation (Chiu et al. 2009). Older people tend to under-report feelings of depression and may not acknowledge being sad or depressed. Hence, non-dysphoric depression is more prominent in late life, presenting with symptoms such as a loss of interest in life, lack of enjoyment in normal activities, fatigue, insomnia, weight loss, thoughts of death, chronic pain, poor concentration, or impaired memory—these may be incorrectly attributed to age, dementia, or physical illness by the older person, family, friends, and doctors (Gallo et al. 1997).

Both ends of the spectrum of depression severity are particularly important in late life. Psychotic depression is more prevalent in older people and generally requires treatment in hospital (Brodaty et al. 1991; Draper and Low 2009). Hypochondriacal and nihilistic delusions may result in presentations that include swallowing difficulties, weight loss, constipation, and conviction about suffering cancer. Milder forms of depression in late life tend to persist and cause considerable disability, possibly because many older people also have physical illness that is contributing to their mood change (Broadhead et al. 1990).

5.1 Aetiology

Early onset depression with recurrent episodes in late life should be distinguished from first episode depression in late life (LO depression). Recurrent EO depression is more likely to have genetic, personality, and adverse early life experience as causal factors, though recent health and psychosocial issues may also be relevant (Brodaty et al. 2001).

LO depression frequently has a close relationship to health problems and genetic predisposition is more likely to occur indirectly; for instance, it may be genetic vulnerability to cerebrovascular disease that increases the risk of depression (Hickie et al. 2001). LO depression may be the first manifestation of neuroendocrine disturbances like hypothyroidism; occult malignancy; vitamin deficiencies; anaemia; and infections. Chronic pain is frequently associated with both depression and suicide. Neurological disorders such as cerebrovascular disease, Alzheimer's disease and PD can predispose more directly to depression through structural brain damage, particularly in fronto-subcortical and hippocampal areas (Chiu et al. 2009). Medications may also cause or exacerbate depression, particularly antihypertensives, steroids, analgesics; similarly psychotropic agents like benzodiazepines (BZDs) and first generation antipsychotics may be associated with depression (Dhondt et al. 2002).

It is essential for anyone who develops depression for the first time in late life to have a thorough physical evaluation, including neuroimaging, particularly when no stressors are apparent (Chiu et al. 2009). Depression risk increases with both, the number of physical illnesses and their severity, but the main predictor is the presence of disability (Prince et al. 1997). The handicap of being unable to perform a usual social role and dependency on others may cause a loss of dignity, a sense of being a burden, and a fear of institutionalisation. Depression may be inappropriately assessed as a normal psychological reaction and left untreated.

The psychological impact of physical illness may also have a causal role in the development of depression (Prince et al. 1997; Vink et al. 2008). The diagnosis of a serious illness such as cancer or dementia may precipitate a depressive disorder in vulnerable individuals. Psychosocial factors that are linked with depression throughout adulthood are also relevant in late life (Vink et al. 2008). Losses such as the deaths of a partner, friends and pets, and the loss of independence, health, home, and lifestyle are cumulative and frequent in older people, although most cope well. Vulnerable older people have often had traumatic early life experiences such as child abuse, alcohol abuse or war-related trauma in a parent (Draper et al. 2008).

Social isolation in combination with physical disablement may result in loneliness and demoralisation, with depression being the result (Prince et al. 1997). Loneliness may relate to inability to maintain adequate social contact and might be resolved by organising social activity. In men, loneliness from the lack of a confidant, say after the death of a spouse, increases the vulnerability to the effects of severe life events (Emmerson et al. 1989). Satisfaction with social support is a significant protector against depression (Jang et al. 2002). Individuals with lower levels of mastery, smaller social networks, and less satisfaction with their support are more likely to become depressed.

5.2 Prognosis

Depression has a considerable range of impacts upon the quality of life of older people that might appear disproportionate to the apparent severity of symptoms. Social and family roles are affected with adverse outcomes such as premature retirement, withdrawal from social activities, and alienation of family and friends. Psychological suffering is associated with chronic loneliness, abuse of alcohol and other substances, and an increased risk of suicide (Chiu et al. 2009).

Poor outcome is associated with physical illness and disability, cognitive impairment, severity of depressive symptoms including psychosis, and inadequate treatment (Chiu et al. 2009). Frontal executive cognitive deficits might develop in vascular depression and merge almost imperceptibly into vascular dementia (Almeida 2008).

Depression in late life is a recurrent disorder (Chiu et al. 2009). Difficulties in coping are often exacerbated by accentuation of physical illness, overuse of medication, reduced self care, and poor nutrition (Braam et al. 2005; Draper and Anstey 1996). Depression also affects the prognosis of many physical illnesses including ischaemic heart disease and stroke (Penninx et al. 1999; Katon and

Ciechanowski 2002; Reynolds et al. 2008). Together these result in an increased use of health and social services as well as higher mortality rates (Unutzer et al. 1997; Penninx et al. 2001; Schulz et al. 2002; Adamson et al. 2005).

5.3 Suicide

In most countries, older people have the highest rates of suicide, but the pattern is variable and over the last 30 years there has been a general decline in late life suicide rates (Draper 2010). Suicidal behaviour in older people has a high risk of a fatal outcome for reasons that include lethality of the method used, high suicidal intent, physical frailty, and difficulties in detection of suicidal intention by health professionals.

Factors that increase suicide risk in older people range from distal early life issues such as childhood adversity, to proximal precipitants in late life such as social isolation, loneliness and physical ill health (Table 2). Suicide risk is increased in the three months after a diagnosis of cancer or dementia (Draper 2010).

Whilst studies consistently show that most late life suicides are clinically depressed, usually suffering Major Depressive Disorder, a significant minority are inflexible individuals who have difficulty coping with the challenges related to ageing including failing health, loss of confidants and demanding life events (Draper 2010).

Table 2 Risk factors for suicide in the elderly

Risk factors
Demographic
Male
Widowed or divorced
Caucasian
Low socio-economic status
Early Life
Childhood adversity
Biological
Recent diagnosis of cancer or dementia
Pain
Psychological
Depressive symptoms
Rigid, less open to experience personality
Substance abuse
Past suicidal behaviour
Social
Social isolation and loneliness
Family discord
Financial problems

6 Bipolar Disorder

Bipolar disorder in late life is considerably less common than late life unipolar depression, with subsequent research limitations such as small sample sizes, lack of longitudinal data, and overemphasis on hospitalised samples that have confounded findings (Depp and Jeste 2004). The prevalence of bipolar disorder declines with age and the vast majority of older sufferers have an EO disorder.

Several studies have shown a long latency between the onset of depression and the index episode of mania, with a mean gap of 15–22 years (Broadhead and Jacoby 1990; Shulman et al. 1992). Although the evidence for a vascular aetiology is less robust than in LO depression, there is a general support for an association between LO mania, vascular risk factors, and cerebrovascular disease (Cassidy and Carroll 2002; de Asis et al. 2006). Bipolar patients who were older at first psychiatric hospitalisation (>50 years) were more likely to present with episodes of psychotic depression and less likely to present with episodes of psychotic mania (Kessing 2006).

A prospective study of acute mania found that clinical symptomatology (behavioural and cognitive acceleration) and outcome did not differ by either chronological age or age of onset (>60 years), although older patients experienced less severe symptoms (Broadhead and Jacoby 1990). Similarly, a study of outpatients showed that, compared to EO (<18 years) cases, the LO (>40 years) group had a less severe form of bipolar disorder with reduction in psychotic features, number of mixed affective episodes and rate of co-morbidity with anxiety disorders (Schurhoff et al. 2000). Nevertheless, the prognosis for older people with bipolar disorder seems foreboding, with the rate of mortality in late life mania being 2.5 times greater than in unipolar depression (Shulman et al. 1992).

One possible reason for the lower rate of mania in late life may be related to the decline in dopamine neurotransmission with age (Kaasinen et al. 2000). The dopamine dysregulation hypothesis of bipolar disorder has posited increased dopaminergic drive in mania and the opposite in depression, for which PD provides a relevant model (Berk et al. 2007). Evidence from PD includes the increased rate of depression, the phenomenon of 'on–off' motor movements associated with apposite mood shifts, and the relatively high rate of hypersexuality and gambling due to dopaminergic agonists like pramipexole (Aiken 2007).

Mania-like syndromes which present with an associated physical aetiology for the first time in late life are more common than true LO mania. A vast number of cerebral and systemic aetiologies have been identified in parallel psychiatric and neurological literatures that have used overlapping terminologies like post-stroke mania, disinhibition syndrome and secondary mania, syndromes that do not always meet full duration and symptom criteria for mania (Shulman 1997). The significance of these reported aetiologies is limited by the tendency for papers to consist of case reports or series.

Management principles have been largely extrapolated from studies of adult cohorts.

7 Psychotic Disorders

7.1 Schizophrenia

Schizophrenia is usually regarded as a disorder of young people, but this popular belief fails to recognise a number of lesser known facts about ageing and schizophrenia: elderly schizophrenics are not uncommon in the population and may be at special risk in relation to their clinical and service needs; schizophrenia may have an onset late and very-late in life; and psychosis is a common concomitant of dementia, the prevalence of which continues to increase in the population.

Very few studies have attempted to estimate the prevalence of schizophrenia in late life. The ECA Study (Goldman and Manderscheid 1987) reported a prevalence of schizophrenia of 0.3% in individuals of age 65 years and over, but this could have been an under-estimate due to a sampling bias against areas likely to have more concentrations of older persons with chronic mental illness. Other investigators have estimated the prevalence from 0.1 to 1% (Copeland et al. 1998; Kua 1992), with the majority (about 85%) of these individuals living in the community. Elderly schizophrenia patients have generally grown old with their illness—90% have an EO disorder (<50 years), 7% onset in the sixth decade and 3% in later decades (Howard et al. 2000). Using a register of contact with psychiatric services, the incidence rate of late-onset schizophrenia (LOS) ($\geq$45 years) was estimated as 12.6 per 100,000 population per year (Copeland et al. 1998).

Since the majority of older sufferers have EOS, how does the disorder change with ageing? The longitudinal course of schizophrenia is quite heterogeneous, but in general, positive symptoms tend to decrease with age (Ciompi 1980). The course of negative symptoms is more variable, with some studies reporting improvement and others worsening with ageing (McGlashan and Fenton 1992; Schultz et al. 1997). This is, however, confounded by factors such as institutionalisation, chronic medication use, demoralisation, and poverty (Davidson et al. 1995), as well as the high rates of depression in this population (Cohen et al. 1996).

Older schizophrenia patients show significant cognitive deficits, which are considered to be a combination of deficits in early life that are part of schizophrenia, compounded by ageing and a lifetime of deprivation. The deficits are not as severe as those seen in Alzheimer's disease, and there is no evidence to suggest that EO patients are more prone to progressive cognitive decline later in life (Heaton et al. 1994). Most older schizophrenic patients continue to show varying degrees of disability because of the cognitive deficits and varying degrees of psychopathology, but some become self-supportive (Hafner et al. 1995).

These patients have high rates of physical illness co-morbidity, in particular cardiovascular disease and diabetes, and their mortality from 'natural' causes is higher than the general population (Simpson and Tsuang 1996). Mortality from non-natural causes (suicide, homicide, or accident) is also higher, although not as high as in younger sufferers. Illicit drug use is less common in older compared to younger schizophrenia patients.

7.2 Very Late Onset Schizophrenia (VLOS)

Onset of schizophrenia after the age of 60 years is called very late onset schizophrenia (VLOS) which accounts for about 3% of all cases of schizophrenia and has received considerable attention in the literature, going back to the 1960s when it was referred to as 'late paraphrenia' (Kay and Roth 1961). The clinical presentation is not dissimilar to EOS (Brodaty et al. 1999), although various authors have been struck by less affective flattening, greater likelihood of paranoid subtype, higher frequency of visual, tactile, or olfactory hallucinations, fewer negative symptoms and less formal thought disorder (Jeste et al. 1995; Howard et al. 1993).

A striking feature of VLOS is the over-representation of women by a factor of 2–10 in various studies. This finding has been difficult to explain although it is consistent with the observation of earlier onset in younger men; estrogen-mediated dopaminergic inhibition has been suggested to be neuroprotective in young women, with estrogen deficiency in late life making a possible contribution to the high rate of VLOS. Patients with VLOS have also been reported to have higher rates of sensory impairment, particularly deafness (Cooper and Curry 1976). While VLOS patients tend to have higher premorbid educational and occupational functioning, they have been reported to have schizoid and paranoid personality traits leading to social isolation and eccentricity as a marker of their premorbid personalities (Kay and Roth 1961; Jeste et al. 1995, Kay et al. 1976).

It has been suggested that VLOS may represent a manifestation of 'organicity', with the implication that it is a mock-up of the 'true' EOS. Familial aggregation of VLOS is less common than in EOS (Howard et al. 1997). Evidence of increased abnormalities on neuroimaging and electroencephalography in VLOS in comparison with EOS has not been consistent, especially when care was taken to exclude patients with known neurological disease (Howard et al. 2000). Structural brain abnormalities in VLOS are similar to those seen in younger patients, with the possible exception of stigmata of small vessel disease (Howard et al. 1995; Sachdev and Brodaty 1999). Hypoperfusion in the frontal and temporal lobes has been reported (Lesser et al. 1993; Dupont et al. 1994), but the adequacy of control samples has been questioned, and the findings are again not qualitatively different from EOS. The cognitive deficits seen in VLOS—in executive function, memory and learning, motor skills, and verbal ability—appear similar to those in EOS, but are qualitatively and quantitatively different from those seen in dementia. A family history of dementia is not higher, and the apolipoprotein E4 genotype is not more frequent (Howard et al. 1995). The evidence, therefore, does not overwhelmingly support the hypothesis that VLOS is an organic disorder.

Another approach taken by investigators to determine organicity is to examine the longitudinal course of VLOS. The results have again been inconsistent, with both stability and decline in cognitive function reported (Holden 1987; Palmer et al. 2003). In a longitudinal study of LOS (>50 years), the LOS and matched EOS subjects did not differ in their cognitive profiles at baseline and 1-year follow-up, but at 5 years a large proportion of the LOS group showed decline and 9/15 were diagnosed with dementia, predominantly Alzheimer's disease (Brodaty et al. 2003a). This raises the possibility that VLOS is a heterogeneous disorder, with a subgroup representing early neurodegenerative disease.

7.3 Delusional Disorder

In delusional disorder, the core feature is the presence of persistent non-bizarre delusions, usually with themes of persecution, grandiosity, jealousy, disease, or erotomania. The onset of the delusions is often gradual and there may be an element of plausibility in the symptoms. Auditory and visual hallucinations, if present, are not prominent. The individual's behavior is not markedly impaired, apart from the impact of the delusion.

The epidemiology of delusional disorder has been inadequately studied. The DSM-IV-TR estimates the population prevalence of delusional disorder at about 0.03%. The onset is in middle to late life, approximately two decades later than schizophrenia. The prevalence is higher in women, and the age of onset in women (60–69 years) is later than in men (40–49 years).

The risk factors for delusional disorder are not well understood. Family and genetic studies suggest that it is distinct from schizophrenia (Kendler 1980). Cerebral insult from a variety of causes, such as brain trauma, stroke, brain tumor, alcohol and substance abuse, increases the risk. Other risk factors include low socioeconomic status, sensory isolation, and immigration status. The course of the disorder is often chronic, especially in the persecutory type. In others, it may fluctuate, and periods of remission may occur.

8 Anxiety Disorders

Anxiety disorders appear to be less common in late life compared with younger adulthood (Flint 1994). Prevalence rates range from 2 to 11.6% (Streiner et al. 2006; Beekman et al. 1998; Forsell and Winbland 1997; Byers et al. 2010).

The National Co-morbidity Survey Replication found a lifetime prevalence of anxiety of 15.3% in those aged over 60 years (Kessler et al. 2005) with specific phobia the most common anxiety disorder (6.5%) followed by social phobia (3.5%). In contrast, the overall lifetime prevalence of anxiety disorders was 28.8% in younger adults, while the 12-month prevalence was found to be 18.1% (Kessler et al. 2005). A Canadian study of community dwelling people aged over 65 years found a 12 month prevalence of anxiety disorder of 5.6% with Specific Phobia (2%), Obsessive Compulsive Disorder (1.5%) and Generalised Anxiety Disorder of (1.2%) being most common (Préville et al. 2008).

A number of possible explanations for the general finding of lower prevalence rates in older adults have been discussed, and it is also possible the screening instruments used may not measure the experience of anxiety in the elderly (Kogan et al. 2000; Jorm 2000). Self-report measures of anxiety have not been validated in this age group (Fuentes and Cox 2000; Kogan et al. 2000). It has been suggested that current diagnostic criteria may under-estimate the rate of anxiety disorders in older people (Palmer et al. 1997). Presentations of anxiety such as tachycardia and hyperventilation may be attributed to physical illness, and avoidance behaviour (such as agoraphobia) to physical disability. Disentangling the relative contribution of physical and psychological factors is difficult, and thorough physical assessment is recommended in LO anxiety.

Distinguishing symptoms of anxiety from those of depression is another challenge, and anxiety symptoms may be attributed to depression and remain untreated. Given the significant rates of co-morbidity for depression and anxiety in elderly people, both disorders should be considered. Screening for cognitive impairment should also take place, as agitation associated with early memory loss may be mistaken for anxiety (Kogan et al. 2000).

9 Substance Use Disorders

The main substance of abuse in the elderly is alcohol, although the relatively low rate of DSM-IV alcohol abuse or dependence (1% in women and 4.8% in men) seems a reflection of more restrictive diagnostic criteria (Grant et al. 2004). In contrast, the respective gender prevalence rates for 'at risk' drinking (two or more drinks at one sitting) of 8 and 13%, and for 'binge' drinking (five or more drinks at one sitting) of 3 and 14% are more relevant to physical and psychiatric morbidity, and mortality (Blazer and Wu 2009). Binge drinking was also associated with nicotine and illicit drug use in both genders, and non-medical use of prescribed medications in females.

There may be two types of older people with alcohol use problems (McGrath et al. 2005). The first, an EO group, account for about 70% of older alcohol abusers. The LO group has been associated with higher income levels and onset of alcohol abuse after a major stressor.

Alcohol may be a marker of mood, anxiety, and impulse control disorders. Recognition of alcohol abuse may be more difficult in older people (McInnes and Powell 1994). Warning signs in younger adults such as erratic behaviour may be attributed to ageing or cognitive impairment, while retirement generally results in fewer cognitive demands and less scrutiny. Other pointers such as poor co-ordination, malnutrition, depressed mood, drowsiness, or incoordination can be similarly explained by co-morbid physical illness. The general problem of patients underestimating the amount of alcohol consumption may be compounded by cognitive impairment.

Treatment of alcohol abuse in the elderly follows the same principles as other adults, and the elderly have been shown to have comparable or better outcomes compared with younger adults (Oslin et al. 2002). A proportion of patients may reduce drinking once the associated psychiatric disorder has been treated, although brief alcohol counselling interventions may further improve outcomes (Oslin 2005). The use of the opioid antagonist, naltrexone, halved the rate of relapse into heavy drinking in a group of 50–70 year old veterans (Oslin et al. 1997).

The misuse of prescribed sedative, anxiolytic, and analgesic medications, in particular the BZD, may be a hidden problem in the elderly. BZD use in the elderly is common, with 7.5–13.7% of persons aged over 65 taking at least one agent in the US (Gleason et al. 1998; Aparasu et al. 2003). Residents of nursing homes have especially high rates of BZD prescription, with 32–41% of psychotropic prescriptions in a US study being for anxiolytics, mainly BZDs (Beardsley et al. 1989; Holmquist et al. 2003). An Austrian sample of 75 year olds showed that the prescription of BZDs was as common as that of antidepressants and often indicated inappropriate treatment of late life depression (Assem-Hilger et al. 2009). The detection of BZD abuse may be even lower than that of alcohol use in the elderly because users tend to be female (Bogunovic and Greenfield 2004).

Alcohol and BZD withdrawal should be considered in cases of delirium (acute confusional state), and the potential for withdrawal noted when older patients are admitted to hospital, regardless of the patient's estimations of their intake. Indicators of alcohol withdrawal include autonomic arousal, and in more severe cases, delirium, and seizures. Elderly patients with BZD withdrawal are more likely to present with confusion than anxiety, insomnia, or perceptual changes (Foy et al. 1986).

Finally, a number of epidemiological surveys have reported low rates of illicit substance (in particular cannabis and stimulant) use disorders in older people (Blazer et al. 1987; Kessler et al. 2005). The National Comorbidity Survey Replication, however, has reported increases in the predicted lifetime prevalence of all psychiatric disorders in more recent cohorts, with the largest increase in illicit substance abuse and dependence (Kessler et al. 2005). Compared to respondents aged 60 years and over at the time of interview, people aged 45–59 were nearly 25 times, and those aged 30–44 were nearly 50 times, more likely to be diagnosed with a substance use disorder.

10 Somatoform Disorders

The somatoform disorders are a particularly troublesome category that overlaps both internal medicine and psychiatry. Their essence is the presentation with a physical (i.e. somatic) symptom(s) for which there is no, or an inadequate, pathophysiological explanation. The DSM classification is heterogenous, consisting of some high threshold, low prevalence disorders such as Somatisation Disorder or Conversion Disorder that are more likely to present in general hospital settings, and disorders such as Hypochondriasis and Body Dysmorphic Disorder which may be better placed within the anxiety disorders (APA 1995).

Only a small proportion of the common somatic symptoms, such as abdominal pain, chest pain, and fatigue, that present to primary care physicians are associated with a clear physical explanation despite thorough and costly investigation (Kroenke and Mangelsdorff 1989). Indeed each sub-speciality of internal medicine is characterised by a common symptom cluster of indeterminate physical aetiology, termed the functional somatic syndromes (FSS) and equivalent to the DSM diagnosis of Undifferentiated Somatoform Disorder—irritable bowel syndrome, chronic fatigue syndrome, fibromyalgia, atypical chest pain, chronic pain syndrome, and so on (Wessely et al. 1999).

Traditionally, somatic symptoms in older people were assumed to be physical in origin, or else indistinguishable from depressive and anxiety (psychological) disorders. Yet the presence of a psychological disorder was a greater predictor of chronic fatigue than physical illness in an older primary care sample (Wijeratne et al. 2007). Further, rather than evolving into a psychological disorder or vice versa, cases of chronic fatigue were longitudinally stable, suggesting that chronic fatigue could be conceptualised as a disorder independent of anxiety and depression. Using exploratory factor analysis, a somatic symptom factor distinct from mood and cognitive symptom factors could also be derived, suggesting the independence of somatic symptoms (Wijeratne et al. 2007; Wijeratne et al. 2006).

Treatment can only be extrapolated from studies of younger adults (Wijeratne et al. 2003). Antidepressants are beneficial in many of the FSS except fatigue syndromes, whilst cognitive behaviour therapy (CBT) and graded exercise therapy are also effective. Prevention consists of using a holistic, bio-psychosocial aetiological model for all presentations, judicious use of physical investigations and specialist referral, and avoiding the use of spurious diagnoses and treatments.

11 Personality Disorders

Personality disorders represent longstanding and persistent dysfunction in an individual's sense of self and functioning within interpersonal relationships, distinct from psychiatric disorders which represent a break in usual function. Personality disorders may present independently, co-exist with, or colour substance use, anxiety, mood, and psychotic disorders.

A diagnosis of personality disorder was a predictor of poorer outcome in elderly inpatients with depressive disorder (Stek et al. 2002). DSM personality disorders were encountered in up to one-third of older people attending a specialist psychiatric clinic, with avoidant, dependent or paranoid sub-types being the most common (Casey and Joyce 1999).

Whether personality becomes less problematic with age is dependent on whether a categorical or dimensional model of personality is used, and the specific characteristics measured. For instance, neuroticism is a dimensional measure of a tendency to be tense and more sensitive to stress but is correlated positively with depression scores (Mulder 2002), so that the decline in neuroticism with age may reflect an apposite decline in depression (Steunenberg et al. 2005). The Temperament and Character Inventory measures four dimensional personality variables—harm avoidance, novelty seeking, reward dependence, and persistence (Cloninger 1994). A study of twins aged 50–96 years showed significant declines with increasing age for all the dimensions except harm avoidance which was stable (Heiman et al. 2003).

Poor recall and the potential lack of a corroborative historian able to describe lifelong difficulties may under-estimate the diagnosis of personality disorder in older people, whilst diagnostic criteria may also be inappropriate. For instance, physical frailty may impede impulsive or reckless behaviours although disturbances of identity and the self may remain (Abrams and Bromberg 2006). Age specific presentations of impulsive (narcissistic, borderline, antisocial) personalities may include refusal to eat or take medication, and splitting of staff in residential care facilities.

The relationship between personality pathology and increasing age may be best represented by a reverse J shaped curve. To again take the example of the impulsive personality type, there may be a decline in distress in middle age because of neurological or psychological maturation, plus certain social and role buffers allowing distress to be contained (Links et al. 1990). There may then be a levelling or even slight increase of personality vulnerability, once old age is reached, with vulnerable elders being unable to emotionally counter losses such as retirement, bereavement, and the individuation of children (Agronin 1994).

12 Management

A seven tiered model of managing psychiatric disorders in older people, incorporating evidence-based interventions that aim to avert individuals from moving up tiers (prevention) and to move individuals down tiers (treatment) has been proposed (Brodaty et al. 2003b). The model embraces population-based prevention of psychiatric disorders in the lower tiers, as well as the institutional care of the most severely disturbed in the higher tiers. It also presents a model that allows for the titration of the intensity of intervention by specialist and primary care services according to the severity of the disorder.

Preventative strategies that have been proposed in late life depression include increasing both the presentation of sufferers in primary care and detection by practitioners via strategies aimed at improving community mental health literacy and practitioner knowledge; public health campaigns encouraging regular physical exercise; and nutritional supplementation such as folate added to grains (Bird and Parslow 2002).

Management per se of each disorder is an individual amalgam of physical, pharmacological, psychological, and social therapies.

The adequate management of physical health is paramount. Addressing risk factors such as hypertension and diabetes mellitus associated with cerebrovascular disease may help prevent onset and delay progression of depressive, bipolar, and psychotic disorders. Similarly, reducing chronic pain, treating urinary incontinence or improving physical mobility can help improve mood and anxiety. Interventions such as exercise groups may have a specific effect on mood (Singh et al. 2001), whilst community groups will reduce social isolation and encourage the formation of supportive relationships.

The psychotropic medications used in older people, apart from the cholinesterase inhibitors used in dementia, are similar to those used in young adults. There are, however, important caveats in their use in older people. Age related pharmacokinetic and pharmacodynamic changes, and polypharmacy in a population with multiple physical illnesses confound drug prescription (Moltke et al. 2005). Drug adherence may be affected by general issues like financial strain and the difficulty in engaging people with severe mental disorders, and by age specific issues such as cognitive impairment.

The long held wisdom of dosing, to start low and go slow, is essential given the greater sensitivity to adverse effects, whether it be the extrapyramidal side-effects associated with antipsychotic drugs, the falls or sedation associated with these and antidepressant drugs. Doses may need to be 1/4 to 1/2 of those used in younger patients, and novel antidepressant and antipsychotic drugs are preferable because of their improved tolerance. Further, medication may be slower to work in older people, so longer trials are required to determine efficacy (Brodaty et al. 1993).

The actual outcomes of psychotropic medication use in older people are unclear. The example of the anxiety disorders, in which medication treatments tend to be extrapolated from studies of younger adults and their efficacy unexamined in older people (Sheikh and Cassidy 2000), is not unusual. Similarly, very few controlled studies have examined treatment response in very LO schizophrenia, but open studies suggest that the response rate to antipsychotics is good, with full remission in 48–61% (Howard et al. 2000).

The usual range of psychotherapeutic treatments may be used in the elderly, either as a primary or adjunctive therapy. The efficacy of CBT has been demonstrated for anxiety disorders in adults (Butler et al. 2006) and has been used in cognitively impaired (Koder 1998) and cognitively intact elderly (Stanley and Novy 2000), although there is a paucity of controlled studies examining efficacy in the elderly.

12.1 Psychiatric Services in Late Life

The provision of psychiatric services to older people shares much in common with delivery of other aspects of healthcare to the elderly. Common features of an effective care system include a single entry point; case management; assessment and involvement of a multidisciplinary team; and use of financial incentives to encourage less expensive community based care (Johri et al. 2003). These features are also mentioned in the World Psychiatric Association's consensus statement on the organisation of care in the psychiatry of the elderly (Wertheimer 1997). The principles of good quality psychiatric care for older people within the consensus statement embrace the following concepts—comprehensive, accessible, responsive, individualised, transdisciplinary, accountable, and systemic (Chiu 2005).

Although there have been no system-wide evaluations, individual components of service delivery have been evaluated (Draper and Low 2004). The strength of the evidence is variable, ranging from strong evidence for community multidisciplinary teams to relatively weak evidence for acute hospital care, but this is mainly due to a lack of controlled studies rather than evidence of ineffectiveness. Integration of acute hospital and community care has been shown to improve outcomes following hospital discharge. The limited evidence also suggests that specialised old age psychiatric services provide more effective psychiatric care to older people than geriatric medicine, adult psychiatric services, and primary care (Draper and Low 2004).

13 Conclusion

The relatively young sub-speciality of geriatric psychiatry has achieved much already, but faces a number of significant challenges. With the baby boomer generation just entering old age, population-based models of service delivery with a public health focus that include health promotion and disease prevention will be required (Cole 2002). Changes in the nature of younger cohorts will also have implications for the epidemiology of psychiatric disorders in older people in the coming decades, in particular for illicit substance abuse and dependence (Kessler et al. 2005). Further refinements in diagnostic criteria are needed to more accurately determine the true prevalence of psychiatric morbidity in older people. Finally, there is a need for formal controlled trials of pharmacological and psychological therapies in this age group.

Acknowledgement Angie Russell assisted with final preparation of references.

References

Abrams R, Bromberg C (2006) Personality disorders in the elderly: a flagging field of inquiry. Int J Geriatr Psychiatry 21:1013–1017

Adamson JA, Price GM, Breeze E, Bulpitt CJ, Fletcher AE (2005) Are older people dying of depression? Findings from the medical research council trial of the assessment and management of older people in the community. J Am Geriatr Soc 53:1128–1132

Agronin M (1994) Personality disorders in the elderly: an overview. J Geriatr Psychiatry 27:151–191

Aiken C (2007) Pramipexole in psychiatry: a systematic review of the literature. J Clin Psychiatry 68:1230–1236

Alexopoulos G, Meyers B, Young R, Mattis S, Kakuma T (1993) The course of geriatric depression with "reversible dementia": a controlled study. Am J Psychiatry 150:1693–1699
Alexopoulos A, Meyers B, Young R, Campbell S, Silbersweig D, Charlson M (1997) "Vascular depression" hypothesis. Arch Gen Psychiatry 54:915–922
Almeida OP (2008) Vascular depression: myth or reality? Int Psychogeriatr 20:645–652
American Psychiatric Association (1995) Diagnostic and statistical manual of mental disorders, 4th edn. American Psychiatric Association, Washington
Andrews G, Henderson S, Hall W (2001) Prevalence, comorbidity, disability and service utilisation. Overview of the Australian National Mental Health Survey. Br J Psychiatry 178:145–153
Aparasu R, Mort J, Brandt H (2003) Psychotropic prescription use by community-dwelling elderly in the United States. J Am Geriatr Soc 51:671–677
Assem-Hilger E, Jungwirth S, Weissgram S, Kirchmeyr W, Fischer P, Barnas C (2009) Benzodiazepine use in the elderly: an indicator for inappropriately treated geriatric depression? Int J Geriatr Psychiatry 24:563–569
Ayuso-Mateos J, Nuevo R, Verdes E, Naidoo N, Chatterji S (2010) From depressive symptoms to depressive disorders: the relevance of thresholds. Br J Psychiatry 196:365–371
Bartels SJ, Clark RE, Peacock WJ et al (2003) Medicare and medicaid costs for schizophrenia patients by age cohort compared with costs for depression, dementia, and medically ill patients. Am J Geriatr Psychiatry 11:648–657
Beardsley RS, Larson DB, Burns BJ, Thompson JW, Kamerow DB (1989) Prescribing of psychotropics in elderly nursing home patients. J Am Geriatr Soc 37:327–330
Beekman A, Deeg D, van Tilburg T, Smit J, Hooijer C, van Tilburg W (1995) Major and minor depression in later life: a study of prevalence and risk factors. J Affect Disord 36:65–75
Beekman A, Bremmer M, Deeg D, Van Balkom A, Smit J, De Beurs E, Van Dyck R, Van Tilburg W (1998) Anxiety disorders in later life: a report from the longitudinal aging study Amsterdam. Int J Geriatr Psychiatry 13:717–726
Beekman ATF, de Beurs E, van Balkom AJ, Deeg DJH, van Dyck R, van Tilburg W (2000) Anxiety and depression in later life: co-occurrence and communality of risk factors. Am J Psychiatry 157:89–95
Berk M, Dodd S, Kauer-Sant'Anna M, Malhi G, Bourin M, Kapczinski F, Norman T (2007) Dopamine dysregulation syndrome: implications for a dopamine hypothesis of bipolar disorder. Acta Psychiatr Scand 116(Suppl 434):41–49
Beyer J, Kuchibhatia M, Looney C, Engstron E, Cassidy F, Krishnan K (2003) Social support in elderly patients with bipolar disorder. Bipolar Disord 5:22–27
Bird M, Parslow R (2002) Potential for community programs to prevent depression in older people. Med J Aust 177(Suppl):S107–S110
Blazer D (2000) Psychiatry and the oldest old. Am J Psychiatry 157:1915–1924
Blazer D, Wu L-T (2009) The epidemiology of at-risk and binge drinking among middle-aged and elderly community adults: national survey on drug use and health. Am J Psychiatry 166:1162–1169
Blazer D, Hughes D, George L (1987) The epidemiology of depression in an elderly community population. Gerontologist 27:281–287
Bogunovic O, Greenfield S (2004) Practical geriatrics: use of benzodiazepines among elderly patients. Psychiatr Serv 55:233–235
Braam AW, Prince MJ, Beekman ATF, Delespaul P, Dewey ME, Geerlings SW et al (2005) Physical health and depressive symptoms in older Europeans. Br J Psychiatry 187:35–42
Broadhead J, Jacoby R (1990) Mania in old age: a first prospective study. Int J Geriatr Psychiatry 5:215–222
Broadhead WE, Blazer DG, George LK, Tse CK (1990) Depression, disability days, and days lost from work in a prospective epidemiological survey. JAMA 264:2524–2528
Brodaty H, Peters K, Boyce P, Hickie I, Parker G, Mitchell P, Wilhelm K (1991) Age and depression. J Affect Disord 23:137–149

Brodaty H, Harris L, Peters K (1993) Prognosis of depression in the elderly: a comparison with younger patients. Br J Psychiatry 163:589–596

Brodaty H, Sachdev P, Rose N, Rylands K, Prenter L (1999) Schizophrenia with onset after age 50 years. I: Phenomenology and risk factors. Br J Psychiatry 175:410–415

Brodaty H, Luscombe G, Parker G, Wilhelm K, Hickie I, Austin MP, Mitchell P (2001) Early and late onset depression in old age: different aetiologies, same phenomenology. J Affect Disord 66:225–236

Brodaty H, Sachdev P, Koschera A, Monk D, Cullen B (2003a) Long-term outcome of late-onset schizophrenia: 5-year follow-up study. Br J Psychiatry 183:213–219

Brodaty H, Draper B, Low LF (2003b) Behavioural and psychological symptoms of dementia: a seven-tiered model of service delivery. Med J Aust 178:231–234

Butler A, Chapman J, Forman E, Beck A (2006) The empirical status of cognitive-behavioural therapy: a review of meta-analyses. Clin Psychol Rev 26:17–31

Byers AL, Yaffe K, Covinsky KE, Friedman MB, Bruce ML (2010) High occurrence of mood and anxiety disorders among older adults: The National Comorbidity Survey Replication. Arch Gen Psychiatry 67:489–496

Casey J, Joyce P (1999) Personality disorder and the temperament and character inventory in the elderly. Acta Psychiatr Scand 100:302–308

Cassidy F, Carroll B (2002) Vascular risk factors in late onset mania. Psychol Med 32:359–362

Chapman DP, Perry GS (2008) Depression as a major component of public health for older adults. Prev Chronic Dis 5:1–9

Chiu E (2005) Principles and best practice model of psychogeriatric service delivery. In: Draper B, Melding P, Brodaty H (eds) Psychogeriatric service delivery: an international perspective. Oxford University Press, Oxford

Chiu E, Ames D, Draper B, Snowdon J (2009) Depressive disorders in the elderly: a review. In: Herrman H, Maj M, Sartorius N (eds) Depressive disorders, 3rd edn. Wiley, Chichester

Chopra M, Zubritsky C, Knott K, Have TT, Hadley T, Coyne J, Oslin D (2005) Importance of subsyndromal symptoms of depression in elderly patients. Am J Geriatr Psychiatry 13:597–606

Ciompi L (1980) Catamnestic long-term study on the course of life and aging of schizophrenics. Schizophr Bull 6:606–618

Cloninger R (1994) The temperament and character inventory (TCI): a guide to its development and use. Center for Psychobiology of Personality, Washington University, St Louis

Cohen CI, Talavera N, Hartung R (1996) Depression among older persons with schizophrenia who live in the community. Psychiatr Serv 47:601–607

Cole MG (2002) Public health models of mental health care for elderly populations. Int Psychogeriatr 14:3–6

Cole MG, Yaffe MJ (1996) Pathway to psychiatric care of the elderly with depression. Int J Geriatr Psychiatry 11:157–161

Cooper AF, Curry AR (1976) The pathology of deafness in the paranoid and affective psychoses of later life. J Psychosom Res 20:107–114

Copeland JRM, Dewey ME, Scott A, Gilmore C, Larkin BA, Cleave N, McCracken CFM, McKibbin PE (1998) Schizophrenia and delusional disorder in older age: community prevalence, incidence, comorbidity and outcome. Schizophr Bull 24:153–161

Davidson M, Harvey PD, Powchik P et al (1995) Severity of symptoms in chronically institutionalized geriatric schizophrenic patients. Am J Psychiatry 152:197–207

de Asis JM, Greenwald B, Alexopoulos G, Kiosses D, Ashtari M, Heo M, Young R (2006) Frontal signal hyperintensities in mania in old age. Am J Geriatr Psychiatry 14:598–604

Depp C, Jeste D (2004) Bipolar disorder in older adults: a critical review. Bipolar Disord 6:343–367

Dhondt TD, Beekman AT, Deeg DJ, Van Tilburg W (2002) Iatrogenic depression in the elderly. Results from a community-based study in the Netherlands. Social Psychiatry Psychiatr Epidemiol 37:393–398

Draper B (2010) Suicidal behaviour. In: Abou-Saleh M, Katona C, Kumar A (eds) Principles and practice of geriatric psychiatry, 3rd edn. Wiley, Chichester

Draper B, Anstey K (1996) Psychosocial stressors, physical illness and the spectrum of depression in elderly inpatients. Aust N Z J Psychiatry 30:567–572

Draper B, Low LF (2004) What is the effectiveness of old age mental health services?. World Health Organization Regional Office for Europe Health Evidence Network, Copenhagen

Draper B, Low LF (2009) Patterns of hospitalisation for depressive and anxiety disorders across the lifespan. Aust J Affect Disord 113:195–200

Draper B, Pfaff J, Pirkis J, Snowdon J, Lautenschlager N, Wilson I, Almeida O (2008) The long-term effects of childhood abuse on the quality of life and health of older people: results from the DEPS-GP project. J Am Geriatr Soc 56:262–271

Dupont RM, Lehr PP, Lamoureaux G, Halpern S, Harris MJ, Jeste DV (1994) Preliminary report: cerebral blood flow abnormalities in older schizophrenic patients. Psychiatry Res 55:121–130

Emmerson JP, Burvill PW, Finlay-Jones R, Hall W (1989) Life events, life difficulties and confiding relationships in the depressed elderly. Br J Psychiatry 155:787–792

Firbank M, Lloyd A, Ferrier N, O'Brien J (2004) A volumetric study of MRI signal hyperintensities in late-life depression. Am J Geriatr Psychiatry 12:606–612

Flint AJ (1994) Epidemiology and comorbidity of anxiety disorders in the elderly. Am J Psychiatry 151:640–649

Forsell Y, Winblad B (1999) Incidence of major depression in a very elderly population. Int J Geriatr Psychiatry 14:368–372

Forsell Y, Winbland B (1997) Anxiety disorders in non-demented and demented elderly patients: prevalence and correlates. J Neuropsychol Neurosurg Psychiatry 62:294–295

Foy A, Drinkwater V, March S, Mearrick P (1986) Confusion after admission to hospital in elderly patients using benzodiazepines. Br Med J Clin Res Ed 293:1072

Fuentes K, Cox B (2000) Prevalence of anxiety disorders in elderly adults: a critical analysis. J Behav Ther Exp Psychiatry 28:269–279

Gallo JJ, Rabins PV, Lyketsos CG et al (1997) Depression without sadness: functional outcomes of nondysphoric depression in later life. J Am Geriatr Soc 45:570–578

Gleason P, Schultz R, Smith N, Newsom J, Kroboth P, Kroboth F, Psaty B (1998) Correlates and prevalence of benzodiazepine use in community-dwelling elderly. J Gen Intern Med 13:243–250

Goldman HH, Manderscheid RW (1987) Epidemiology of chronic mental disorder. In: Menninger W, Hannah G (eds) The chronic mental patient, vol II. American Psychiatric Press, Washington

Grant B, Dawson D, Stinson F, Chou S, Dufour M, Pickering R (2004) The 12-month prevalence and trends in DSM-IV alcohol abuse and dependence: United States, 1991–1992 and 2001–2002. Drug Alcohol Depend 74:223–234

Hafner H, Nowotny B, Loffler W et al (1995) When and how does schizophrenia produce social deficits? Eur Arch Psychiatry Clin Neurosci 246:17–28

Heaton R, Paulsen JS, McAdams LA et al (1994) Neuropsychological deficits in schizophrenics. Arch Gen Psychiatry 51:469–476

Heiman N, Stallings M, Hofer S, Hewitt J (2003) Investigating age differences in the genetic and environmental structure of the tridimensional personality questionnaire in later adulthood. Behav Genet 33:171–180

Henderson A (1994) Does ageing protect against depression? Soc Psychiatry Psychiatr Epidemiol 29:107–109

Henderson AS, Korten AE, Levings C, Jorm AF, Christensen H, Jacomb PA, Rodgers B (1998) Psychotic symptoms in the elderly: a prospective study in a population sample. Int J Geriatr Psychiatry 13:484–492

Hickie I, Scott E, Naismith S, Ward PB, Turner K, Parker G, Mitchell P, Wilhelm K (2001) Late-onset depression: genetic, vascular and clinical contributions. Psychol Med 31:1403–1412

Hickie I, Naismith S, Ward P et al (2007) Serotonin transporter gene status predicts caudate nucleus but not amygdala or hippocampal volumes in older persons with major depression. J Affect Disord 98:137–142

Holden NL (1987) Late paraphrenia or the paraphrenias? A descriptive study with a 10-year followup. Br J Psychiatry 150:635–639

Holmquist IB, Svensson B, Hoglund P (2003) Psychotropic drugs in nursing and old age homes: relationships between needs of care and mental health status. Eur J Clin Pharmacol 59:669–676

Howard R, Castle D, Wessely S, Murray R (1993) A comparative study of 470 cases of early-onset and late-onset schizophrenia. Br J Psychiatry 163:352–357

Howard R, Dennehey J, Lovestone S, Birkett J, Sham P, Powell J, Castle D, Murray R, Levy R (1995) Apolipoprotein E genotype and late paraphrenia. Int J Geriatr Psychiatry 10:147–150

Howard R, Graham C, Sham P, Dennehey J, Castle DJ, Levy R, Murray R (1997) A controlled family study of late-onset non-affective psychosis (late paraphrenia). Br J Psychiatry 170:511–514

Howard R, Rabins PV, Seeman MV, Jeste DV, the International Late-Onset Schizophrenia Group (2000) Late-onset schizophrenia and very-late-onset schizophrenia-like psychosis: an international consensus. Am J Psychiatry 157:172–178

Jang Y, Haley WE, Small BJ, Mortimer JA (2002) The role of mastery and social resources in the associations between disability and depression in later life. Gerontologist 42:807–813

Jeste DV, Harris MJ, Krull A, Kuck J, McAdams LA, Heaton R (1995) Clinical and neuropsychological characteristics of patients with late-onset schizophrenia. Am J Psychiatry 152:722–730

Jeste D, Twamley E, Eyler Z et al (2003) Aging and outcome in schizophrenia. Acta Psychiatr Scand 107:336–343

Jeste D, Blazer D, First M (2005) Aging-related diagnostic variations: need for diagnostic criteria appropriate for elderly psychiatric patients. Biol Psychiatry 58:265–271

Johri M, Beland F, Bergman H (2003) International experiments in integrated care for the elderly: a synthesis of the evidence. Int J Geriatr Psychiatry 18:222–235

Jorm A (2000) Does old age reduce the risk of anxiety and depression? A review of epidemiological studies across the life span. Psychol Med 30:11–22

Kaasinen V, Vilkmanb H, Hietalac J, Någrend K, Heleniuse H, Olssonf H, Fardef L, Rinne J (2000) Age-related dopamine D2/D3 receptor loss in extrastriatal regions of the human brain. Neurobiol Aging 21:683–688

Kales H, Maixne D, Mellow A (2005) Cerebrovascular disease and late-life depression. Am J Geriatr Psychiatry 13:88–98

Katon W, Ciechanowski P (2002) Impact of major depression on chronic medical illness. J Psychosom Res 53:859–863

Kay DWK, Roth M (1961) Environmental and hereditary factors in the schizophrenias of old age ("late paraphrenia") and their bearing on the general problem of causation in schizophrenia. J Ment Sci 107:649–686

Kay DWK, Cooper AF, Garside RF, Roth M (1976) The differentiation of paranoid from affective psychoses by patients' premorbid characteristics. Br J Psychiatry 129:207–215

Kendler KS (1980) The nosological validity of paranoia. Arch Gen Psychiatry 37:699–706

Kessing L (2006) Diagnostic subtypes of bipolar disorder in older versus younger adults. Bipolar Disord 8:56–64

Kessler R, Berglund P, Demler O, Jin R, Walters E (2005) Lifetime prevalence and age-of-onset distributions of DSM-IV disorders in the National Comorbidity Survey Replication. Arch Gen Psychiatry 62:593–602

Kinsella K, Velkoff V (2001) U.S. Census Bureau. An aging world. U.S. Government Printing Office, Washington

Koder D (1998) Treatment of anxiety in the cognitively impaired elderly: can cognitive-behaviour therapy help? Int Psychogeriatr 10:173–182

Kogan J, Edelstein B, McKee D (2000) Assessment of anxiety in older adults: current status. J Anxiety Disord 14:109–132

Kroenke K, Mangelsdorff D (1989) Common symptoms in ambulatory care: incidence, evaluation, therapy and outcome. Am J Med 86:262–266

Kua EHA (1992) Community study of mental disorders in elderly Singaporean Chinese using the GMS-AGECAT package. Aust N Z J Psychiatry 26:502–506

Lawrence V, Banerjee S, Bhugra D, Sangha K, Turner S, Murray J (2006) Coping with depression in later life: a qualitative study of help-seeking in three ethnic groups. Psychol Med 36:1375–1383

Lenze E, Mulsant BH, Shear MK, Schulberg HC, Dew MA, Begley AE, Pollock BG, Reynolds CF III (2000) Comorbid anxiety disorders in depressed elderly patients. Am J Psychiatry 157:722–728

Lesser IM, Miller BL, Swartz JR, Boone KB, Mehringer CM, Mena I (1993) Brain imaging in late-life schizophrenia and related psychoses. Schizophr Bull 19:773–782

Lindesay J (1991) Phobic disorders in the elderly. Br J Psychiatry 159:531–541

Links P, Mitton J, Steiner M (1990) Predicting outcome for borderline personality disorder. Compr Psychiatry 31:490–498

Lloyd A, Ferrier N, Barber R, Gholkar A, Young A, O'Brien J (2004) Hippocampal volume change in depression: late- and early-onset illness compared. Br J Psychiatry 184:488–495

McGlashan TH, Fenton WS (1992) The positive–negative distinction in schizophrenia: review of natural history validators. Arch Gen Psychiatry 49:63–72

McGrath A, Crome PI, Crome B (2005) Substance misuse in the older population. Postgrad Med J 81:228–231

McHugh P (2005) Striving for coherence. Psychiatry's efforts over classification. JAMA 293:2526–2528

McInnes E, Powell J (1994) Drug and alcohol referrals: are elderly substance abuse diagnoses and referrals being missed? Br Med J 308:444–446

Moltke L, Abernethy D, Greenblatt D (2005) Kinetics and dynamics of psychotropic drugs in the elderly. In: Salzman C (ed) Clinical geriatric psycho-pharmacology. Lippincott Williams and Wilkins, Philadelphia

Mossey J, Knott K, Craik R (1990) The effects of persistent depressive symptoms on hip fracture recovery. J Gerontol 45:M163–M168

Mulder R (2002) Personality pathology and treatment outcome in major depression: a review. Am J Psychiatry 159:359–371

Murphy E (1982) Social origins of depression in old age. Br J Psychiatry 141:135–142

O'Brien J, Lloyd A, McKeith I, Gholkar A, Ferrier N (2004) A longitudinal study of hippocampal volume, cortisol levels, and cognition in older depressed subjects. Am J Psychiatry 161:2081–2090

Oldehinkel A, Ormel J, Brilman E et al (2003) Psychosocial and vascular risk factors of depression in later life. J Affect Disord 74:237–246

Oslin D (2005) Evidence-based treatment of geriatric substance abuse. Psychiatr Clin North Am 28:897–911

Oslin D, Liberto J, O'Brien J, Krois S, Norbeck J (1997) Naltrexone as an adjunctive treatment for older patients with alcohol dependence. Am J Geriatr Psychiatry 5:324–332

Oslin DW, Pettinati H, Volpicelli JR (2002) Alcoholism treatment adherence. Am J Geriatr Psychiatry 10:740–747

Ostling S, Skoog I (2002) Psychotic symptoms and paranoid ideation in a nondemented population-based sample of the very old. Arch Gen Psychiatry 59:53–59

Ostling S, Palsson SP, Skoog I (2002) The incidence of first-onset psychotic symptoms and paranoid ideation in a representative population sample followed from age 70–90 years. Relation to mortality and later development of dementia 2007. Int J Geriatr Psychiatry 22:520–528

Palmer BW, Jeste DV, Sheikh JI (1997) Anxiety disorders in the elderly: DSM-IV and other barriers to diagnosis and treatment. J Affect Disord 46:183–190

Palmer BW, Bondi MW, Twamley EW et al (2003) Are late-onset schizophrenia spectrum disorders neurodegenerative conditions? Annual rates of change on two dementia measures. J Neuropsychiatry Clin Neurosci 15:45–52

Pálsson SP, Östling S, Skoog I (2001) The incidence of first-onset depression in a population followed from the age of 70–85. Psychol Med 31:1159–1168

Penninx BW, Leveille S, Ferrucci L et al (1999) Exploring the effect of depression on physical disability: longitudinal evidence from the established populations for epidemiologic studies of the elderly. Am J Public Health 89:1346–1352

Penninx BW, Beekman AT, Honig A et al (2001) Depression and cardiac mortality: results from a community-based longitudinal study. Arch Gen Psychiatry 58:221–227

Penter C, Other-Gee B (2005) Scoping project: older people and depression. Office for Seniors Interests and Volunteering, Government of Western Australia, Perth

Préville M, Boyer R, Grenier S, Dubé M, Voyer P, Ounti R, Baril M-C, Streiner D, Cairney J, Brassard J (2008) The epidemiology of psychiatric disorders in Quebec's older adult population. Can J Psychiatry 53:822–832

Prince MJ, Harwood RH, Blizard RA et al (1997) Social support deficits, loneliness and life events as risk factors for depression in old age. The Gospel Oak Project VI. Psychol Med 27:323–332

Rajii T, Ismail Z, Mulsant B (2009) Age at onset and cognition in schizophrenia: meta-analysis. Br J Psychiatry 195:286–293

Regier D, Boyd J, Burke J, Rae D, Myers J, Kramer M, Robins L, George L, Karno M, Locke B (1988) One month prevalence of mental disorders in the United States based on five epidemiologic catchment area sites. Arch Gen Psychiatry 45:977–986

Reynolds SL, Haley WE, Kozlenko N (2008) The impact of depressive symptoms and chronic diseases on active life expectancy in older Americans. Am J Geriatr Psychiatry 16:425–432

Rovner B, German P, Broadhead J, Morriss R, Brant L, Blaustein J, Folstein M (1990) The prevalence and management of dementia and other psychiatric disorders in nursing homes. Int Psychogeriatr 2:13–24

Sachdev P, Brodaty H (1999) Quantitative study of signal hyperintensities on T2-weighted magnetic resonance imagining in late-onset-schizophrenia. Am J Psychiatry 156:1958–1967

Sachdev P, Brodaty H, Rose N, Cathcart S (1999) Schizophrenia with onset after age 50 years. 2: Neurological, neuropsychological and MRI investigation. Br J Psychiatry 175:416–421

Sandanger I, Nygard JF, Ingebrigtsen G et al (1999) Prevalence, incidence and age at onset of psychiatric disorders in Norway. Soc Psychiatry Psychiatr Epidemiol 34:570–579

Schultz SK, Miller DD, Oliver SE (1997) The life course of schizophrenia: age and symptom dimensions. Schizophr Res 23:15–23

Schulz R, Drayer RA, Rollman BL (2002) Depression as a risk factor for non-suicide mortality in the elderly. Biol Psychiatry 52:205–225

Schurhoff F, Bellivier F, Jouvent R, Mouren-Simeoni M, Bouvard M, Allilaire J, Leboyer M (2000) Early and late onset bipolar disorders: two different forms of manic-depressive illness? J Affect Disord 58:215–221

Seitz D, Purandare N, Conn D (2010) Prevalence of psychiatric disorders among older adults in long-term care homes: a systematic review. Int Psychogeriatr. doi:10.1017/S1041610210000608

Shah S, Harris M (1997) A survey of general practitioners' confidence in their management of elderly patients. Aust Fam Physician 26:S12–S17

Sheikh J, Cassidy E (2000) Treatment of anxiety disorders in the elderly: issues and strategies. J Anxiety Disord 14:173–190

Shulman K (1997) Disinhibition syndromes, secondary mania and bipolar disorder in old age. J Affect Disord 46:175–182

Shulman K, Tohen M, Satlin A, Mallya G, Kalunian D (1992) Mania compared with unipolar depression in old age. Am J Psychiatry 149:341–345

Simpson JC, Tsuang MT (1996) Mortality among patients with schizophrenia. Schizophr Bull 22:485–499

Simpson S, Jackson A, Baldwin R et al (1998) Subcortical hyperintensities in late-life depression: acute response to treatment and neuropsychological impairment. Int Psychogeriatr 9:257–275

Singh N, Clements K, Singh MF (2001) The efficacy of exercise as a long-term antidepressant in elderly subjects. A randomized, controlled trial. J Gerontol Biol Sci 56:M497–M504

Stanley M, Novy D (2000) Cognitive-behavior therapy for generalized anxiety in late life: an evaluative overview. J Anxiety Disord 14:191–207

Steffans D, Pieper C, Bosworth H, MacFall J, Provenzale J, Payne M et al (2005) Biological and social predictors of long-term geriatric depression outcome. Int Psychogeriatr 17:41–56

Stek M, Exel EV, Tilburg WV, Westendorp R, Beekman A (2002) The prognosis of depression in old age: outcome six to eight years after clinical treatment. Aging Ment Health 6:282–285

Steunenberg B, Twisk J, Beekman A et al (2005) Stability and change of neuroticism in aging. J Gerontol 60B:27–33

Streiner DL, Cairney J, Veldhuizen S (2006) The epidemiology of psychological problems in the elderly. Can J Psychiatry 51:185–191

Taylor W, Steffens D, MacFall J, McQuoid D, Payne M, Provenzale J, Krishnan K (2003) White matter hyperintensity progression and late-life depression outcomes. Arch Gen Psychiatry 60:1090–1096

Taylor W, Steffens D, Payne M et al (2005) Influence of serotonin transporter promoter region polymorphisms on hippocampal volumes in late-life depression. Arch Gen Psychiatry 62:537–544

Trollor J, Anderson T, Sachdev P, Brodaty H, Andrews G (2007) Prevalence of mental disorders in the elderly: the Australian national mental health and well-being survey. Am J Geriatr Psychiatry 15:455–466

Unutzer J, Patrick DL, Simon G et al (1997) Depressive symptoms and the cost of health services in HMO patients aged 65 years and older: a 4-year prospective study. JAMA 277:1618–1623

Vink D, Aartsen MJ, Schoevers RA (2008) Risk factors for anxiety and depression in the elderly: a review. J Affect Disord 106:29–44

Wertheimer J (1997) Psychiatry of the elderly: a consensus statement. Int J Geriatr Psychiatry 12:432–435

Wessely S, Nimnuan C, Sharpe M (1999) Functional somatic syndromes: one or many? Lancet 354:936–939

Wijeratne C, Brodaty H, Hickie I (2003) The neglect of somatoform disorders by old age psychiatry: some explanations and suggestions for future research. Int J Geriatr Psychiatry 18:812–819

Wijeratne C, Hickie I, Davenport T (2006) Is there an independent somatic symptom dimension in older people? J Psychosom Res 61:197–204

Wijeratne C, Hickie I, Brodaty H (2007) The characteristics of fatigue in an older primary care sample. J Psychosom Res 62:153–158

Wilhelm K, Kotze B, Waterhouse M, Hadzi-Pavlovic D, Parker G (2004) Screening for depression the medically ill. A comparison of self-report measures, clinician judgement, and DSM-IV diagnoses. Psychosomatics 45:461–469

World Health Organisation (1992) The ICD-10 classification of mental and behavioural disorders: clinical descriptions and diagnostic guidelines. World Health Organisation, Geneva

Part V
Modifiers of Brain Aging

The Impact of Physical and Mental Activity on Cognitive Aging

Amy J. Jak

Abstract With the aging of the population, there is continued emphasis on finding interventions that prevent or delay onset of cognitive disorders of aging. Pharmacological interventions have proven less effective than hoped in this capacity and a greater emphasis has therefore been placed on understanding behavioral interventions that will positively impact dementia risk. Building on a robust animal literature, a substantial volume of research has emerged, particularly over the last 5 years, to suggest that modifiable behaviors impact brain plasticity in both humans and animals. This chapter aims to provide a critical summary of this ever growing body of research, focusing specifically on participation in physical and cognitive activities among older adults and their impact on cognition, the brain, and cognitive aging outcomes. The animal literature on activity and cognition provides a series of hypotheses as to how exercise exerts its cognitive and brain benefits. Research in animals is briefly reviewed in the context of these hypotheses as it provides the groundwork for investigations in humans. The literature on physical and cognitive activity benefits to brain and cognition in humans is reviewed in more detail. The largely positive impact of physical and cognitive activities on cognition and brain health documented in epidemiological, cross sectional, and prospective randomized controlled studies are summarized. While most studies have targeted older adults in general, the implications of exercise and cognitive interventions in individuals with Alzheimer's disease (AD) or mild

A. J. Jak (✉)
Department of Psychology Service,
Veteran's Affairs San Diego Healthcare System,
San Diego, CA, USA
e-mail: ajak@ucsd.edu

A. J. Jak
Department of Psychiatry School of Medicine,
University of California, San Diego, CA, USA

Curr Topics Behav Neurosci (2012) 10: 273–291
DOI: 10.1007/7854_2011_141

Published Online: 5 August 2011

cognitive impairment (MCI) are also described as is the evidence supporting the ability for physical activity to modify genetic risk. The connection between activity levels and brain volume, white matter integrity, and improved functionality is reviewed. Practical recommendations regarding the nature, duration, intensity and age of onset of physical or mental activity necessary to reap cognitive and brain benefits are also detailed. Most studies have investigated a singular behavioral factor or intervention, but there is some research detailing the impact of combining both mental and physical activity to boost brain health; this emerging literature is also reviewed. Finally, we comment on the limitations of the extant literature and directions for future research, in particular the need for prospective trials of activity interventions in older adults.

Contents

1 The Impact of Physical and Mental Activity on Cognitive Aging 274
2 Physical Activity 275
2.1 Animal Studies 275
2.2 Human Studies 276
3 Cognitive Activity 282
3.1 Combining Physical and Cognitive Activities 284
3.2 Conclusions, Limitations, and Recommendations 285
References 287

1 The Impact of Physical and Mental Activity on Cognitive Aging

Alzheimer's disease (AD) is the leading cause of dementia in the elderly. If current trends continue, by the year 2050, 14 million older Americans are expected to have AD. Even a modest delay in the onset of dementia would substantially reduce the number of dementia cases and result in numerous benefits to quality of life as well as in healthcare cost reduction. Pharmacological interventions have proven less effective than hoped in delaying or arresting the onset of dementia or other disorders of cognitive aging and a greater emphasis has therefore been placed on understanding behavioral interventions that will positively impact dementia risk. Behavioral factors investigated have centered primarily on physical exercise, cognitively stimulating activities, social interactions, and diet and a substantial literature has emerged to suggest that these modifiable behaviors impact brain plasticity in both animals and humans. This chapter focuses specifically on participation in physical and cognitive activities among older adults and their impact on cognition, the brain, and cognitive aging outcomes. Commentary will also be made on directions for future research, in particular the need for prospective trials of activity interventions in older adults.

2 Physical Activity

2.1 Animal Studies

A substantial animal literature documents the cognitive benefits of exercise and, more recently, the literature has expanded to confirm similar benefits in humans. There are several hypotheses as to why "what is good for the heart is good for the brain" and the mechanisms by which exercise exerts its cognitive and brain benefits. One hypothesis is that exercise is neuroprotective, possibly via its ability to reduce oxidative stress and insulate the brain from injury and neurodegenerative diseases (Cotman et al. 2007; Kiraly and Kiraly 2005). In rats, exercise reduced age-related oxidative damage, particularly lipid oxidation in the cerebellum (Cui et al. 2009). Another possibility is that exercise appears to facilitate learning via improved long term potentiation (O'Callaghan et al. 2007). Physical activity may also result in neurogenesis within the hippocampus (van Praag et al. 2005). Van Praag et al. (2005) studied aged mice that were predominantly sedentary until provided with a running wheel. Even the older mice that were allowed activity demonstrated enhanced learning (as determined by performance on the Morris water maze) and reversed age-related decline in hippocampal neurogenesis by 50%. Further studies confirmed the benefits of exercise in older rats, benefits that included not only improved memory and neurogenesis but also reduced apoptosis in the hippocampus (Kim et al. 2010). Another hypothesis is that exercise upregulates growth factors, such as brain-derived neurotrophic factor (BDNF) (Knaepen et al. 2010), associated with energy metabolism and homeostasis, which in turn lead to cognitive benefits. BDNF in particular may be integral to inducing neurogenesis (Churchill et al. 2002), and has been linked to hippocampal plasticity, learning, and memory in animal models (Lu and Gottschalk 2000). In addition to growth factors, serotonin levels may increase with exercise while also reducing levels of corticosteroids (Colcombe et al. 2004a). Alternatively, exercise may exert its positive cognitive influence more indirectly via impact on other risk factors such as diabetes and/or hypertension, which may also be mediated by the above mentioned upregulation of growth factors and ultimately reduction in inflammation (Cotman et al. 2007). Finally, exercise may also counteract some genetic risk factors for dementia. Wheel running resulted in improvements in accuracy of completion of a water maze task in mice carrying the risk allele of apolipoprotein (APOE ε4), enough to make their performance comparable to mice not carrying the risk allele (APOE ε3) (Nichol et al. 2009).

Hypotheses about the mechanisms by which cognitively stimulating activities impart brain and cognitive benefits are in many ways similar to those proposed for physical activity. One hypothesis as to why mentally stimulating activities may protect against cognitive decline also centers around neuroprotective actions (Kramer et al. 2004). There is also speculation that cognitively stimulating activities provide reserve capacity and/or result in enhanced neural networks that may delay cognitive decline even in the presence of structural brain changes with age (Kramer et al. 2004). Studies have shown that complex environments alter brain structure in rodents and nonhuman primates, including increased dendritic branching, synaptogenesis, alterations in glial cells, enhancement of capillary network, and neurogenesis (Kramer et al. 2004). Furthermore, environmental complexity correlates positively with cortical volume/thickness, perhaps by either neurogenesis or by augmentation of existing neuropil (Churchill et al. 2002).

2.2 Human Studies

2.2.1 Epidemiological and/or Cross Sectional Studies

The extant literature on the relationship between exercise and brain and cognitive changes in humans is largely retrospective, either examining these variables epidemiologically or by assessing one's current cardiorespiratory fitness and its relationship to cognitive functioning and brain integrity. Nonetheless, this literature has generally supported the link between exercise and brain health (Lytle et al. 2004). One large-scale population-based study, the Canadian Study of Health and Aging, found reduced risk for AD with exercise, among other lifestyle factors such as wine and coffee consumption (Lindsay et al. 2002) and reported physical activity as a means to reduce risk for cognitive decline and dementia (Laurin et al. 2001). Larson et al. (2006) followed 1,740 older adults over a mean follow-up period of 6.2 years and found that those who exercised three or more times per week had significantly lower incidence rates of dementia than those who exercised less (Larson et al. 2006). Data from the Nurses' Health study revealed that, with increasing amounts of long term physical activity, cognitive performance, including verbal fluency, memory, attention, and global cognition, improved significantly; adjusting for cardiovascular risk factors did not significantly attenuate these relationships (Weuve et al. 2004). Physical activity was also associated with less cognitive decline over time in this sample. In practical terms, exercise resulted in a 20% reduction in risk for cognitive impairment or the equivalent taking 3 years off your age (Weuve et al. 2004).

Epidemiological studies have also provided information about when in one's lifetime exercise needs to begin to reap ongoing cognitive benefits. A recent longitudinal examination of over 13,000 individuals at least 70 years of age in the Nurses' Health Study revealed that mid-life physical activity rates were significantly associated with higher survival rates (Sun et al. 2010). While walking increased the odds of successful aging in this study, increased pace of walking also resulted in a significant increase in the odds of successful aging, regardless of the amount of walking. Physical activity improved survival rates regardless of body mass index (BMI), although women who had low BMI and were active had the highest successful survival odds as compared to women who were more sedentary with higher BMI (Sun et al. 2010). Moderate exercise done either in mid- or late-life reduced the likelihood of MCI, though neither light nor vigorous exercise were associated with risk reduction (Geda et al. 2010). Those who engaged in physical activity at any point in life (teenage, 30s, midlife, and late life) were less likely to have cognitive impairment as older adults (Middleton et al. 2010). For women, exercise during the teenage years had the strongest likelihood of impacting late life cognitive changes, but even those who did not exercise as teenagers but introduced exercise later in life experienced cognitive benefits (Middleton et al. 2010). Other population-based studies have also found an association between early-life physical activity and cognition. Dik et al. (2003) queried 1,241 older adults retrospectively about their physical activity when they were 15 and 25 years old and found men, but not women, who exercised more in early life had stronger information processing speed than those who exercised less early in life (Dik et al. 2003). In a large-scale investigation of the physical and leisure activity of 1919 individuals at several points in middle adulthood, both physical and mental activities were significantly and positively associated with memory performance at midlife (after controlling for sex, education, socioeconomic status, IQ, and health/mental health conditions) (Richards et al. 2003). Furthermore, physical exercise was related to a significantly slower rate of decline in memory from ages 43 to 53 (Richards et al. 2003). Finally, in addition to evidence that mid-life exercise is beneficial to later-life cognitive functioning, there is evidence that it is never too late to begin exercising. Sumic et al. (2007), investigated the impact of physical activity on cognitive impairment in a sample of adults over the age of 85. They found that at least 4 hours of exercise per week was protective against cognitive impairment in the oldest-old and was particularly important for older women (Sumic et al. 2007).

Meta-analyses have also offered strong evidence that exercise does benefit cognition in older adults. In a meta-analysis of exercise trials by Colcombe and Kramer (2003), they reported effect sizes of exercise on cognition of 0.16 for control groups and 0.48 for exercise groups. The data suggest some improvement on cognitive tasks in both groups, however, the control group improved only by about 0.125 standard deviations on cognitive measures while the exercise groups gained 0.50 standard deviations. Furthermore, exercise appeared to have the largest effect size related to improvements in executive functioning, followed by mental control, spatial tasks, and psychomotor speed (Colcombe and Kramer 2003).

However, not all studies have been favorable in support of exercise as a mechanism to prevent cognitive decline. Sturman et al. (2005) studied 4,055 adults 65 and older and found that exercise was associated with slowing the rate of cognitive decline, but the result was no longer significant after adjusting for participation in cognitive activities, depression, vascular disease, and likely preclinical dementia (Sturman et al. 2005). Other studies examining both physical and mental activities have reported benefits to cognitive aging only for the participation in mentally stimulating activities and not physical activity and are reported in more detail below (see "Combined Physical and Cognitive Activities" section).

2.2.2 Studies in Impaired Populations

Once individuals have progressed to AD, there is evidence that exercise still is beneficial. Scarmeas et al. (2010) found that higher levels of physical activity prior to onset of AD led to longer rates of survival, though did not impact cognitive changes or functional decline. Even small amounts of physical activity were beneficial to reducing mortality risk by up to 75% (Scarmeas et al. 2010). While the physical activity did lead to fewer medical comorbidities in this sample, the protective effect of exercise was still present even after adjusting for APOE, smoking, medical, and cognitive status suggesting that the benefits did not arise from reduction of medical risk factors alone. In those who are already diagnosed with Alzheimer's disease, there is also evidence that physical activity can improve nutritional status, MMSE scores, reduce fall risks, and improve behavioral problems (Rolland et al. 2002).

While most emphasis in the literature has been placed on the role of exercise in reducing risk for dementia in general, and often AD specifically, moderate exercise has also been shown to lower risk for vascular cognitive impairment in women (Middleton et al. 2008b). Ravaglia et al. (2008) found that physical activity (dichotomously rated as either present or absent and self-reported) positively impacted risk for vascular dementia but not AD (Ravaglia et al. 2008).

2.2.3 Exercise and Genetic Risk Factors

Building on results obtained from animal studies, research is also emerging on the interaction between genes and environment in humans, particularly the ability of exercise to modify genetic risk factors for Alzheimer's disease, such as the ε4 allele of the apolipoprotein E gene (APOE). In humans, there is evidence that exercise is particularly beneficial to carriers of the APOE ε4 allele. In a sample of middle aged adults, ε4 carriers had improved reaction times with physical activity while non-carriers did not (Deeny et al. 2008). Etnier et al. (2007) also concluded that exercise is particularly beneficial in female ε4 homozygotes; those with higher levels of aerobic fitness (VO_2 max) performed better on tests of verbal memory (Affective Auditory Verbal Learning Test; AAVLT), the complex figure test, and the Paced Auditory Serial Addition Test (PASAT) (Etnier et al. 2007). Conversely, a study of 1,449 people at year 21 of follow-up found that lack of physical activity, high fat intake, alcohol, and smoking in middle age were associated with increased dementia risk especially in ε4 carriers (Kivipelto et al. 2008). In contrast, others have found greater benefits of exercise in those not at genetic risk. Schuit et al. (2001) found that higher levels of physical activity reduced dementia risk in those who were APOE ε4 negative but not in those who were ε4 positive; however they also found that inactivity in ε4 carriers was associated with greater risk of cognitive decline (Schuit et al. 2001).

2.2.4 Prospective Physical Activity Interventions

Prospective activity interventions with older adults in the form of randomized controlled trials (RCTs) are the gold standard for research but are more limited,

Table 1 Summary of randomized controlled trials of the impact of exercise on cognition in older adults

Study	Type of exercise	Control group	Duration	Outcome
Baker et al. (2010)	"Aerobic exercise"—treadmill, stationary bicycle, or elliptical trainer	Stretching and balance exercises	6 months	Improved executive functioning
Lautenschlager et al. (2008)	"Moderate-intensity physical activity" (most frequently walking)	Health education	6 months	Improved delayed recall
Anderson-Hanley et al. (2010)	Combination of chair and standing exercises with small weights	Wait-list	4 weeks	Improvements in working memory and executive functioning
van Uffelen et al. (2008b)	Supervised aerobic walking	Trainer-led postural/flexibility classes	1 year	Improved memory and executive functioning in a subgroup
Erickson et al. (2011)	Trainer-led aerobic walking	Trainer-led stretching/toning	1 year	Increased hippocampal volume

though emerging, in the literature (see Table 1). In a recent trial, 33 sedentary adults were randomized to either 6 months of supervised aerobic exercise (3–4 times per week for 45–60 min) or a stretching control group. Those in the exercise group showed improved executive functioning, with women experiencing broader improvements than men (Baker et al. 2010). In contrast, women in the stretching control group actually experienced cognitive declines over the intervention period. Lautenschlager et al. (2008) randomized 170 older adults with subjective or objective mild cognitive impairment to either a 6 month home-based exercise program (3–4 50 min weekly sessions, primarily walking) or a control group (care as usual). Global cognition (as measured by the ADAS-cog) and delayed recall were significantly better in the exercise as compared to the control group, benefits that were maintained for a follow-up period of 1 year (Lautenschlager et al. 2008). In contrast, van Uffelen et al. (2008b) found no benefit of either walking (1 year group supervised walking two times per week for 1 h) or vitamin B on cognition in a study of 179 older adults. However, for the subgroup with the best compliance with the walking intervention, modest memory improvements were noted (van Uffelen et al. 2008b).

In a meta-analysis of randomized controlled trials of exercise interventions for cognitive health, an overall effect size of 0.48 was found in those assigned to exercise conditions and only 0.16 for control groups (Colcombe and Kramer 2003), however the studies included did not include longitudinal follow-up. Other systematic reviews of RCTs of exercise suggested that, in cognitively healthy populations, participation in exercise results in benefits to memory, processing speed, and executive functioning (van Ufflelen et al. 2008a). In those with cognitive decline already present, cognitive benefits were in the domains of general cognitive functioning and executive functioning (van Ufflelen et al. 2008b). Exercise appears to be particularly beneficial in those who already evidenced cognitive decline; the review by van Ufflelen et al. (2008a) reports cognitive benefits in one-third of the studies with participants with no cognitive decline but found benefits in two-thirds of the studies in which participants had cognitive decline.

2.2.5 Neuroimaging Correlates of Exercise

Examining structural and functional MRI variables has revealed positive effects on brain integrity in older adults who exercise. In one of the earliest studies showing positive brain changes with exercise in humans, Colcombe et al. (2003) showed that adults over age 55 with higher levels of cardiovascular fitness had less age-related volume loss in the frontal, parietal, and temporal lobes (Colcombe et al. 2003). Colcombe et al. (2006) conducted a prospective study in which older adults ages 60–79 were randomized to either a 6-month aerobic activity condition or a toning and stretching control group. Those in the exercise group had significantly larger gray and white matter volumes (specifically anterior white matter, inferior frontal gyrus, anterior cingulate, and superior temporal gyrus gray matter) after the intervention than did the control group. Other investigators have also found positive structural brain changes related to exercise. Older adults ages 55–79 with higher rates of exercise over the last 10 years had larger superior frontal lobes and reduced rate of medial temporal lobe atrophy (Bugg and Head 2009). Closely paralleling findings in the animal literature, Erickson et al. (2009) found that higher levels of exercise in older adults resulted in larger hippocampal volumes bilaterally (Erickson et al. 2009). A recent randomized trial of exercise training demonstrated the ability to increase hippocampal volumes in humans by 2% (Erickson et al. 2011). Objective measures of fitness are also associated with structural brain changes in older adults. VO_2 max levels are associated with whole brain and white matter volumes (Burns et al. 2008). However others have found reduced brain atrophy rates with exercise in AD but not in normal aging (Burns et al. 2008).

Functional brain activation is also affected by exercise. Highly fit older adults showed greater activation during selective attention tasks in prefrontal and parietal cortices, activations more similar to younger adults (Colcombe et al. 2004b) and suggests higher levels of fitness leads to more efficient functioning of the prefrontal cortex.

There may be some benefit of exercise to brain vessel health, resulting in reduced tortuosity and increased small vessels (Bullit et al. 2009). Marks et al. (2010) also found a relationship between white matter and exercise. In older adults, higher aerobic fitness, as measured by VO_2 max, was associated with increased fractional anisotropy (FA) in the left middle cingulum whereas BMI and abdominal adiposity were associated with lower FA in the right posterior cingulum (Marks et al. 2010). A 12 week prospective walking exercise intervention in individuals with amnestic MCI also revealed improvements in indices of white matter microstructural integrity in the medial temporal and frontal lobes (Smith et al. 2011).

2.2.6 How Much, How Often, and What Kind of Exercise?

The vast majority of the existing literature supports the positive cognitive effects of exercise in older adults. But these positive results lead to practical questions, such as how much exercise is enough for cognitive benefits? Exercise at least as intense as walking a minimum of three times per week was more likely to result in maintenance of cognition than was participation in lower amounts of physical activity (Middleton et al. 2008a). Participating in a number of different activities may be particularly important in reducing dementia risk (Schuit et al. 2001) and participating in three or more activities provided the greatest reduction in dementia risk (Podewils et al. 2005). Even relatively brief training programs provided cognitive benefit, though slightly less than longer-term exercise programs (Colcombe and Kramer 2003). However, exercise sessions less than 30 minutes may have limited impact on cognition (Colcombe and Kramer 2003). Maintaining consistent duration and intensity of exercise over time may also be a key to unlocking cognitive benefits. In a large-scale study, van Gelder et al. (2004) found that the duration and intensity of activities declined over the 10 years of the study. Those whose activity duration declined by more than 60 min per day over the 10 years had a two and a half times greater decline in cognitive functioning (MMSE) than those with stable duration of activity (van Gelder et al. 2004). They reported similar findings for decrease in duration of activity but found no cognitive decline in men who actually increased their activity duration (van Gelder et al. 2004).

Another practical question about how to achieve the cognitive benefits of exercise is what specific activity should one participate in? Most older adults participate in 2–3 different activities, and walking is one of the most common (Podewils et al. 2005). Given its popularity and ease of implementation, it is one of the more widely investigated physical activities. Abbott and colleagues found that men who walked the least (averaged less than 0.25 miles/day) had a 1.8 times greater increase in dementia as opposed to those who walked the most (more than 2 miles per day) (Abbott et al. 2004). Walking was also associated with cognitive benefits in older women. Five thousand nine hundred twenty five women followed-up after 6–8 years revealed that those who reported walking more were less likely to experience cognitive decline (Yaffe et al. 2001). The median distance walked was 49 city blocks per week resulting in a 37% reduction in the chance of cognitive decline (Yaffe et al. 2001). This relationship held true even after adjusting for age, education, health factors, depression, cerebrovascular risk factors, and estrogen use. Our own pilot data are also consistent with cognitive benefits stemming from daily increases in walking. In fifteen sedentary but healthy older adults (ages 65–80) who participated in a 12 week trial in which half progressively increased their daily step counts and half did not, we found the walking group had significant improvements on tests of executive functioning after 1 and 3 months from baseline and were maintained 3 months after the intervention ceased (Jak et al. 2011).

Less attention has been paid to the benefit of non-aerobic activities, such as strength-training, on cognitive functioning. A small sample of 16 older adults participated in a trial of a strength program for 1 month and subsequently performed better than controls on digits backward task and the Stroop task (Anderson-Hanley et al. 2010). Yoga has also been studied and while it appears to have quality of life benefits, its cognitive benefits have not been demonstrated (Oken et al. 2006). Meta-analytic data suggests that those who participate in a combination of strength and aerobic exercise had more reliable improvements in cognition than those only completing aerobic activity (Colcombe and Kramer 2003) so combinations of strength and aerobic activities may be particularly beneficial.

Even day-to-day physical activity appears to have a positive effect on cognition. Barnes et al. (2008) measured daily activity via actigraphy and found that those women who had the highest daytime activity had significantly better cognitive functioning (as measured by Trails B and MMSE). This general daily activity was only moderately correlated with volitional exercise and still remained significant even when purposeful exercise was controlled for, suggesting there is additional benefit from simply moving more during the day (Barnes et al. 2008).

3 Cognitive Activity

In addition to the growing research support for exercise as a mechanism to impact cognitive aging, multiple avenues of evidence suggest that learning and mental activity positively impact cognitive function and may also be protective against the ills of aging. Occupational complexity (Schooler et al. 1999), cognitively complex leisure activities [for example, reading, hobbies, etc. (Kramer et al. 2004)], and practicing an engaged lifestyle [(e.g., learning a new language or a new card game; (Hultsch et al. 1999)] all positively influence cognitive functioning and may be associated with lower risk for Alzheimer's disease (Wilson et al. 2002a). A meta-analysis integrating data from over 29,000 participants revealed almost a 50% reduction in risk for cognitive decline from participation in cognitively stimulating activities; however, the effects of education and occupational complexity were similar in their rates of risk reduction (Valenzuela and Sachdev 2009). Even early life indices such as 'idea density' measured in written autobiographical records have been associated with later-life cognitive functioning. Specifically, high idea density in young adulthood is associated with lowered rates of cognitive impairment in older age as well as less brain structural change identified upon autopsy (Riley et al. 2005). Practice at specific cognitive skills has also been demonstrated to alter brain structure. In a study of London taxi drivers, Maguire et al. 2003 found that taxi drivers had larger gray matter volumes in the posterior hippocampus than did age-matched non-taxi drivers, a volume that was positively correlated with the amount of time spent driving (Maguire et al. 2003).

In one hallmark study of cognitive activity and its impact on cognitive aging, (Wilson et al. 2002b) found that cognitive activity was significantly related to reduced AD risk and carried the same benefits for both black and white individuals, for both genders, and across age and education levels, suggesting that cognitive activity mediates the impact of education and occupation. Furthermore, they found that those with limited cognitive activity were twice as likely to have AD as those who were more cognitively active; they did not find exercise to be related to AD risk (Wilson et al. 2002a). Similarly, Verghese et al. (2006) found that participation in cognitive activities lowered risk for developing amnestic MCI (Verghese et al. 2006) and dementia (Verghese et al. 2003) but they did not find the same protective effect for physical activity. Scarmeas et al. (2001) also found that participating in leisure activities resulted in reduced relative risk of dementia even after controlling for health factors, cerebrovascular disease, and depression (Scarmeas et al. 2001). For those who do go on to develop dementia, there is evidence that participation in cognitive activities delays the onset of accelerated memory decline, though unfortunately, eventual rapid decline was not eliminated (Hall et al. 2009).

Cognitive training in most modalities appears to benefit the trained skill, particularly if the follow-up time frame is short (Valenzuela and Sachdev 2009). As with investigations of physical activity, randomized controlled trials investigating the impact of cognitive training on aging outcomes, particularly with longitudinal follow-up have been more limited. In one RCT investigating training programs targeting individual cognitive domains (reasoning, memory, and processing speed), the authors reported improvements in IADL's following the reasoning training, though not following training in processing speed or memory training. These reasoning skills learned were maintained even 5 years later (Willis et al. 2006). Ball et al. (2002) also found similarly persistent results in that older adults who received training in a particular domain exhibited cognitive gains in that domain that persisted for 2 years (Ball et al. 2002). In one of the only RCTs of cognitive training to also include neuroimaging, a group of older and middle aged adults underwent memory training for 8 weeks, resulting in improvements in source memory but also in increase in cortical thickness, as compared to controls who did not receive the memory training (Engvig et al. 2010).

However, in some meta-analyses, no significant benefits of cognitive training in older adults were found; studies showing large effect sizes often had outcome measures directly related to the trained task, but with no effect of generalizability (Papp et al. 2009).

While the general results documenting cognitive benefits of participating in mentally stimulating activities have been positive, many still pose the practical question of which specific activity should be encouraged for older adults? While all leisure activities appear to impart some cognitive benefit, traveling, odd jobs, knitting, and gardening have evidence of being particularly beneficial; participating in 2–3 activities was the most beneficial while participating in only one leisure activity did not significantly reduce dementia risk (Fabrigoule et al. 1995). Reading, playing board games, dancing, and playing musical instruments have also been specifically related to reduced dementia risk (Verghese et al. 2003). A study of bridge players suggested cognitive benefits; bridge players performed better than those who did not play bridge on measures of working memory and reasoning (Clarkson-Smith and Hartley 1990). However, with correlation data, it is impossible to determine causality and it is plausible that those with better working memory and reasoning self-select to play a complex game like bridge.

Another practical question involves determining how much cognitive training is enough? A meta-analysis of existing controlled trials of cognitive interventions suggested that 2–3 months of regular training results in cognitive effects that persist over time and translated into clinical gains of 1.2 points on the MMSE for cognitively healthy elders and 2.6 points for those with MCI (Valenzuela and Sachdev 2009).

To attempt to capitalize on the cognitive benefits of participation in mentally stimulating activities, numerous programs and games have been developed to enhance cognition in older adults. The drawback to most cognitive training programs is that individuals improve on the trained task only, with no generalizability to broader tasks or cognition. There have been some, however, that have demonstrated better transfer of learning to real-world situations. The generalizability of processing speed, in particular, has been more robust. For transfer of cognitive skills other than processing speed, it appears that complexity of the training program is essential. For example, there is evidence that training older adults in a complex computer based strategy game called Rise of Nations for 23.5 hours over 4–5 weeks imparts benefits to memory, reasoning, and multitasking (Basak et al. 2008). A randomized controlled trial of a cognitive training program, with several different exercises over time, showed improvement on the trained tasks as well improvements on neuropsychological testing in the trained group but not in the control group (Mahncke et al. 2006). Multi-domain cognitive training also appears to have benefits in mild Alzheimer's disease leading to cognitive benefits and delay of disease progression (Gates and Valenzuela 2010). In MCI, training that involves multiple cognitive domains also appears to provide more benefit than solely memory strategy training (Gates and Valenzuela 2010).

3.1 Combining Physical and Cognitive Activities

Combination approaches may ultimately hold the most promise for making the largest impact on cognitive aging. For example, one study documented that combining cognitive training with acetylcholinesterase inhibitors provides a more significant benefit than either treatment in isolation (Rozzini et al. 2007). Specifically, Rozzini et al. (2007) conducted a small trial comparing three groups of individuals with MCI: a cognitive training plus cholinesterase inhibitor group, a cholinesterase inhibitor without training group, and a no treatment group. Those receiving both cognitive training and cholinesterase inhibitors was the only group to reap cognitive benefits (Rozzini et al. 2007). Combinations of both physical and mental activities also have positive results. A study in mice revealed that both wheel running and environmental enrichment independently stimulated neurogenesis, however, sequential exercise and environmental enrichment led to even greater levels of neurogenesis than either intervention in isolation (Fabel et al. 2009). The authors hypothesized that the exercise prepares the potential for neurogenesis that can be realized to a greater extent if also exposed to cognitive enrichment. Fabre et al. (2002) reported similar findings in humans; combined mental and physical activity programs reap greater benefits than either in isolation. Memory (as measured by story memory and paired associate learning) was significantly improved in the cognitive training, physical training, and combined training groups as compared to the control group, however the combined training group improvement in memory was significantly better than either of the other training groups (Fabre et al. 2002). Using the modified Mini-Mental State Examination measured in over 2,500 older adults at three points in time over 8 years, Yaffe et al. (2009) created three groups: those with stable cognition, minor declines, and major declines. They found that age, white race, minimum of high school education, not smoking, and weekly moderate or vigorous exercise were factors that distinguished the stable group from the minor decliners (Yaffe et al. 2009). They also found that those who worked/volunteered, had social support, did not live alone, were also more likely to maintain cognition.

3.2 Conclusions, Limitations, and Recommendations

The overwhelming majority of evidence indicates that both physical exercise and participation in cognitively stimulating activities are both important for maintaining cognitive and brain integrity with age and also in lowering risk for cognitive decline and dementia. Exercise at midlife does translate into reduced risk of dementia but RCTs and other research also support the idea that it is never too late to begin physical activity to reap its cognitive benefits. Practical recommendations for the type and amount of activity also emerge from the research. Aerobic exercise of some sort seems necessary for cognitive benefits; strength training and yoga may have other mind/body benefits but in isolation do not appear to aid cognition in the same way aerobic activity does. Physical activity, at least as intense as walking, for at least 30 min bouts, at least three times a week is supported in the literature as a reasonable amount of exercise to achieve cognitive benefit. Once exercising, maintaining a consistent intensity and duration of exercise over time also appears to be important for maintaining cognition and reducing dementia risk. Participating in a variety of physical and mental activities also appears to be key.

Regarding mental activities, cognitive complexity emerges as one consistently necessary factor for impacting cognitive aging outcomes. Activities like traveling, knitting, games, and playing musical instruments, in part because of their cognitive complexity, reduce dementia risk. Steady participation in 3–4 different activities for durations of 2–3 months emerges from the literature as a threshold for achieving cognitive benefits. With cognitive training programs, multi-domain training programs have the greatest generalizability, though most all training programs reap benefits at least in the trained task.

Although the existing literature on exercise, mental activities, and cognitive health has grown significantly over the last decade, it is still a literature with limitations. The number of RCTs are growing but there is still a relative paucity of prospective studies of exercise or cognitive interventions. There is also heavy reliance on self-report of both exercise and cognitive activity and less objective measurement of these variables. However, studies that do have objective measures of cardiovascular fitness and aerobic capacity report results consistent with those derived from self-report data. For example, higher levels of aerobic fitness (as measured by VO_2 max or maximal oxygen consumption) are associated with better cognition, particularly global cognition, executive functioning, and attention, after a follow-up period of 6 years (Barnes et al. 2003).

There is also often a lack of sufficient follow-up to determine either effectiveness of interventions or persistence of gains. Outcome measures that address daily functioning and the degree to which exercise or other activities impact functional skills are also a limitation of most studies to date. Often cognitive outcomes are measured by global cognitive measures, such as the MMSE; broader cognitive outcomes would help clarify the specific nature of the benefits of the intervention, i.e., does one activity benefit memory while another might benefit attention. There are exceptionally few imaging studies examining the impact on brain integrity of cognitively stimulating activities or cognitive training. Imaging studies in this area are essential in determining the nature of brain changes that may result from mental activity and afford a better understanding of the mechanisms by which mentally stimulating activities exert any benefits.

Some recommendations can be drawn from the existing literature about how much, how long, and what kinds of activity impart cognitive benefits. However, RCTs directly investigating these factors need to be undertaken. No study has systematically and prospectively varied the time and/or duration of exercise to determine, for example, whether 60 min of activity provides significantly more cognitive benefits than 30 min or whether swimming provides greater cognitive or brain benefits than walking. There is support that exercise in middle age aids cognition over the lifespan and that even the oldest-old can gain benefits from exercise, but there is no definitive data about the age of initiation of physical activity required to reap the greatest cognitive and brain benefits. With better understanding of these factors, physical and mental activity interventions can then be more clearly targeted for future use for reducing or delaying the onset of negative cognitive outcomes with age or as combination therapies. The potential additive impact of medications currently used to slow memory decline paired with activity interventions need to be more fully explored as they may produce the most robust benefits.

It does appear that behavioral factors can serve a substantial protective function against future cognitive decline or dementia and no known study has found that aerobic exercise or mental stimulation makes cognition worse. Behavioral modifications hold minimal risk for harm for most older adults and there is strong research support that exercise and cognitive activity are beneficial from a longitudinal cognitive perspective. While no amount of mental or physical activity can guarantee protection from dementia or cognitive decline, the evidence does suggest that older adults should be encouraged to pursue/maintain regular physical activity and mental stimulation to increase the odds of positive cognitive aging outcomes.

References

Abbott RD, White LR, Ross GW, Masaki KH, Curb JD, Petrovitch H (2004) Walking and dementia in physically capable elderly men. JAMA: J Am Med Assoc 292(12):1447–1453

Anderson-Hanley C, Nimon JP, Westen SC (2010) Cognitive health benefits of strengthening exercise for community-dwelling older adults. J Clin Exp Neuropsychol 32(9):996–1001

Baker LD, Frank LL, Foster-Schubert K, Green PS, Wilkinson CW, McTiernan A et al (2010) Effects of aerobic exercise on mild cognitive impairment: a controlled trial. Arch Neurol 67(1):71–79

Ball K, Berch DB, Helmers KF, Jobe JB, Leveck MD, Marsiske M et al (2002) Effects of cognitive training interventions with older adults: a randomized controlled trial. J Am Med Assoc 288(18):2271–2281

Barnes DE, Yaffe K, Satariano WA, Tager IB (2003) A longitudinal study of cardiorespiratory fitness and cognitive function in healthy older adults. J Am Geriatr Soc 51(4):459–465

Barnes DE, Blackwell T, Stone KL, Goldman SE, Hillier T, Yaffe K (2008) Cognition in older women: the importance of daytime movement. J Am Geriatr Soc 56(9):1658–1664

Basak C, Boot WR, Voss MW, Kramer AF (2008) Can training in real-time strategy video game attenuate cognitive decline in older adults? Psychol Aging 23:765–777

Bugg JM, Head D (2009) Exercise moderates age related atrophy of the medial temporal lobe. Neurobiol Aging. doi:10.1016/j.neurobiolaging.2009.03.008

Bullit E, Rahman FN, Smith JK, Kim E, Zeng D, Katz LM et al (2009) The effect of exercise on the cerebral vasculature of healthy and aged subjects as visualized by MR angiography. Am J Neuroradiol 30:1857–1863

Burns JM, Cronk BB, Anderson HS, Donnelly JE, Thomas GP, Harsha A et al (2008) Cardiorespiratory fitness and brain atrophy in early Alzheimer disease. Neurology 71(3):210–216

Churchill JD, Galvez R, Colcombe S, Swain RA, Kramer AF, Greenought WT (2002) Exercise, experience, and the aging brain. Neurobiol Aging 23:941–955

Clarkson-Smith L, Hartley AA (1990) The game of bridge as an exercise in working memory and reasoning. J Gerontol 45(6):P233–P238

Colcombe SJ, Kramer AF (2003) Fitness effects on the cognitive function of older adults. Psychol Sci 14(2):125–130

Colcombe SJ, Erickson KI, Raz N, Webb A, Cohen NJ, McAuley E et al (2003) Aerobic fitness reduces brain tissue loss in aging humans. J Gerontol: Med Sci 58A(2):176–180

Colcombe SJ, Kramer AF, Erickson KI, Scalf P, McAuley E, Cohen NJ et al (2004a) Cardiovascular fitness, cortical plasticity, and aging. Proc Natl Acad Sci 101(9):3316–3321

Colcombe SJ, Kramer AF, McAuley E, Erickson KI, Scalf P (2004b) Neurocognitive aging and cardiovascular fitness. J Mol Neurosci 24:9–14

Colcombe SJ, Erickson KI, Scalf PE, Kim JS, Prakash R, McAuley E, Elavsky S, Marquez DX, Hu L, Kramer AF (2006) Aerobic exercise training increases brain volume in aging humans. J Gerontol A Biol Sci Med Sci 61(11):1166–1170

Cotman CW, Berchtold NC, Christie LA (2007) Exercise build brain health: key roles of growth factor cascades and inflammation. Trends Neurosci 30(9):464–472
Cui L, Hofer T, Rani A, Leeuwenburgh C, Foster TC (2009) Comparison of lifelong and late life exercise on oxidative stress in the cerebellum. Neurobiol Aging 30(6):903–909
Deeny SP, Poeppei D, Zimmerman JB, Roth SM, Brandauer J, Witkowski S et al (2008) Exercise, APOE, and working memory: MEG and behavioral evidence for benefit of exercise in epsilon4 carriers. Biol Psychol 78:179–187
Dik MG, Deeg DJH, Visser M, Jonker C (2003) Early life physical activity and cognition at old age. J Clin Exp Neuropsychol 25(5):643–653
Engvig A, Fjell AM, Westlye LT, Moberget T, Sundseth O, Larsen VA et al (2010) Effects of memory training on cortical thickness in the elderly. Neuroimage 52:1667–1676
Erickson KI, Prakash RS, Voss MW, Chaddock L, Hu L, Morris KS et al (2009) Aerobic fitness is associated with hippocampal volume in elderly humans. Hippocampus 19(10):1030–1039
Erickson KI, Voss MW, Prakash RS, Basak C, Szabo A, Chaddock L et al (2011) Exercise training increases size of hippocampus and improves memory. Proc Natl Acad Sci 108(7): 3017–3022
Etnier JL, Caselli RJ, Reiman EM, Alexander GE, Sibley BA, Tessier D et al (2007) Cognitive performance in older women relative to APOE-e4 genotype and aerobic fitness. Med Sci Sports Exerc 39(1):199–207
Fabel K, Wolf SA, Ehninger D, Babu H, Leal-Galicia P, Kempermann G (2009) Additive effects of physical exercise and environmental enrichment on adult hippocampal neurogenesis in mice. Frontiers Neurosci 3:1–7
Fabre C, Chamari K, Mucci P, Masse-Biron J, Prefaut C (2002) Improvement of cognitive function by mental and/or individualized aerobic training in healthy elderly subjects. Int J Sports Med 23(6):415–421
Fabrigoule C, Letenneur L, Dartigues JF, Zarrouck M, Commenges D, Barberger-Gateau P (1995) Social and leisure activities and risk of dementia: a prospective longitudinal study. J Am Geriatr Soc 43(5):485–490
Gates N, Valenzuela MJ (2010) Cognitive exercise and its role in cognitive function in older adults. Curr Psychiatry Rep 12:20–27
Geda YE, Roberts RO, Knopman DS, Christianson TJ, Pankratz VS, Ivnik RJ et al (2010) Physical exercise, aging, and mild cognitive impairment: a population-based study. Arch Neurol 67(1):80–86
Hall CB, Lipton RB, Sliwinski M, Katz MJ, Derby CA, Verghese J (2009) Cognitive activities delay onset of memory decline in persons who develop dementia. Neurology 73:356–361
Hultsch DF, Hertzog C, Small BJ, Dixon RA (1999) Use it or lose it: engage lifestyle as a buffer of cognitive decline in aging. Psychol Aging 14(2):245–263
Jak AJ, McCauley A, Gravno J, Jurick S, Bondi MW (2011) Impact of a walking intervention on executive functioning in older adults. Paper presented at the international neuropsychological society annual meeting
Kim S, Ko I, Kim B, Shin M, Cho S, Kim C et al (2010) Treadmill exercise prevents aging-induced failure of memory through an increase in neurogenesis and suppression of apoptosis in rat hippocampus. Exp Gerontol 45:357–365
Kiraly MA, Kiraly SJ (2005) The effect of exercise on hippocampal integrity: review of recent research. Int J Psychiatry Med 35(1):75–89
Kivipelto M, Rovio S, Ngandu T, Kareholt I, Eskelinen M, Winblad B et al (2008) Apolipoprotein E e4 magnifies lifestyle risks for dementia: a population-based study. J Cell Mol Med 12:2762–2771
Knaepen K, Goekint M, Heyman EM, Meeusen R (2010) Neuroplasticity: exercise-induced response of peripheral brain-derived neurotrophic factor. A systematic review of experimental studies in human subjects. Sports Med 40(9):765–801
Kramer AF, Bherer L, Colcombe SJ, Dong W, Greenought WT (2004) Environmental influences on cognitive and brain plasticity during aging. J Gerontol: Med Sci 59a(9):940–957

Larson EB, Wang L, Bowen JD, McCormick WC, Teri L, Crane PK et al (2006) Exercise is associated with reduced risk for incident dementia among persons 65 years of age and older. Ann Intern Med 144(2):73–82

Laurin D, Verreault R, Lindsay J, MacPherson K, Rockwood K (2001) Physical activity and risk of cognitive impairment and dementia in elderly persons. Arch Neurol 58:498–504

Lautenschlager NT, Cox KL, Flicker L, Foster JK, van Bockxmeer FM, Xiao J et al (2008) Effect of physical activity on cognitive function in older adults at risk for alzheimer disease. Am Med Assoc 300(9):1027–1037

Lindsay J, Laurin D, Verreault R, Herbert R, Helliwell B, Hill GB et al (2002) Risk factors for Alzheimer's disease: a prospective analysis from the Canadian study of health and aging. Am J Epidemiol 156(5):445–453

Lu B, Gottschalk W (2000) Modulation of hippocampal synaptic transmission and plasticity by neurotrophins. Prog Brain Res 128:231–241

Lytle ME, Vanderbilt J, Pandav RS, Dodge HH, Ganguli M (2004) Exercise level and cognitive decline. Alzheimer Dis Assoc Disord 18:57–64

Maguire EA, Spiers HJ, Good CD, Hartly T, Frackowiak R, Burgess N (2003) Navigation expertise and the human hippocampus: a structural brain imaging analysis. Hippocampus 13:250–259

Mahncke HW, Connor BB, Appelman J, Ahsanuddin ON, Hardy J, Wood RA et al (2006) Memory enhancement in healthy older adults using a brain plasticity-based training program: a randomized, controlled study. PNAS 103(33):12523–12528

Marks BL, Katz LM, Styner M, Smith JK (2010) Aerobic fitness and obesity: relationship to cerebral white matter integrity in the brain of active and sedentary older adults. British J Sports Med. doi:10.1136/bjsm.2009.068114

Middleton LE, Kirkland SA, Rockwood K (2008a) Prevention of CIND by physical activity: different impact on VCI-NO compared with MCI. J Neurol Sci 269:80–84

Middleton LE, Mitnitski A, Fallah N, Kirkland SA, Rockwood K (2008b) Changes in cognition and mortality in relation to exercise in late life: a population-based study. Plos ONE 3(9):1–7

Middleton LE, Barnes DE, Li-Yung L, Yaffe K (2010) Physical activity over the life course and its association with cognitive performance and impairment in old age. J Am Geriatr Soc 58(7):1322–1326

Nichol K, Deeny SP, Seif J, Camaclang K, Cotman CW (2009) Exercise improves cognition and hippocampal plasticity in APOE epsilon4 mice. Alzheimer's Dementia 5(4):287–294

O'Callaghan RM, Ohle R, Kelly AM (2007) The effects of forced exercise on hippocampal plasticity in the rat: a comparison of LTP, spatial and non-spatial learning. Behav Brain Res 176(2):362–366

Oken BS, Zajdel D, Kishiyama S, Flegal K, Dehen C, Haas M et al (2006) Randomized, controlled, six-month trial of yoga in healthy seniors: effects on cognition and quality of life. Altern Ther Health Med 12(1):40–47

Papp KV, Walsh SJ, Snyder PJ (2009) Immediate and delayed effects of cognitive interventions in healthy elderly: a review of current literature and future directions. Alzheimer's Dementia 5:50–60

Podewils L, Guallar E, Kuller LH, Fried LP, Lopez OL, Carlson M et al (2005) Physical activity, APOE genotype, and dementia risk: findings from the cardiovascular health cognition study. Am J Epidemiol 161(7):639–651

Ravaglia G, Forti P, Lucicesare A, Pisacane N, Rietti E, Bianchin M et al (2008) Physical activity and dementia risk in the elderly: findings from a prospective Italian study. Neurology 70:1786–1794

Richards M, Hardy R, Wadsworth MEJ (2003) Does active leisure protect cognition? Evidence from a national birth cohort. Soc Sci Med 56:785–792

Riley KP, Snowdon DA, Desrosiers MF, Markesbery WR (2005) Early life linguistic ability, late life cognitive function, and neuropathology: findings from the Nun study. Neurobiol Aging 26:341–347

Rolland Y, Rival L, Pillard F, Lafont C, Rivere D, Albarede J et al (2002) Feasibility of regular physical exercise for patients with moderate to severe Alzheimer disease. J Nutr, Health Aging 4(2):109–113

Rozzini L, Costardi D, Chilovi BV, Franzoni S, Trabucchi M, Padovani A (2007) Efficacy of cognitive rehabilitation in patients with mild cognitive impairment treated with cholinesterase inhibitors. Int J Geriatr Psychiatry 22(4):356–360

Scarmeas N, Levy G, Tang M, Manly JJ, Stern Y (2001) Influence of leisure activity on the incidence of Alzheimer's disease. Neurology 57:2236–2242

Scarmeas N, Luchsinger JA, Brickman AM, Cosentino S, Schupf N, Xin-Tang M et al (2010) Physical activity and Alzheimer disease course. Am J Geriatr Psychiatry 19(5):471–481

Schooler C, Mulatu MS, Oates G (1999) The continuing effects of substantively complex work on the intellectual functioning of older workers. Psychol Aging 14(3):483–506

Schuit AJ, Feskens EJ, Launer LJ, Kromhout D (2001) Physical activity and cognitive decline, the role of the apolipoprotein e4 allele. Med Sci Sports Exerc 33(5):772–777

Smith J, Verber MD, Nielson KA, Antuono P, Hason RJ, Mattes AJ et al (2011) Effects of walking exercise on white matter integrity in amnestic mild cognitive impairment. Paper presented at the international neuropsychological society annual meeting, Boston, Ma

Sturman MT, Morris MC, Mendes de Leon CF, Bienias JL, Wilson RS, Evans DA (2005) Physical activity, cognitive activity, and congnitive decline in a biracial community population. Arch Neurlogy 62:1750–1754

Sumic A, Michael YL, Carlson NE, Howieson DB, Kaye JA (2007) Physical activity and the risk of dementia in oldest old. J Aging Health 19(2):242–259

Sun Q, Townsend MK, Okerke OI, Franco OH, Hu FB, Grodstein F (2010) Physical activity at midlife in relation to successful survival in women at age 70 years or older. Arch Intern Med 170(2):124–125

Valenzuela MJ, Sachdev P (2009) Can cognitive exercise prevent the onset of dementia? Systematic review of randomized clinical trials with longitudinal follow-up. Am J Geriatr Psychiatry 17(3):179–187

van Gelder BM, Tijhuis MA, Kalminjn S, Giampaoli S, Nissinen A, Kromhout D (2004) Physical activity in relation to cognitive decline in elderly men: the FINE study. Neurology 63(12):2316–2321

van Praag H, Shubert T, Zhao C, Gage FH (2005) Exercise enhances learning and hippocampal neurogenesis in aged mice. J Neurosci 25(38):8680–8685

van Uffelen J, Chinapaw MJM, Hopman-Rock M, van Mechelen W (2008a) The effects of exercise on cognition in older adults with and without cognitive decline: a systematic review. Clin J Sport Med 18(6):486–500

van Uffelen J, Chinapaw MJM, van Mechelen W, Hopman-Rock M (2008b) Walking or vitamin B for cognition in older adults with mild cognitive impairment? A randomised controlled trial. British J Sports Med 42:344–351

Verghese J, Lipton RB, Katz MJ, Hall C, Derby CA, Kuslansky G et al (2003) Leisure activities and the risk of dementia in the elderly. New Engl J Med 348:2508–2516

Verghese J, LeValley A, Derby C, Kuslansky G, Katz M, Hall C et al (2006) Leisure activities and the risk of amnestic mild cognitive impairment in the elderly. Neurology 66(6):821–827

Weuve J, Kang J, Manson JE, Breteler MMB, Ware JH, Grodstein F (2004) Physical activity, including walking, and cognitive function in older women. JAMA 292(12):1454–1461

Willis SL, Tennstedt SL, Markiske M, Ball K, Elias J, Koepke KM et al (2006) Long-term effects of cognitive training on everyday functional outcomes in older adults. J Am Med Assoc 296(23):2805–2814

Wilson RS, Bennett DA, Bienias JL, Aggarwal NT, Mendes de Leon CF, Morris MC et al (2002a) Cognitive activity and incident AD in a population-based sample of older persons. Neurology 59:1910–1914

Wilson RS, Mendes de Leon CF, Barnes LL, Schneider JA, Bienias JL, Evans DA et al (2002b) Participation in cognitively stimulating activities and risk of incident alzheimer disease. J Am Med Assoc 287(6):742–748

Yaffe K, Barnes D, Nevitt M, Lui L, Covinsky K (2001) A prospective study of physical activity and cognitive decline in elderly women: women who walk. Arch Intern Med 161(14): 1703–1708

Yaffe K, Fiocco AJ, Lindquist K, Vittinghoff E, Simonsick EM, Newman AB et al (2009) Predictors of maintaining cognitive function in older adults: the Health ABC study. Neurology 9(72):23

Potential Benefits and Limitations of Enriched Environments and Cognitive Activity on Age-Related Behavioural Decline

Rosa Redolat and Patricia Mesa-Gresa

Abstract The main aim of this chapter is to review preclinical studies that have evaluated interventions which may aid in preventing or delaying age-related behavioural decline. Animal models of Environmental Enrichment (EE) are useful for evaluating the influence of cognitive, physical and social stimulation in mitigating cognitive decline at different ages. The EE paradigm has been proposed as a non-invasive treatment for alleviating age-related memory impairment and neurodegenerative diseases. While in this complex environment, rodents can be stimulated at different levels (physical, social, cognitive and sensorial), although a synergism between all these components is likely to play an important role. We will summarize available data relating to EE as a potential therapeutic strategy that slows down or counteracts age-related cognitive and behavioural changes. EE also alters physiological responses and induces neurobiological changes such as stimulation of neurogenesis and neural plasticity. At the behavioural level, EE improves learning and memory tasks and reduces anxiety. Several variables seem to influence the behavioural and cognitive benefits induced by EE, including the age at which animals are first exposed to EE, total period during which animals are submitted to EE, gender, the cognitive task evaluated, the drug administered and individual factors. Cognitive and physical stimulation of animals in enriched experimental environments may lead to a better understanding of factors that promote the formation of cognitive reserve (CR) and a healthier life in humans. In the present chapter we review the potential benefits of EE in aged rodents and in animal models of Alzheimer Disease (AD). Results obtained in preclinical models

R. Redolat (✉) · P. Mesa-Gresa
Departamento de Psicobiología, Universitat de València, Blasco Ibáñez 21, 46010 Valencia, Spain
e-mail: Rosa.redolat@uv.es

P. Mesa-Gresa
e-mail: Patricia.mesa@uv.es

Curr Topics Behav Neurosci (2012) 10: 293–316
DOI: 10.1007/7854_2011_134

Published Online: 4 June 2011

of EE may be relevant to future research into mental and neurodegenerative diseases, stress, aging and development of enviromimetics. Finally, we outline the main limitations of EE studies (variability between laboratories, difficulty of separating the different components of EE, gender of experimental subjects, individual differences in the response to EE), evaluating the potential benefits of enriched environments and the neurobiological mechanisms that underlie them. We conclude that there are experimental data which demonstrate the cognitive benefits of rearing rodents in enriched environments and discuss their implication for clarifying which variables contribute to the formation of the CR.

Keywords Age-related Behavioural Decline · Aging · Alzheimer's Disease · Cognitive Reserve · Enviromimetics · Individual Differences

List of Abbreviations

AD: Alzheimer's disease
Aβ: β-Amyloid deposition
CR: Cognitive reserve
EE: Environmental enrichment
LTP: Long-term potentiation

Contents

1 Introduction 294
2 The Study of Age-Related Behavioural Decline in Animal Models 296
3 How Lifestyle Factors May Have an Impact on Age-Related Behavioural Decline 296
4 The Environmental Enrichment Paradigm 297
5 Benefits of Environmental Enrichment and Cognitive Activity on Age-Related Behavioural Decline 302
5.1 Benefits of Environmental Enrichment in Aged Rodents 304
5.2 Benefits of Environmental Enrichment in Mouse Models of Alzheimer's Disease 304
5.3 Other Benefits 305
6 Can We Establish an Analogy Between "Enriched Environments" in Rodents and "Active Lifstyle" in Humans: Some Reasons for Caution 306
7 Main Limitations of the Studies Regarding Effects of Environmental Enrichment on Age-Related Behavioural Decline 308
8 Conclusions and Future Directions 309
References 311

1 Introduction

In the developed world of the twenty first century, the elderly represent a continuously increasing percentage of the population. Demographical studies suggest that this trend will be maintained in the coming years (Christensen et al. 2009).

These demographical changes are sure to have political, socio/economical and biological/medical consequences (Fratiglioni and Qiu 2009; Partridge 2010). One of the most important implications is a rise in the rate of age-related cognitive decline, Alzheimer's disease (AD) and other neurodegenerative diseases (Bishop et al. 2010; O'Callaghan et al. 2009).

These changes drive us to evaluate different interventions (either behavioural/environmental or pharmacological) which may aid in preventing or delaying age-related behavioural decline (Fratiglioni and Qiu 2009; Middleton and Yaffe 2010). The development and use of cognitive enhancers poses ethical and practical problems and is a subject of current debate (Darby 2010). Therefore, the application of environmental strategies could be a more fruitful approach (Petrosini et al. 2009).

There is growing evidence that maintaining an active lifestyle delays age-related behavioural decline and reduces the risk of developing AD and other neurodegenerative diseases (Lewejohann et al. 2009; Nichol et al. 2009). The extraordinary life of Rita Levi-Montalcini, the scientist who won the Nobel Prize in 1947 for her identification and characterization of the Nerve Growth Factor, is one example of this general rule. She has maintained an astonishing level of activity and at her 100 years of age continues to work in her research laboratory and foundation for the education of African girls (Chao 2010).

Epidemiological studies have shown that different lifestyle factors (such as educational achievement, occupational attainment, social activity, leisure activities and engagement in mentally stimulating activities) are correlated with the delayed onset of mild cognitive impairment and a lower incidence of AD (Daffner 2010; Fratiglioni and Qiu 2009; Gates and Valenzuela 2010; Papp et al. 2009). Many of these factors are modifiable and can have a protective effect even at middle or advanced ages, leading to a more successful cognitive aging (Plassman et al. 2010). Other factors, such as healthy diet (Scarmeas et al. 2009; van Praag 2009) and moderate alcohol consumption (Brust 2010), may also play a role.

There are, however, many variables which obstruct the interpretation of retrospective and epidemiological studies in humans. For that reason, animal models have been proposed in order to evaluate the beneficial effects of cognitive activity throughout life on age-related cognitive decline (Curtis and Nelson 2003; Petrosini et al. 2009). As Daffner (2010) has recently emphasized *"basic science and animal studies may aid in identifying which factors promote successful cognitive aging"*. Studies have reported a range of effects, among which some potential benefits have emerged (Cotel et al. 2010). In general, animals maintained in enriched environments display a slower rate of cognitive decline and neurobiological changes indicative of increased plasticity (Nithianantharajah and Hannan 2006).

In the current chapter, the preclinical studies that have helped to obtain a better understanding of the potential benefits of enriched environments and the neurobiological mechanisms that underlie them are discussed. In addition, the limitations of such studies are outlined and approaches to help overcome them are proposed.

2 The Study of Age-Related Behavioural Decline in Animal Models

Aging has been associated with a decline in different behavioural functions in humans and animals. Rodents have been used as models of human aging in order to evaluate changes in body composition and locomotor, sensory and cognitive decline. The need to develop animal models of aging that include a complete battery of behavioural tests has been emphasized (Fahlström et al. 2009). In general, aging has been better characterized in the rat than in the mouse, although the latter species is increasingly used, particularly knockout strains as preclinical animal models of AD (Morrissette et al. 2009; Philipson et al. 2010).

One of the most widely explored age-related changes is the impairment in learning and memory processes in different mice strains and using different learning tasks (both spatial, such as the Morris water-maze and eight-arm radial arm maze and non-spatial) (Gresack et al. 2007a, b; O'Callaghan et al. 2009; Pawlowski et al. 2009). Fahlström et al. (2009) have recently suggested that, when interpreting alterations in cognitive parameters, a poorer performance could be explained by a reduction in exploratory drive or changes in sensorimotor functions. However, in C57Bl/6 mice the lower levels of locomotion observed from the age of 10 months are independent of the cognitive status of each mouse (Fouquet et al. 2009).

Several neurobiological and neurochemical changes may correlate with these alterations of behaviour. One of the structures that is most vulnerable to age-related degeneration is the hippocampus. In this context, impairments in long-term potentiation (LTP) in the dentate gyrus, a decline in neurogenesis, reductions in growth factors such as insulin-like growth factor 1, impairment in some measures of plasticity and increased oxidative stress have been related to age (Lister and Barnes 2009; O'Callaghan et al. 2009). However, it is still unclear which mechanisms control neurogenesis in the hippocampus (Kuzumaki et al. 2010).

It is important to emphasize individual differences with respect to the rate of aging; not all subjects age equally, and inter-individual variability increases with age (Stranahan et al. 2011). Fouquet et al. (2009) have developed an experimental method whose aim is the early detection of age-related decline in cognitive function. Using a spatial navigation task (the starmaze), they identified sub-groups of cognitively impaired C57BL/6 male mice among middle-aged (10 month) and aged (17 month) animals that displayed an alteration of late-LTP maintenance. With age, there are also increases in individual differences in exploration and acquisition of motor skills.

3 How Lifestyle Factors May Have an Impact on Age-Related Behavioural Decline

The amount and form of cognitive enrichment that a person experiences in the course of his/her life may have profound consequences on cognitive ability and mental function (Daffner 2010; Nithianantharajah and Hannan 2009). Until recently, few efforts had been made to identify the environmental factors that can contribute to a healthy aging. However, this has changed thanks to recent studies that have identified several variables (activity participation, frequent contacts in a large social network, emotional support) that are neuroprotective against cognitive aging (Bielak 2009; Plassman et al. 2010).

Since the maintenance of an active lifestyle seems to contribute to the prevention of dementia, it is logical to ask which elements are critical for maintaining a healthy brain. In humans, it is difficult to separate the social, cognitive, physical and other components of an active lifestyle. Cognitive training seems to be important in maintaining cognitive health and has attracted public awareness and media interest in recent years. However, experimental evidence in young (Owen et al. 2010) and aged subjects (Gates and Valenzuela 2010) has caused the benefits of this type of training to be questioned, especially with respect to their application to real life and long-term effects. Social engagement, physical activity and implication in productive activities have also been proposed as effective methods for improving health both directly and indirectly (Hillman et al. 2008; Qiu et al. 2010; van Praag 2009).

Animal models are useful for evaluating the influence of cognitive, physical and social enrichment on age-related decline, and can throw light on the mechanisms implicated in the beneficial effects of a complex environment on mitigating cognitive decline at different ages. Preclinical studies in which researchers have evaluated how the brain responds to changing experiences have often employed the "Environmental Enrichment" (EE) paradigm (Nithianantharajah and Hannan 2009; van Praag et al. 2000). EE supplements the experience that a subject receives in standard conditions and can be considered as "*a richness that cures*" (Pizzorusso et al. 2007). It has been proposed as a non-invasive treatment for alleviating age-related memory impairment and neurodegenerative diseases (Bennett et al. 2006). However, how it cures and whether its benefits are similar to those of cognitive-enhancing drugs is still to be clarified (Pizzorusso et al. 2007; Qiu et al. 2010).

4 The Environmental Enrichment Paradigm

The potential benefits of enriched environments were underscored for the first time in the twenty-fourth century by Charles Darwin. The modern application of the experimental paradigm of EE was established by D. O. Hebb, although the introduction of EE as a scientific paradigm is generally attributed to M. Rosenzweig and colleagues (Diamond 2001; Rosenzweig et al. 1962; Solinas et al. 2010; van Praag et al. 2000). In the 1940s, Hebb (1947) reported positive effects of EE on animal learning and memory. As he described in his seminal paper, "*The effects of early experience on problem solving at maturity*", he took rats from his laboratory as pets for his children, allowing them to play and run free in the family's home. When these animals were compared with those that had remained in

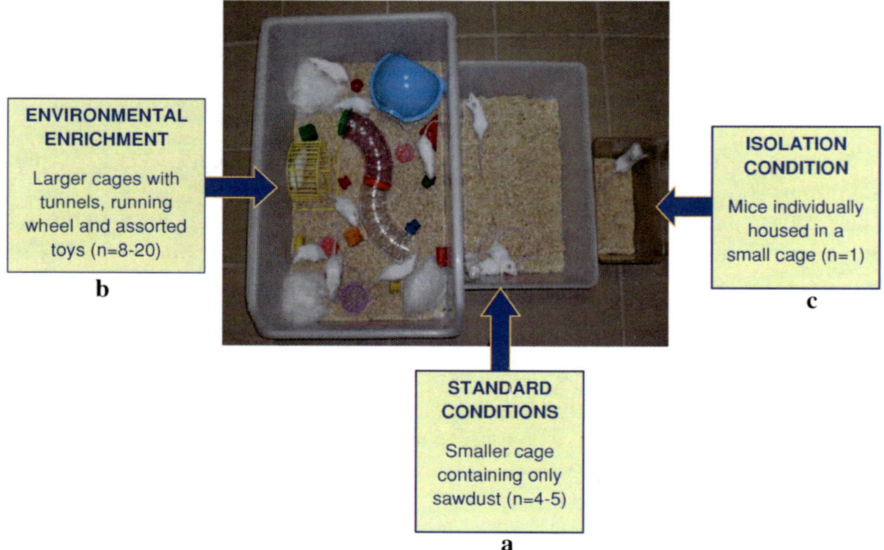

Fig. 1 Comparison between different forms of housing usually applied in rodent studies: (*a*) Standard Social Housing, (*b*) Environmental Enrichment, (*c*) Isolation Housing

standard laboratory conditions, he observed a general improvement in the performance of enriched rats in the Hebb-Williams maze. Since these groundbreaking studies, a great deal of research has confirmed that rodents that have been kept in complex enriched environments display anatomical, neurochemical, physiological and behavioural alterations (Hughes and Collins 2010). A more detailed characterization of these changes is now possible thanks to the sophisticated techniques applied in recent years, which are broadening our knowledge of the benefits of enriched experience (Petrosini et al. 2009; Sztainberg et al. 2010).

Different experimental procedures have been applied in order to evaluate the effects of environmental factors on neurobehavioural plasticity; these include EE, communal nesting, stress, physical activity and cognitive training. In the enriched environment, mice or rats are maintained in higher cages and larger groups (8–12 animals per cage, even 20 per cage) than would be usual. In these environments, different materials (colourful toys, plastic houses, igloos, ladders, tunnels, ropes, nesting material…) are introduced with the aim of encouraging exploratory activity (see Fig. 1). Furthermore, toys and other objects (and often food location) are changed or re-arranged frequently (usually daily or twice a week) in order to increase exploration. EE may offer enhanced social interaction, sensory and cognitive stimulation and problem-solving opportunities (Nithianantharajah and Hannan 2006, 2009; van Praag et al. 2000). In such environments, rodents increase play and curiosity, exhibiting highly motivated behaviours. In response to this stimulation, the brain undergoes a broad range of changes and general remodelling in order to cope with the environmental demands (Leggio et al. 2005; Mandolesi

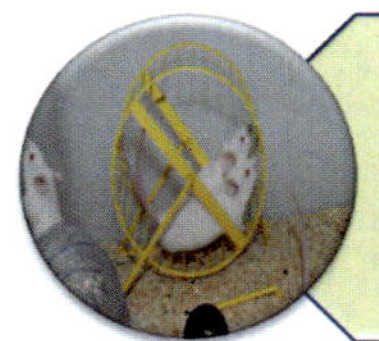

Physical activity stimulated by exploration of larger cages, voluntary exercise in running wheels and manipulation of different objects in the cage (ropes, balls, house, tunnels, ladders…).

Cognitive stimulation induced by the complex environment and stimulated by novelty, complexity, exploration and manipulation through continuous changes of the objects during the enrichment period.

Sensorial stimulation throughout different stimuli (houses, ladders, balls, tunnels, colourful and textured toys) that may offer oportunities for somatosensory and visual stimulation.

Social interaction is greater in enriched environments because of the higher number of conspecifics in each cage (8-20 animals). These conspecifics may also provide social support in stressful or complex situations.

Fig. 2 Main components included in the Environmental Enrichment paradigm

et al. 2008). As Sale et al. (2009) emphasized in a recent paper, "*enriching the environment empowers the brain*".

Experimental evidence indicates that rodents can be stimulated at different levels while in this complex environment: (1) *Physical*: The greater space offered by larger cages encourages rodents to be more active and explorative. Different objects present in the cage (running wheels ropes, balls, platforms, toys, houses, tunnels, ladders…) also stimulate physical activity, (2) *Socia*l: Higher number of conspecifics in each cage increases social interaction, (3) *Cognitive*: Learning opportunities may be offered by different objects and toys or nesting materials present in the environment, (4) *Sensorial*: Different stimuli (houses, ladders, balls, tunnels, colourful and textured toys) may also offer opportunities for somatosensory and visual stimulation (see Fig. 2). It cannot be denied the synergism between these components is likely to play an important role in determining the outcome of a continuous exposure to an EE. In support of this hypothesis, research in humans suggests that it is important to maintain an active lifestyle at all levels (social, intellectual and physical) in order to maintain cognitive functions and counteract age-related cognitive decline (Daffner 2010). Recently, the interaction between environmental enrichment, genetics and pharmacological manipulations has been the focus of attention (Pamplona et al. 2009) Fig. 2.

Many studies have been published regarding the impact of EE on age-related changes in behaviour and physiology in rodents. It is not within the scope of this chapter to provide a comprehensive description of these changes, but we will summarize the results of most interest in relation to EE as a potential therapeutic strategy that slows down or counteracts age-related cognitive and behavioural changes. At the behavioural level, the majority of studies emphasize an improvement in the ability to learn both spatial and non-spatial tasks (Diniz et al. 2010; Hansalik et al. 2006; Leggio et al. 2005; Viola et al. 2010) and faster adaptation to novel environments both in young (Brenes et al. 2008; Zimmermann et al. 2001) and aged animals (Hughes and Collins 2010). Rodents exposed to enriched environments have also been described as less fearful (Chapillon et al. 1999). In the forced swimming test, a predictive model of the efficacy of antidepressant drugs, EE reduces immobility behaviour, which may reflect an antidepressive-like effect (Cryan et al. 2005), though results seem to depend on the duration of EE (Brenes and Fornaguera 2008).

EE also alters physiological responses and induces neurobiological changes. Structural changes (including increases of weight, dendritic branching and number of synapses) have been reported in several brain regions, especially in the cortex and hippocampus (Nithianantharajah and Hannan 2006). An enriched environment stimulates neurogenesis and promotes neural plasticity in the hippocampus in both young and aged rodents. It also produces an increase in the number of neurotransmitter receptors, and an elevated presence of neurotrophic and growth factors in different areas of the brain (Kempermann et al. 2002). Long-term changes induced by EE involve alterations in gene expression profiles in the neurons and glial cells, especially in regions of the brain related to different neurodegenerative diseases, such as the striatum or hippocampus (Li et al. 2007; Thiriet et al. 2008). All these neurobiological changes could be related to the formation of a cognitive reserve (CR) (Petrosini et al. 2009). This poses the question of how the neuronal changes induced by an EE are translated into the changes in behaviour that are subsequently observed (Pizzorusso et al. 2007). This question can be responded in several ways based on the experimental evidence available and the hypotheses of different research groups. Curtis and Nelson (2003) suggest that complex environments induce the formation of new networks, leading to a more efficient processing of the environment and greater flexibility in problem solving.

If we compare different studies that have applied the paradigm of EE, several variables seem to influence the behavioural and cognitive benefits that are induced (Hu et al. 2010). Below we outline some of these variables, emphasizing those may have greater impact in aging studies:

(1) *Age at which animals are first exposed to EE* Is of great relevance in determining its effects. During some critical periods (for example, peri-adolescence and adolescence) the effects of an EE may be more evident

(Pietropaolo et al. 2008). In preclinical studies in mice, EE has been initiated during adolescence (Solinas et al. 2009), adulthood (Madroñal et al. 2010) or at an advanced age (Diamond, 2001), while in other studies it has been introduced during the pre-weaning period or even "in utero" (Lonetti et al. 2010; Sparling et al. 2010). At early postnatal ages, the effects of EE may be mediated by maternal care; pups exposed to enriched environments receive a higher level of stimulation through licking, grooming and physical contact (Maruoka et al. 2009).

(2) *The total period during which animals are submitted to EE* Some authors have hypothesized that changes in adulthood are sufficient for building a CR to be spent at a more advanced age (Petrosini et al. 2009). However, experimental evidence suggests that if a complex environment is present early in life and maintained throughout the life span, these changes will be more robust (Kempermann et al. 2002). The total period during which animals are exposed to EE differs among studies. In some cases, rodents are exposed for long periods (ranging from weeks to months, or even years), whereas, in others, exposure is limited to short periods (a few hours per day or repeated exposures over a few days). Some authors have suggested that 1 month exposure is enough to induce significant changes in rats with respect to animals maintained in isolation or standard housing (Brenes et al. 2008). Even shorter periods may be sufficient for producing marked behavioural changes in rats (Brenes et al. 2008) or mice (Redolat et al. 2009a), though longer periods obviously induce more robust effects. Changes in the open field have been observed after only 1 week of exposure whereas, in more complex tasks, such as the forced swimming test and Morris water-maze, longer periods of exposure are needed (Brenes and Fornaguera 2008; Leggio et al. 2005). Kobayashi et al. (2002) demonstrated that EE had beneficial effects in both adult (11 month-old) and aged (22 month-old) rats. In adult rats the effects of short-term (3 months) or long-term (24 months) enrichment were similar, though in aged rats the beneficial effects of long-term exposure were more pronounced, suggesting that aged animals also exhibit plasticity in their cognitive functions (Kobayashi et al. 2002). Further studies are necessary to respond to the question of whether an active lifestyle must be maintained during the whole life span if cognitive benefits are to be obtained, or whether exposure during critical periods is sufficient to prevent or delay some diseases (Van Dellen et al. 2008).

(3) *Gender* Few studies have focused on the differential impact of EE on males and females. At a neurobiological level, increases have been reported in dendritic density in the occipital cortex of males and in the somatosensory cortex of females (Diamond 2001). At a behavioural level, an increase in motor activity has been observed in the open field test among males, whereas a decrease has been reported among females (Elliott and Grunberg 2005). There are data that reveal differences between rats subjected to experimental brain injury and housed in enriched situations; performance in tasks of spatial memory has been shown to improve in males but not in females (Peña et al. 2006). The effects of EE on the elevated plus-maze seem to be

gender-dependent in social interaction tasks but not in the anxiety response (Peña et al. 2006). Few data have been published in aged rodents. Female rats are more susceptible to the negative effects of stress and respond differently to emotional tasks (Peña et al. 2009), although results are inconclusive. A recent study has shown that exposure to EE has an anxiolytic effect in male mice and an anxiogenic effect in females (Lin et al. 2010). Experimental evidence also suggests that lifelong enrichment can alter the effects of some hormones on behaviour and memory. Gresack et al. (2007a) reported significant differences in memory task performance between enriched young, middle-aged and old female mice treated with estradiol. The performance of aged mice in an object recognition task was enhanced after EE, whereas estradiol had no effect. Further studies are needed in order to evaluate gender-dependent behavioural and neurobiological differences in the effects of EE.

5 Benefits of Environmental Enrichment and Cognitive Activity on Age-Related Behavioural Decline

In humans, complex experiences induced by exposure to enriched environments influence changes in synaptic neurotransmission and plasticity that may provide the basis for the construction of a CR. The beneficial effects of physical activity and mental stimulation have been explained within the framework of this hypothesis. A subject that has built up a rich CR will have more alternatives on which to rely in the case of suffering a brain pathology: he/she will become more resilient and will be able to better apply available brain resources, thus maintaining his/her level of cognitive performance as age advances (Mandolesi et al. 2008; Petrosini et al. 2009). Numerous studies support the role of different factors that may favour the construction of a more "enriched" environment for humans (early education, complex social relationships, health behaviours, participation in leisure activities, bilingualism or multilingualism, creativity, occupational attainment and a generally active lifestyle) (McFadden and Basting 2010; Middleton and Yaffe 2010; Petrosini et al. 2009). It should also be taken into account that, in addition to a "cognitive" reserve, a "motivational" reserve can also be built up (Forstmeier and Maercker 2008).

Different authors have proposed that cognitive and physical stimulation in the EE paradigm is a useful model of active aging and may help to obtain a better understanding of factors that promote a healthier life (Frick and Benoit 2010; Petrosini et al. 2009). There is increasing experimental evidence suggesting that cognitive and social stimulation and physical activity aid in preventing age-related cognitive decline and reducing the risk of neurodegenerative disease (La Rue 2010). However, in humans, it is difficult to separate the contribution of each factor to CR and to evaluate the molecular mechanisms that underlie the beneficial effects of enriched environments. Animal models cannot parallel human factors such as level of education, intelligence, literacy, health care and cognitive, social and physical stimulation (Frick and Benoit 2010; Petrosini et al. 2009). Nevertheless, animal experimental models may be of use in obtaining a better control of variables and, in turn, demonstrating the neuroprotective effects offered by complex environments. Moreover, the evaluation of animals in complex environments allows us to study brain plasticity and resilience for coping with brain damage (Mandolesi et al. 2008). Nithianantharajah and Hannan (2010) have recently published an extensive review of the literature in which they evaluate the cognitive and neurobiological effects of environmental enrichment and physical activity in different animal models of Alzheimer's disease and brain CR. We must take into account that enriched environments are very complex experimental paradigms in which it is extremely difficult to dissociate the contribution of the different factors encountered (novelty, complexity, physical activity, cognitive stimulation, social behaviour) (Kempermann et al. 2010). Furthermore, there is an enormous heterogeneity among EE models (van Praag et al. 2000; Nithianantharajah and Hannan 2009, 2010). In spite of these challenges, there is agreement in the literature regarding human and animal studies that nutrition, social engagement, cognitive complexity and physical activity can favour the health of the brain and lead to a more successful aging. It would be of interest to evaluate in more detail combinations of different environmental variables (Reichman et al. 2010). There is increasing interest in developing and testing programmes that potentiate cognitive fitness, for which more scientific studies are needed (La Rue 2010). Results obtained in animal models are a possible indication of the direction that future research should take (Frick and Benoit 2010; Mandolesis 2009).

In this context, experimental evidence suggests that, in addition to pharmacological interventions (such as cognitive enhancement), other strategies can be implemented (such as physical and cognitive enrichment) in order to counteract age-related cognitive decline (Petrosini et al. 2009). The concept of cognitive enrichment is not applied only to aged subjects, since, in order to obtain greater benefits, it should be initiated early in life (Roe et al. 2007). The biological basis of the benefits induced by cognitive training is unknown. Changes in the density of cortical dopamine D1 receptors after intensive training for 5 weeks in a working memory task have been reported, although more studies are needed in order to correlate cognitive training with neurobiological changes (McNab et al. 2009).

Animal models of EE may help to confirm the theory of CR. The following sections summarize the results of preclinical studies that throw light on the benefits of EE in aged or transgenic mice used as models of AD. In a broad sense, EE in rodents may be applied as an experimental paradigm that has some parallelism with an active lifestyle in humans: (a) encouragement to perform physical exercise (running wheels, objects, tunnels and toys for climbing; changing objects frequently); (b) exposure to a visual, auditory, olfactory, tactile and taste-based stimuli (odours, different textures, sounds, toys and colourful objects); and (c) increased social contact (Curtis and Nelson 2003; Pamplona et al. 2009).

5.1 Benefits of Environmental Enrichment in Aged Rodents

Preclinical studies can help to respond to the question of whether it is possible to construct a CR in aged rodents. Some authors have suggested that an enriched environment is stimulating for young but not for old animals, whereas others have provided evidence that extends the well-known benefits of EE and physical activity to all ages (Frick and Fernandez 2003; Kempermann et al. 2002). Nevertheless, we must take into account that results may depend on the task and strain/species evaluated, the age at which training begins, and other factors (see Sect. 4). I t has been hypothesized that more complex enrichment protocols or a more long-term enrichment are needed in order to obtain beneficial effects in aging rodents (Bennett et al. 2006).

One of the variables that seems to have an impact on the effects of EE is the age at which environmental enrichment begins (Kobayashi et al. 2002; Segovia et al. 2008a, b). However, there are few studies regarding the effects of exposure to EE when initiated at an advanced age. Surprisingly, some studies have found that the behavioural changes and up-regulation of neurogenesis observed following exposure to EE are more significant in old animals that in young subjects, although beneficial effects may also depend on the type of enrichment (Kempermann et al. 1998). Bennett and co-workers (2006) compared the effects of continuous (10 weeks) versus daily enrichment (3 h per day) on spatial memory in aged male mice, and observed that continuous, but not daily, enrichment counteracted the age-related decline in spatial memory.

5.2 Benefits of Environmental Enrichment in Mouse Models of Alzheimer's Disease

As previously discussed, EE is thought to represent a therapeutic intervention that prevents the development of AD. In recent years, the potential protective effect of enriched environments and stimulating cognitive and physical activities has been evaluated in different types of transgenic mice *(CRND, APP23, AD11, PDAPP + PS1...)* used as models of AD. At a cognitive level, the results reported are generally consistent and indicate the benefits of exposure to EE with respect to the different changes associated with neurodegenerative diseases. After long-term exposure to EE (9 months), *APP23* mice displayed an improved acquisition of a spatial task (Wolf et al. 2006). Berardi et al. (2007) found that EE prevented both visual and spatial memory deficits in *AD11* mice for up to at least 12 months. Lifelong EE seems to offer cognitive protection in a variety of behavioural tasks in male and female *PDAPP + PS1* transgenic mice, thus counteracting age-related cognitive decline (Costa et al. 2007). More recently, Herring et al. (2008) observed an increase in hippocampal neurogenesis and structural plasticity when EE (4 months) was initiated at 30 days of age in CRND mice that already exhibited β-amyloid deposition (Aβ) pathology and cognitive deficits at 3 months of age. Voluntary wheel running may also delay the manifestation of AD-like behavioural symptoms in a triple transgenic mouse model of AD (3×TG-AD) (Pietropaolo et al. 2008).

Other authors, however, have not confirmed the beneficial effects of EE in preclinical models of AD. For example, Levi and Michaelson (2007) found that, after exposure to an enriched environment, APOE ε4 transgenic mice did not display any cognitive improvement. More recently, Cotel et al. (2010) confirmed that a combination of continuous EE and physical activity paradigms for 4 months beginning at 2 months of age did not improve working performance in memory tasks or levels of anxiety in *APP/PS1* KI mice. Görtz et al. (2008) also observed that EE did not counteract the deficits of transgenic TgCRND mice in two learning tasks (the Barnes maze and an object recognition task), though it did reduce anxiety levels.

The mechanisms by which EE may enhance learning and memory in different mouse models are largely unknown. It has been argued that enhanced cognition after enrichment may improve LTP and/or neurogenesis (Hu et al. 2010; Wolf et al. 2006). Results regarding the effects of EE on amyloid pathology are contradictory: some studies have confirmed that EE reduces the extent of amyloid deposition in transgenic mice (Berardi et al. 2007; Costa et al. 2007; Cracchiolo et al. 2007; Hu et al. 2010; Lazarov et al. 2005), while others have found no effects on plaque load (Arendash et al. 2004; Wolf et al. 2006; Cotel et al. 2010). Recently, Hu et al. (2010) demonstrated for the first time that EE attenuates the hyperphosphorylation of Tau protein, the other neuropathological hallmark of AD. More complete mouse models of AD are needed in order to better evaluate treatment strategies for this disease (Zahs and Ashe 2010).

Although the mechanisms implicated may be independent of brain Aβ (Cracchiolo et al. 2007), the majority of experimental results suggest that the stimulation offered by EE protects against AD-like cognitive impairment (Costa et al. 2007). In some cases, this protection is detectable even before symptoms appear (Herring et al. 2008). Some authors have stressed the need for EE to be continuous, although benefits may depend on the type of activity (Pietropaolo et al. 2008). In order to explain why activity and experience can improve the outcome of AD, Kempermann and other authors have proposed what they call the "*neurogenic reserve hypothesis*", which suggests that EE and activity promote neurogenesis and allow the aged hippocampus to cope better with environmental demands (Kempermann 2008). The type of EE paradigm or the mouse model of AD applied in each study may influence the results obtained, although the data generally confirm that the brain can also display considerable plasticity during aging (Laviola et al. 2008).

5.3 Other Benefits

The beneficial effects of an EE on behavioural recuperation after stroke in young rats are well-documented, though few studies have been carried out in aged rodents. Buchhold et al. (2007) demonstrated that EE (consisting of social groups of eight rats, maintained in larger cages with a running wheel, catwalk, playing tools and a tunnel) improved both behavioural recovery and neuropathological hallmarks (slower infarct development, fewer proliferating astrocytes and smaller glial scar) in aged rats that had suffered a stroke. The required recovery period depended on the task evaluated, since a longer exposure to EE was required in the more complex tasks (rotarod, labyrinth, radial arm maze, inclined plane). The delay period was shorter for young than for aged rats, although spontaneous motor activity was not fully recovered in either group. The cellular and molecular mechanisms underlying the beneficial effects of EE post-stroke are not well understood, although they may seem to be related to neurogenesis (Buchhold et al. 2007). More recently, Buga et al. (2009) confirmed that exposure to enriched environments may also aid in the functional recovery after brain ischemia in aged rats.

EE is an intervention which also helps in the recovery from different types of brain lesions (Mandolesi et al. 2008). Rearing rats in enriched environments may attenuate the impairment in cognitive flexibility induced by basal forebrain lesions (De Bartolo et al. 2008) and improve spatial learning in rats with subicular lesion-induced neurodegeneration (Dhanushkodi et al. 2007). Long-term EE housing up-regulates gene expression and enhances behavioural performance (cognitive flexibility and spatial memory) in non-lesioned and lesioned rats which display cholinergic damage (Paban et al. 2009). More recently, Sozda et al. (2010) confirmed that exposure to a typical enriched environment, in comparison with an atypical one, produced optimal benefits in the recuperation of traumatic brain injury.

Recent experimental evidence suggests that enriched environments, in combination with genetic factors, may be a non-invasive strategy for combating some types of cancer (Cao et al. 2010), which opens the door to a whole new world of possibilities. The need of performing similar studies in aged mice has been underscored (Pang and Hannan 2010).

6 Can We Establish an Analogy Between "Enriched Environments" in Rodents and "Active Lifstyle" in Humans: Some Reasons for Caution

A parallelism, with certain limitations, can be established between the procedure used in rodents and the way an enriching experience can promote the construction of a CR in humans that can counter or delay cognitive deficits or neurodegenerative diseases in aging. However, there are some limitations to enriched environments that may represent obstacles to this extrapolation: (1) the intensity and duration of exposure to an EE is difficult to control in human studies; (2) the age of onset of enrichment and timing of exposure may influence the results obtained; (3) a synergism between the different components of EE seems to be necessary in order to obtain beneficial effects in both rodents and human studies; (4) some human traits (education, occupation, emotion, language, literacy, complex behaviours, cognition, social relationships) cannot be adequately mimicked in animal experiments (Curtis and Nelson 2003; Laviola et al. 2008; Mandolesi et al. 2008; Morrissette et al. 2009). The use of the EE paradigm and the extrapolation to humans of the results obtained with it can be applied in areas such as aging, stress, neurodegenerative and psychological diseases. In the following paragraphs a brief summary of some of these applications is provided:

1. Evaluation of the contribution of environmental variables to different psychiatric disorders and neurodegenerative diseases. EE has been proposed as a useful procedure for modelling protective and risk factors for depression (Laviola et al. 2008), anxiety (Chapillon et al.1999), stress (Schloesser et al. 2010), recovery from cholinergic lesions (Paban et al. 2009) and stroke (Buchhold et al. 2007). Furthermore, hippocampal neurogenesis is currently under evaluation as a target for the treatment of different mental illness, such as major depressive disorder, post-traumatic stress disorder or AD (DeCarolis and Eisch 2010). Of special interest in the aging research is the study of EE as an intervention which may impact vulnerability to the cognitive effects of recurrent depression which usually results in neural insult and increases the risk of AD (Dotson et al. 2010).
2. Stress and aging: Different studies have been performed in order to evaluate whether or not EE ameliorates the behavioural and physiological effects induced by different stressors in rodents (Schloesser et al. 2010). Stress may have long-term effects causing a decrease of plasticity and increasing the risk of cognitive decline and AD at advanced ages (Sterlemann et al. 2010; Pardon and Rattray 2008). Previous studies have shown that the characteristics of the postnatal environment may modulate the emotional and neuroendocrine response to stress in mammals (Coutellier and Würbel 2009). It has been suggested that one of the benefits induced by EE is an enhancement of the ability to cope with stressors, which leads to lower levels of reactivity to stressful situations (Schloesser et al. 2010; Segovia et al. 2008b). This neuroprotective response could be related to the increased plasticity and neurogenesis induced in the adult hippocampus by an enriched environment (Llorens Martin et al. 2007; Schloesser et al. 2010). Experimental evidence indicates that EE diminishes the reactivity to stress of the prefrontal dopaminergic and cholinergic systems in the rat (Segovia et al. 2008a). Future studies should explore in depth the interaction between stress and aging and the impact of exposure to enriched environments in order to increase resilience to stress (Bloss et al. 2010; Segovia et al. 2008b).
3. Search of preventive strategies and non-pharmacological interventions for delaying age-related cognitive changes and AD: EE represents a useful model for evaluating the relevance of cognitive activity and physical exercise in constructing CR or preventing AD. The combination of cognitive stimulation and physical exercise seems to be a useful strategy for achieving this aim (Harburger et al. 2007; La Rue 2010).

4. Development of enviromimetic drugs: Based on experimental evidence obtained with EE, some authors hypothesize that it would be possible to develop pharmacological agents that either mimic or enhance the benefits induced by EE (Hannan 2004; Herring et al. 2008; McOmish and Hannan 2007).

In conclusion, although the extrapolation of results from animal models to human conditions is complicated (Laviola et al. 2008; Solinas et al. 2010), research about the potential benefits of enriched environments can be of great help in evaluating the potential efficacy of different environmental interventions in a wide range of situations.

7 Main Limitations of the Studies Regarding Effects of Environmental Enrichment on Age-Related Behavioural Decline

There are evident limitations when interpreting the results obtained in preclinical studies of EE and extrapolating them to human subjects. The most important of these limitations and their implications for studies performed in aged subjects are explored below:

1. Variability between laboratories: Detailed examination of the research published about this topic reveals an enormous variability between laboratories (size of cage, complexity and duration of learning and social situations, access to running wheels, exercise paradigms, number of cagemates, frequency of object-changing and experimental handling). All of these factors may play a role in the behavioural and physiological effects observed after exposure to EE at different ages (Van de Weerd et al. 1997, 2002). One possible solution with respect to reducing this variability and increasing the reproducibility of results is the standarization of environmental conditions (for example with standard enriched cages, such as Crial Two® or Marlau® enriched cages). Furthermore, in much research, authors do not evaluate the enrichment programme used (for example, whether or not animals use the objects for enrichment).
2. Normal versus impoverished environments: It has been suggested that standard housing in laboratory conditions can represent a "poor" or "impoverished" environment. Other studies, however, emphasize that we must distinguish between "normal" and "impoverished" environments. Some authors prefer the use of the term "environmental complexity", since "enriched environments" are in general more similar to the natural habitat of animals than normal laboratory conditions.
3. Difficulty of separating the different components in an EE: The great majority of the EE paradigms includes access to running wheels, although the specific impact of exercise is generally not evaluated. Furthermore, when rodents are housed in groups it is not possible to quantify the physical activity of each

animal. On the other hand, some complex environments do not include running wheels, which means the effects of exercise cannot be assessed. Furthermore, it is difficult to distinguish between cognitive stimulation and other types of enrichment.

4. Gender of experimental subjects: The majority of research about the effects of EE has been performed in male rodents. Although some previous studies have confirmed the existence of sex differences in the behavioural and/or neurobiological effects of EE (see Sect. 4), there is a scarcity of information regarding this aspect. Some recent studies performed in both normal aged (Gresack et al. 2007a, b) and transgenic mouse models of AD have addressed this question in females (Görtz et al. 2008; Mirochnic et al. 2009). A problem encountered with male mice is the increase of aggressive behaviour when they are maintained in the laboratory for long periods, as occurs in longitudinal studies.
5. Behavioural processes evaluated: The majority of studies about the neurobiological consequences of EE have focused on learning and memory and in the brain structures (such the hippocampus) involved in these behavioural functions. For instance, few studies have evaluated alterations in emotional levels (Lin et al. 2010). Furthermore, the majority of the research to have demonstrated significant effects of EE in learning and memory has been performed in aging rodents or in animals used as models of AD or other neurodegenerative diseases. It has been suggested that, in rodents without cognitive impairment, a more intense enrichment protocol (increasing novelty by changing objects daily or other types of enrichment) is necessary in order to obtain benefits.
6. Individual differences in the response to EE: Although the topic of individual differences in rodents has recently attracted great scientific interest (Redolat et al. 2009b), few studies have considered how individual rodents interact with their enriched environments. Greater individual differences have been reported in aged rodents, especially an increase in the variability of the hippocampus and its functions, and in the performance of spatial tasks.

8 Conclusions and Future Directions

In conclusion, EE is an experimental paradigm of a cognitively and physically active lifestyle in humans (Nithianantharajah and Hannan 2006). Its application may allow us to broaden our knowledge of experience and training-induced plasticity (Mirochnic et al. 2009) and of the mechanisms involved in the benefits of physical, social and cognitive enrichment in humans, thus contributing to advances in preventive non-invasive interventions for age-related neurodegenerative and other diseases (Curtis and Nelson 2003; Papp et al. 2009). A better understanding of the effects of EE on aging and AD will hopefully lead to the development of novel clinical approaches to the prevention and treatment of these disorders (Hannan 2004; McOmish and Hannan 2007).

The results of preclinical studies in rodents that point towards the potential benefits of EE vary depending on factors, including (a) the type of enrichment constructed, (b) the cognitive task evaluated, (c) the age at which exposure to EE begins (and whether or not it includes a critical period), (d) the total period of exposure, (e) individual factors (sex and age of the animals, species/strain), and (f) the drugs administered. However, in spite of this disparity, there are experimental data that would seem to be conclusive in demonstrating the cognitive benefits derived from rearing in enriched environments.

Studies about EE in animal models may have implications for evaluating which variables contribute to the formation of the CR. Nevertheless, in spite of the numerous epidemiological data that support this idea of a continuous cognitive and physical activity as the main factor in healthier aging, the neurobiological mechanisms that underlie these benefits are still unclear (La Rue 2010; Wolf et al. 2006).

The neurobiological, cognitive and emotional changes induced by EE as an experimental paradigm are currently under evaluation. In this way, the knowledge about the environmental factors related to conditions such as aging, AD and other neurodegenerative diseases (Herring et al. 2008; Laviola et al. 2008; Lazarov et al. 2005), recurrent depression (Laviola et al. 2008), anxiety (Costa et al. 2007) or stress (Segovia et al. 2008a, b) may be improved. They also have implications for animal welfare in laboratory settings (Sparling et al. 2010).

There are several interesting future directions for research about the impact of EE on age-related cognitive decline: (1) Evaluation of the interaction between EE and physical activity in aged animals (Kannangara et al. 2010); (2) Development and application of broader batteries of cognitive tests, including cognitively demanding tasks (Leggio et al. 2005); (3) A clearer distinction between positive (larger CR, better performance in cognitive tests, lower anxiety) and negative (increase of aggressive behaviour, stress) effects of exposure to EE; (4) The mechanisms that can potentiate increases in neural plasticity when the brain faces age-related or neurodegenerative changes (Paban et al. 2009); or (5) The evaluation of the combination of different interventions (for example, simultaneous application of EE and cognitive enhancers) with possible additional efficacy (Pamplona et al. 2009).

There are, however, many questions that are still to be answered: When must the exposure to complex environments begin in order for maximal benefits to be obtained? Which is the main component of the enriched environment related to the construction of the CR? As Petrosini et al. (2009) emphasizes, "*the answer to these questions will help us to better define the target populations for future interventions and consequently to better delineate preventive and therapeutic strategies*". All the knowledge we obtain could be then applied to develop better therapeutic strategies to delay or prevent age-related behavioural decline, AD and other neurodegenerative diseases (Hu et al. 2010; Morrissette et al. 2009; Pizzorusso et al. 2007).

Acknowledgments Supported by a grant from "Ministerio de Ciencia e Innovación" (Spain) and Plan E (Grant number: PSI2009-10410) and "Conselleria d'Educació i Ciéncia" from Generalitat Valenciana (Spain) (Grant number: GVACOMP2010-273). We would like to thank Asunción Pérez-Martínez for her help in photography and Mr. Brian Normanly for his help in revising the manuscript.

References

Arendash GW, Garcia MF, Costa DA, Cracchiolo JR, Wefes IM, Potter H (2004) Environmental enrichment improves cognition in aged Alzheimer's transgenic mice despite stable beta-amyloid deposition. Neuroreport 15:1751–1754

Bennett JC, McRae PA, Levy LJ, Frick KM (2006) Long-term continuous, but not daily, environmental enrichment reduces spatial memory decline in aged male mice. Neurobiol Learn Mem 85:139–152

Berardi N, Braschi C, Capsoni S, Cattaneo A, Maffei L (2007) Environmental enrichment delays the onset of memory deficits and reduces neuropathological hallmarks in a mouse model of Alzheimer-like neurodegeneration. J Alzheimers Dis 11:359–370

Bielak AA (2009) How can we not 'lose it' if we still don't understand how to 'use it'? Unanswered questions about the influence of activity participation on cognitive performance in older age—a mini-review. Gerontology 56:507–519

Bishop NA, Lu T, Yankner BA (2010) Neural mechanisms of ageing and cognitive decline. Nature 464:529–535

Bloss EB, Janssen WG, McEwen BS, Morrison JH (2010) Interactive effects of stress and aging on structural plasticity in the prefrontal cortex. J Neurosci 30:6726–6731

Brenes JC, Fornaguera J (2008) Effects of environmental enrichment and social isolation on sucrose consumption and preference: associations with depressive-like behavior and ventral striatum dopamine. Neurosci Lett 436:278–282

Brenes JC, Rodriguez O, Fornaguera J (2008) Differential effect of environment enrichment and social isolation on depressive-like behavior, spontaneous activity and serotonin and norepinephrine concentration in prefrontal cortex and ventral striatum. Pharmacol Biochem Behav 89:85–93

Brust JC (2010) Ethanol and cognition: indirect effects, neurotoxicity and neuroprotection: a review. Int J Environ Res Public Health 7:1540–1557

Buchhold B, Mogoanta L, Suofu Y, Hamm A, Walker L, Kessler C, Popa-Wagner A (2007) Environmental enrichment improves functional and neuropathological indices following stroke in young and aged rats. Restor Neurol Neurosci 25:467–484

Buga AM, Balseanu A, Popa-Wagner A, Mogoanta L (2009) Strategies to improve post-stroke behavioral recovery in aged subjects. Rom J Morphol Embryol 50:559–582

Cao L, Liu X, Lin EJ, Wang C, Choi EY, Riban V, Lin B, During MJ (2010) Environmental and genetic activation of a brain-adipocyte BDNF/leptin axis causes cancer remission and inhibition. Cell 142:52–64

Chao MV (2010) A conversation with Rita Levi-Montalcini. Annu Rev Physiol 72:1–13

Chapillon P, Manneche C, Belzung C, Caston J (1999) Rearing environmental enrichment in two inbred strains of mice: 1. Effects on emotional reactivity. Behav Genet 29:41–46

Christensen K, Doblhammer G, Rau R, Vaupel JW (2009) Ageing populations: the challenges ahead. Lancet 374:1196–1208

Costa DA, Cracchiolo JR, Bachstetter AD, Hughes TF, Bales KR, Paul SM, Mervis RF, Arendash GW, Potter H (2007) Enrichment improves cognition in AD mice by amyloid-related and unrelated mechanisms. Neurobiol Aging 28:831–844

Cotel MC, Jawhar S, Christensen DZ, Bayer TA, Wirths O (2010) Environmental enrichment fails to rescue working memory deficits, neuron loss, and neurogenesis in APP/PS1KI mice. Neurobiol Aging, in press

Coutellier L, Würbel H (2009) Early environmental cues affect object recognition memory in adult female but not male C57BL/6 mice. Behav Brain Res 203:312–315

Cracchiolo JR, Mori T, Nazian SJ, Tan J, Potter H, Arendash GW (2007) Enhanced cognitive activity-over and above social or physical activity-is required to protect Alzheimer's mice against cognitive impairment, reduce Abeta deposition, and increase synaptic immunoreactivity. Neurobiol Learn Mem 88:277–294

Cryan JF, Mombereau C, Vassout A (2005) The tail suspension test as a model for assessing antidepressant activity: review of pharmacological and genetic studies in mice. Neurosci Biobehav Rev 29:571–625

Curtis WJ, Nelson CA (2003) Toward building a better brain: Neurobehavioral outcomes, mechanisms, and processes of environmental enrichment. In S. Luthar (Ed.), Resilience and Vulnerability: Adaptation in the context of childhood adversities. In: Anonymous Cambridge University Press., New York, pp 463–488

Daffner KR (2010) Promoting successful cognitive aging: a comprehensive review. J Alzheimers Dis 19:1101–1122

Darby R (2010) Ethical issues in the use of cognitive enhancement. Pharos Alpha Omega Alpha Honor Med Soc 73:16–22

De Bartolo P, Leggio MG, Mandolesi L, Foti F, Gelfo F, Ferlazzo F, Petrosini L (2008) Environmental enrichment mitigates the effects of basal forebrain lesions on cognitive flexibility. Neuroscience 154:444–453

DeCarolis NA, Eisch AJ (2010) Hippocampal neurogenesis as a target for the treatment of mental illness: a critical evaluation. Neuropharmacology 58:884–893

Dhanushkodi A, Bindu B, Raju TR, Kutty BM (2007) Exposure to enriched environment improves spatial learning performances and enhances cell density but not choline acetyltransferase activity in the hippocampus of ventral subicular-lesioned rats. Behav Neurosci 121:491–500

Diamond MC (2001) Response of the brain to enrichment. An Acad Bras Cienc 73:211–220

Diniz DG, Foro CA, Rego CM, Gloria DA, de Oliveira FR, Paes JM, de Sousa AA, Tokuhashi TP, Trindade LS, Turiel MC, Vasconcelos EG, Torres JB, Cunnigham C, Perry VH, da Costa Vasconcelos PF, Diniz CW (2010) Environmental impoverishment and aging alter object recognition, spatial learning, and dentate gyrus astrocytes. Eur J Neurosci 32: 509–519

Dotson VM, Beydoun MA, Zonderman AB (2010) Recurrent depressive symptoms and the incidence of dementia and mild cognitive impairment. Neurology 75:27–34

Elliott BM, Grunberg NE (2005) Effects of social and physical enrichment on open field activity differ in male and female Sprague-Dawley rats. Behav Brain Res 165:187–196

Fahlström A, Yu Q, Ulfhake B (2009) Behavioral changes in aging female C57BL/6 mice. Neurobiol Aging, in press

Forstmeier S, Maercker A (2008) Motivational reserve: lifetime motivational abilities contribute to cognitive and emotional health in old age. Psychol Aging 23:886–899

Fouquet C, Petit GH, Auffret A, Gaillard E, Rovira C, Mariani J, Rondi-Reig L (2009) Early detection of age-related memory deficits in individual mice. Neurobiol Aging, in press

Fratiglioni L, Qiu C (2009) Prevention of common neurodegenerative disorders in the elderly. Exp Gerontol 44:46–50

Frick KM, Benoit JD (2010) Use it or lose it: Environmental enrichment as a means to promote successful cognitive aging. Sci World J 10:1129–1141

Frick KM, Fernandez SM (2003) Enrichment enhances spatial memory and increases synaptophysin levels in aged female mice. Neurobiol Aging 24:615–626

Gates N, Valenzuela M (2010) Cognitive exercise and its role in cognitive function in older adults. Curr Psychiatry Rep 12:20–27

Görtz N, Lewejohann L, Tomm M, Ambree O, Keyvani K, Paulus W, Sachser N (2008) Effects of environmental enrichment on exploration, anxiety, and memory in female TgCRND8 Alzheimer mice. Behav Brain Res 191:43–48

Gresack JE, Kerr KM, Frick KM (2007a) Life-long environmental enrichment differentially affects the mnemonic response to estrogen in young, middle-aged, and aged female mice. Neurobiol Learn Mem 88:393–408

Gresack JE, Kerr KM, Frick KM (2007b) Short-term environmental enrichment decreases the mnemonic response to estrogen in young, but not aged, female mice. Brain Res 1160:91–101

Hannan AJ (2004) Molecular mediators, environmental modulators and experience-dependent synaptic dysfunction in Huntington's disease. Acta Biochim Pol 51:415–430

Hansalik M, Skalicky M, Viidik A (2006) Impairment of water maze behaviour with ageing is counteracted by maze learning earlier in life but not by physical exercise, food restriction or housing conditions. Exp Gerontol 41:169–174

Harburger LL, Lambert TJ, Frick KM (2007) Age-dependent effects of environmental enrichment on spatial reference memory in male mice. Behav Brain Res 185:43–48

Hebb DO (1947) The effects of early experience on problem solving at maturity. Am Psychol 2:306–307

Herring A, Ambree O, Tomm M, Habermann H, Sachser N, Paulus W, Keyvani K (2008) Environmental enrichment enhances cellular plasticity in transgenic mice with Alzheimer-like pathology. Exp Neurol 216:184–192

Hillman CH, Erickson KI, Kramer AF (2008) Be smart, exercise your heart: exercise effects on brain and cognition. Nat Rev Neurosci 9:58–65

Hu YS, Xu P, Pigino G, Brady ST, Larson J, Lazarov O (2010) Complex environment experience rescues impaired neurogenesis, enhances synaptic plasticity, and attenuates neuropathology in familial Alzheimer's disease-linked APPswe/PS1DeltaE9 mice. FASEB J 24:1667–1681

Hughes RN, Collins MA (2010) Enhanced habituation and decreased anxiety by environmental enrichment and possible attenuation of these effects by chronic alpha-tocopherol (vitamin E) in aging male and female rats. Pharmacol Biochem Behav 94:534–542

Kannangara TS, Lucero MJ, Gil-Mohapel J, Drapala RJ, Simpson JM, Christie BR, van Praag H (2010) Running reduces stress and enhances cell genesis in aged mice. Neurobiol Aging, in press

Kempermann G (2008) The neurogenic reserve hypothesis: what is adult hippocampal neurogenesis good for? Trends Neurosci 31:163–169

Kempermann G, Kuhn HG, Gage FH (1998) Experience-induced neurogenesis in the senescent dentate gyrus. J Neurosci 18:3206–3212

Kempermann G, Gast D, Gage FH (2002) Neuroplasticity in old age: sustained fivefold induction of hippocampal neurogenesis by long-term environmental enrichment. Ann Neurol 52: 135–143

Kempermann G, Fabel K, Ehninger D, Babu H, Leal-Galicia P, Garthe A, Wolf SA (2010) Why and how physical activity promotes experience-induced brain plasticity. Front Neurosci 4:1-9

Kobayashi S, Ohashi Y, Ando S (2002) Effects of enriched environments with different durations and starting times on learning capacity during aging in rats assessed by a refined procedure of the Hebb-Williams maze task. J Neurosci Res 70:340–346

Kuzumaki N, Ikegami D, Tamura R, Hareyama N, Imai S, Narita M, Torigoe K, Niikura K, Takeshima H, Ando T, Igarashi K, Kanno J, Ushijima T, Suzuki T, Narita M (2010) Hippocampal epigenetic modification at the brain-derived neurotrophic factor gene induced by an enriched environment. Hippocampus, in press

La Rue A (2010) Healthy brain aging: role of cognitive reserve, cognitive stimulation, and cognitive exercises. Clin Geriatr Med 26:99–111

Laviola G, Hannan AJ, Macri S, Solinas M, Jaber M (2008) Effects of enriched environment on animal models of neurodegenerative diseases and psychiatric disorders. Neurobiol Dis 31:159–168

Lazarov O, Robinson J, Tang YP, Hairston IS, Korade-Mirnics Z, Lee VM, Hersh LB, Sapolsky RM, Mirnics K, Sisodia SS (2005) Environmental enrichment reduces Abeta levels and amyloid deposition in transgenic mice. Cell 120:701–713

Leggio MG, Mandolesi L, Federico F, Spirito F, Ricci B, Gelfo F, Petrosini L (2005) Environmental enrichment promotes improved spatial abilities and enhanced dendritic growth in the rat. Behav Brain Res 163:78–90

Levi O, Michaelson DM (2007) Environmental enrichment stimulates neurogenesis in apolipoprotein E3 and neuronal apoptosis in apolipoprotein E4 transgenic mice. J Neurochem 100:202–210

Lewejohann L, Hoppmann AM, Kegel P, Kritzler M, Kruger A, Sachser N (2009) Behavioral phenotyping of a murine model of Alzheimer's disease in a seminaturalistic environment using RFID tracking. Behav Res Methods 41:850–856

Li C, Niu W, Jiang CH, Hu Y (2007) Effects of enriched environment on gene expression and signal pathways in cortex of hippocampal CA1 specific NMDAR1 knockout mice. Brain Res Bull 71:568–577

Lin ED, Choi E, Liu X, Martin A, During MJ (2010) Environmental enrichment exerts sex-specific effects on emotionality in C57BL/6 J mice. Behav Brain Res, in press

Lister JP, Barnes CA (2009) Neurobiological changes in the hippocampus during normative aging. Arch Neurol 66:829–833

Llorens-Martin MV, Rueda N, Martinez-Cue C, Torres-Aleman I, Florez J, Trejo JL (2007) Both increases in immature dentate neuron number and decreases of immobility time in the forced swim test occurred in parallel after environmental enrichment of mice. Neuroscience 147:631–638

Lonetti G, Angelucci A, Morando L, Boggio EM, Giustetto M, Pizzorusso T (2010) Early environmental enrichment moderates the behavioral and synaptic phenotype of MeCP2 null mice. Biol Psychiatry 67:657–665

Madroñal N, Lopez-Aracil C, Rangel A, del Rio JA, Delgado-Garcia JM, Gruart A (2010) Effects of enriched physical and social environments on motor performance, associative learning, and hippocampal neurogenesis in mice. PLoS One 5:e11130

Mandolesi L, De Bartolo P, Foti F, Gelfo F, Federico F, Leggio MG, Petrosini L (2008) Environmental enrichment provides a cognitive reserve to be spent in the case of brain lesion. J Alzheimers Dis 15:11–28

Maruoka T, Kodomari I, Yamauchi R, Wada E, Wada K (2009) Maternal enrichment affects prenatal hippocampal proliferation and open-field behaviors in female offspring mice. Neurosci Lett 454:28–32

McFadden SH, Basting AD (2010) Healthy aging persons and their brains: promoting resilience through creative engagement. Clin Geriatr Med 26:149–161

McNab F, Varrone A, Farde L, Jucaite A, Bystritsky P, Forssberg H, Klingberg T (2009) Changes in cortical dopamine D1 receptor binding associated with cognitive training. Science 323:800–802

McOmish CE, Hannan AJ (2007) Enviromimetics: exploring gene environment interactions to identify therapeutic targets for brain disorders. Expert Opin Ther Targets 11:899–913

Middleton LE, Yaffe K (2010) Targets for the prevention of dementia. J Alzheimers Dis 20: 915–924

Mirochnic S, Wolf S, Staufenbiel M, Kempermann G (2009) Age effects on the regulation of adult hippocampal neurogenesis by physical activity and environmental enrichment in the APP23 mouse model of Alzheimer disease. Hippocampus 19:1008–1018

Morrissette DA, Parachikova A, Green KN, LaFerla FM (2009) Relevance of transgenic mouse models to human Alzheimer disease. J Biol Chem 284:6033–6037

Nichol K, Deeny SP, Seif J, Camaclang K, Cotman CW (2009) Exercise improves cognition and hippocampal plasticity in APOE epsilon4 mice. Alzheimers Dement 5:287–294

Nithianantharajah J, Hannan AJ (2006) Enriched environments, experience-dependent plasticity and disorders of the nervous system. Nat Rev Neurosci 7:697–709

Nithianantharajah J, Hannan AJ (2009) The neurobiology of brain and cognitive reserve: mental and physical activity as modulators of brain disorders. Prog Neurobiol 89:369–382

Nithianantharajah J, Hannan AJ (2010) Mechanisms mediating brain and cognitive reserve: Experience-dependent neuroprotection and functional compensation in animal models of neurodegenerative diseases. Prog Neuropsychopharmacol Biol Psychiatry, In press

O'Callaghan RM, Griffin EW, Kelly AM (2009) Long-term treadmill exposure protects against age-related neurodegenerative change in the rat hippocampus. Hippocampus 19:1019–1029

Owen AM, Hampshire A, Grahn JA, Stenton R, Dajani S, Burns AS, Howard RJ, Ballard CG (2010) Putting brain training to the test. Nature 465:775–778

Paban V, Chambon C, Manrique C, Touzet C, Alescio-Lautier B (2009) Neurotrophic signaling molecules associated with cholinergic damage in young and aged rats Environmental enrichment as potential therapeutic agent. Neurobiol Aging, in press

Pamplona FA, Pandolfo P, Savoldi R, Prediger RD, Takahashi RN (2009) Environmental enrichment improves cognitive deficits in Spontaneously Hypertensive Rats (SHR): relevance for Attention Deficit/Hyperactivity Disorder (ADHD). Prog Neuropsychopharmacol Biol Psychiatry 33:1153–1160

Pang TYC, Hannan AJ (2010) Environmental enrichment: a cure for cancer? It's all in the mind. J Mol Cell Biol 2:302–304

Papp KV, Walsh SJ, Snyder PJ (2009) Immediate and delayed effects of cognitive interventions in healthy elderly: a review of current literature and future directions. Alzheimers Dement 5:50–60

Pardon MC, Rattray I (2008) What do we know about the long-term consequences of stress on ageing and the progression of age-related neurodegenerative disorders? Neurosci Biobehav Rev 32:1103–1120

Partridge L (2010) The new biology of ageing. Philos Trans R Soc Lond B Biol Sci 365:147–154

Pawlowski TL, Bellush LL, Wright AW, Walker JP, Colvin RA, Huentelman MJ (2009) Hippocampal gene expression changes during age-related cognitive decline. Brain Res 1256: 101–110

Peña Y, Prunell M, Dimitsantos V, Nadal R, Escorihuela RM (2006) Environmental enrichment effects in social investigation in rats are gender dependent. Behav Brain Res 174:181–187

Peña Y, Prunell M, Rotllant D, Armario A, Escorihuela RM (2009) Enduring effects of environmental enrichment from weaning to adulthood on pituitary-adrenal function, pre-pulse inhibition and learning in male and female rats. Psychoneuroendocrinology 34:1390–1404

Petrosini L, De Bartolo P, Foti F, Gelfo F, Cutuli D, Leggio MG, Mandolesi L (2009) On whether the environmental enrichment may provide cognitive and brain reserves. Brain Res Rev 61:221–239

Philipson O, Lord A, Gumucio A, O'Callaghan P, Lannfelt L, Nilsson LN (2010) Animal models of amyloid-beta-related pathologies in Alzheimer's disease. FEBS J 277:1389–1409

Pietropaolo S, Sun Y, Li R, Brana C, Feldon J, Yee BK (2008) The impact of voluntary exercise on mental health in rodents: a neuroplasticity perspective. Behav Brain Res 192:42–60

Pizzorusso T, Berardi N, Maffei L (2007) A richness that cures. Neuron 54:508–510

Plassman BL, Williams JW Jr, Burke JR, Holsinger T, Benjamin S (2010) Systematic review: NIH state of the science conference: factors associated with risk for and possible prevention of cognitive decline in later life. Ann Intern Med 153:182–193

Qiu C, Xu W, Fratiglioni L (2010) Vascular and psychosocial factors in Alzheimer's disease: epidemiological evidence toward intervention. J Alzheimers Dis 20:689–697

Redolat R, Mesa P, Perez-Martinez A (2009a) Influence of environmental enrichment and nicotine administration on exploratory behavior in mice. Abstract presented at XIXth IAGG World Congress of Gerontology and Geriatrics

Redolat R, Perez-Martinez A, Carrasco MC, Mesa P (2009b) Individual differences in novelty-seeking and behavioral responses to nicotine: a review of animal studies. Curr Drug Abuse Rev 2:230–242

Reichman WE, Fiocco AJ, Rose NS (2010) Exercising the brain to avoid cognitive decline: examining the evidence. Aging Health 6:565–584

Roe CM, Xiong C, Miller JP, Morris JC (2007) Education and Alzheimer disease without dementia: support for the cognitive reserve hypothesis. Neurology 68:223–228

Rosenzweig MR, Krech D, Bennett EL, Zolman JF (1962) Variation in environmental complexity and brain measures. J Comp Physiol Psychol 55:1092–1095

Sale A, Berardi N, Maffei L (2009) Enrich the environment to empower the brain. Trends Neurosci 32:233–239

Scarmeas N, Luchsinger JA, Schupf N, Brickman AM, Cosentino S, Tang MX, Stern Y (2009) Physical activity, diet, and risk of Alzheimer disease. JAMA 302:627–637

Schloesser RJ, Lehmann M, Martinowich K, Manji HK, Herkenham M (2010) Environmental enrichment requires adult neurogenesis to facilitate the recovery from psychosocial stress. Mol Psychiatry 15:1152–1163

Segovia G, Del Arco A, de Blas M, Garrido P, Mora F (2008a) Effects of an enriched environment on the release of dopamine in the prefrontal cortex produced by stress and on working memory during aging in the awake rat. Behav Brain Res 187:304–311

Segovia G, Del Arco A, Garrido P, de Blas M, Mora F (2008b) Environmental enrichment reduces the response to stress of the cholinergic system in the prefrontal cortex during aging. Neurochem Int 52:1198–1203

Solinas M, Thiriet N, El Rawas R, Lardeux V, Jaber M (2009) Environmental enrichment during early stages of life reduces the behavioral, neurochemical, and molecular effects of cocaine. Neuropsychopharmacology 34:1102–1111

Solinas M, Thiriet N, Chauvet C, jaber M (2010) Prevention and treatment of drug addiction by environmental enrichment. Prog Neurobiol 92:572–592

Sozda CN, Hoffman AN, Olsen AS, Cheng JP, Zafonte RD, Kline AE (2010) Empirical comparison of typical and atypical environmental enrichment paradigms on functional and histological outcome after experimental traumatic brain injury. J Neurotrauma 27:1047–1057

Sparling JE, Mahoney M, Baker S, Bielajew C (2010) The effects of gestational and postpartum environmental enrichment on the mother rat: a preliminary investigation. Behav Brain Res 208:213–223

Sterlemann V, Rammes G, Wolf M, Liebi C, Ganea K, Müller MB, Schmidt MV (2010) Chronic social stress during adolescence induces cognitive impairment in aged mice. Hippocampus 20:540–549

Stranahan AM, Haberman RP, Gallagher M (2011) Cognitive decline is associated with reduced reelin expression in the entorhinal cortex of aged rats. Cereb Cortex 21:392–400

Sztainberg Y, Kuperman Y, Tsoory M, Lebow M, Chen A (2010) The anxiolytic effect of environmental enrichment is mediated via amygdalar CRF receptor type 1. Mol Psychiatry 15:905–917

Thiriet N, Amar L, Toussay X, Lardeux V, Ladenheim B, Becker KG, Cadet JL, Solinas M, Jaber M (2008) Environmental enrichment during adolescence regulates gene expression in the striatum of mice. Brain Res 1222:31–41

Van de Weerd HA, Van Loo PL, Van Zutphen LF, Koolhaas JM, Baumans V (1997) Nesting material as environmental enrichment has no adverse effects on behavior and physiology of laboratory mice. Physiol Behav 62:1019–1028

Van de Weerd HA, Aarsen EL, Mulder A, Kruitwagen CL, Hendriksen CF, Baumans V (2002) Effects of environmental enrichment for mice: variation in experimental results. J Appl Anim Welf Sci 5:87–109

van Dellen A, Cordery PM, Spires TL, Blakemore C, Hannan AJ (2008) Wheel running from a juvenile age delays onset of specific motor deficits but does not alter protein aggregate density in a mouse model of Huntington's disease. BMC Neurosci 9:34

van Praag H (2009) Exercise and the brain: something to chew on. Trends Neurosci 32:283–290

van Praag H, Kempermann G, Gage FH (2000) Neural consequences of environmental enrichment. Nat Rev Neurosci 1:191–198

Viola GG, Botton PH, Moreira JD, Ardais AP, Oses JP, Souza DO (2010) Influence of environmental enrichment on an object recognition task in CF1 mice. Physiol Behav 99:17–21

Wolf SA, Kronenberg G, Lehmann K, Blankenship A, Overall R, Staufenbiel M, Kempermann G (2006) Cognitive and physical activity differently modulate disease progression in the amyloid precursor protein (APP)-23 model of Alzheimer's disease. Biol Psychiatry 60:1314–1323

Zahs KR, Ashe KH (2010) 'Too much good news'—are Alzheimer mouse models trying to tell us how to prevent, not cure, Alzheimer's disease? Trends Neurosci 33:381–389

Zimmermann A, Stauffacher M, Langhans W, Wurbel H (2001) Enrichment-dependent differences in novelty exploration in rats can be explained by habituation. Behav Brain Res 121:11–20

Reproductive Experience may Positively Adjust the Trajectory of Senescence

Craig Howard Kinsley, R. Adam Franssen and Elizabeth Amory Meyer

Abstract Although aging is inexorable, aging well is not. From the perspective of research in rats and complementary models, reproductive experience has significant effects; indeed, benefits, which include better-than-average cognitive skills, a slowing of the slope of decline, and a healthier brain and/or nervous system well later into life. Work from our lab and others has suggested that the events of pregnancy and parturition, collectively referred to as reproductive experience—an amalgam of hormone exposure, sensory stimulation, and offspring behavioral experience and interaction—may summate to flatten the degree of decline normally associated with aging. Mimicking the effects of an enriched environment, reproductive experience has been shown to: enhance/protect cognition and decrease anxiety well out to two-plus years; result in fewer hippocampal deposits of the Alzheimer's disease herald, amyloid precursor protein (APP); and, in general, lead to a healthier biology. Based on a suite of recent work in organisms as diverse as nematodes, flies, and mammals, the ubiquitous hormone insulin and its large family of related substances and receptors may play a major role in mediating some of the effects of RE on the parameters of aging studied thus far. We will discuss the current set of data that suggest mechanisms for successful biological and neurobiological aging, and the implications for understanding aging and senescence in their broadest terms.

C. H. Kinsley (✉) · E. A. Meyer
Department of Psychology, Center for Neuroscience,
Gottwald Science Center and 116 Richmond Hall,
University of Richmond, B-326/328, 28 Westhampton Way,
Richmond VA 23173, USA
e-mail: ckinsley@richmond.edu

R. A. Franssen
Department of Biology, Longwood University, Farmville, VA, USA

Curr Topics Behav Neurosci (2012) 10: 317–345
DOI: 10.1007/7854_2011_123

Published Online: 25 May 2011

Keywords Amyloid precursor protein (APP) · Enriched environment · Hippocampus · Lactation · Learning and memory · Medial preoptic area · Oxytocin · Parity effects · Pregnancy · Reproductive experience · Senescence · Steroid hormones

Contents

1 Introduction 318
2 Aging, in General, and Brain Aging and Consequences for Behavior and Physiological Function 318
3 Evidence for Reproductive Experiential Mediation of Nervous System Development.... 324
4 Mechanisms and Likely Connections to Disparate Species and Systems 330
5 Summary, Conclusions, and Future Directions 337
References 338

1 Introduction

Life leads inexorably to death. The path, however, may be smooth and clear of obstacles, or it can be strewn with impediments. The form that aging takes is subject to regulation and mediation through genetics, experience, diet, and a host of interwoven factors. Given the inevitability of the body's descent into decrepitude, the focus is on extending the quality and/or quantity of life: modifying aging and the preserving of vitality for as long as possible. Of the many physiological factors that may control aging are those that, ironically, produce and protect life, namely, reproductive experience. This section will address the facets of aging that may be affected by the experiences associated with reproduction. Understanding these natural conditions and their effect on the aging process may elucidate principles that can be exploited to aid in successful aging.

2 Aging, in General, and Brain Aging and Consequences for Behavior and Physiological Function

Scientists studying a wide array of species, from worms to humans, have examined the factors that drive aging in an effort to better understand the process. Whether the motive is to discover the "fountain of youth", or simply to identify why/how we age, recent research provides fascinating details of this process. Along with bodily changes that occur in humans with aging (i.e., metabolism slowing, decrease in heart and cardiovascular function, weakening of immune function, etc.) there are a host of cognitive changes occurring as well. Further, with Alzheimer's disease and dementia rates on the rise, cognitive decline may be the greatest fear associated with aging. Descartes' famous maxim, "I think, therefore,

I am" is an appropriate descriptor for most people. That is, once one's mind departs—they are no longer "themselves". This section of the chapter will highlight recent work done in the field of the aging brain while emphasizing that aging is not a "fixed" process but one of plasticity and malleability.

Throughout the following chapter, signaling pathways and gene expression will be discussed as they are key players influencing the rate at which organisms age. Changes to intracellular organelles and the currency of neural activity, glucose utilization, form the very foundation for quality of aging. The "elderly" cell or neuron presents a complex and fascinating object for study. Among the factors that have attracted attention are those organelles that, for the entire life of the cell, fuel its activity, the mitochondrion. That, these change with age has attracted attention as primary sites of the initiation of aging in the neuron. Reduced expression of genes regulating mitochondrial function during aging is highly conserved across species (reviewed in Bishop et al. 2010). Specifically, brain and muscle are highly susceptible to faulty mitochondrial function (Bishop et al. 2010). Mitochondrial function begins to decline, and "longevity pathways" are likely activated as a compensatory mechanism to increase stress resistance (Bishop et al. 2010). As we discuss in more detail, stress has profound effects on the brain and cognition, but can exert its effects through multiple pathways. Mitochondrial integrity is an important bulwark against the many insults that can hasten the aging process. For the neuron, mitochondrial health and activity are crucial for defending against neurodegeneration. Factors that promote mitochondrial health would, therefore, be expected to facilitate longevity. (Interestingly, our laboratory has preliminary data showing enhanced mitochondrial activity—demonstrated by quantifying Mitotracker Green© activity in medial preoptic and hippocampal tissues—in lactating females compared to virgins [Morgan, Kinsley et al. unpublished observations]).

The insulin/IGF-1 pathway has been implicated in many aspects of aging. It functions as a nutrient sensor and has been demonstrated to regulate the DAF-2 gene, which in turn regulates reproductive development, resistance to oxidative stress, and autophagy among other things (reviewed in Kenyon 2010; see below). In the mammalian brain, decreased signaling of the insulin/IGF-1 pathway promotes a decrease in the brain pathology of Alzheimer's disease, however, increased signaling may be neuroprotective (reviewed in Kenyon 2010). In mammals, insulin and IGF-1 have been shown to support learning and memory and promote neuronal survival through inhibition of apoptosis (Van der Heide et al. 2006). Further, as most pathways involved in aging, the insulin/IGF-1 pathway is plastic. For example, it has been shown that pregnancy-associated plasma protein (PAPPA) raises IGF-1 levels, thus activating a downstream signaling pathway shown to positively affect lifespan (Kenyon 2010). We explore more deeply the nervous system connections later-on in this chapter, but it should be kept in mind that a lifetime of battering via stress, nutrient challenges, injury, etc., takes its toll on the nervous system. Even normal activities, such as breathing, carry a price.

The phenomenon of oxidative stress refers to various pathologic changes observed in living organisms in response to excessive levels of cytotoxic oxidants and free radicals in the environment. Aging increases reactive oxygen species

(ROS) such as free radicals that lead to cell and tissue damage. So what drives this change? An upregulation of oxidative stress-response genes has been identified in worms, flies, mice, rats, chimpanzees, and humans (reviewed in Bishop et al. 2010; Yankner et al. 2008). These data suggest that oxidative damage is conserved across species and is a mechanism of age related functional decline (Muller et al. 2007), and that such damage may be a fixed aspect of aging; other data indicate aging to be plastic, indeed. It may be surprising to some that changes in their daily routine may affect gene expression, as Fischer et al. (2007) demonstrated in the mouse. Here, the authors employed a mouse model capable of inducing expression of the p25 gene, which elicits such events as neurodegeneration, hippocampal synaptic degradation, and memory loss. They reported that environmental enrichment promoted the recovery of lost memories in these animals in addition to synaptic plasticity and induction of histone acetylation marks. Further, administration of a pharmacological histone deacetylase inhibitor evoked the same effect as environmental enrichment, including neuronal plasticity and recovery of memory (Fischer et al. 2007; reviewed in Bishop et al. 2010). Again, aging may not be a fixed process but one of many levels of malleability. As we will see later, the events associated with reproductive experience may be akin to a significant form of environmental enrichment.

Intra-cellular events, including intra-neuronal, regulate tissue aging responses. Autophagy, the cell's ability to cannibalize its own internal organelles when an inadequate supply of nutrients exists, like the body raiding its fat stores in times of famine, is crucial to regulating the aging process (Bishop et al. 2010). In worms (Melendez et al. 2003), flies (Simonsen et al. 2008), and mice (Hara et al. 2006), increased autophagy extends the lifespan. The enhanced autophagy effect may be related to its reduction of insulin-like signaling (Melendez et al. 2003), as well as to the prevention/removal of the buildup of protein aggregates like those observed in Huntington's or Alzheimer's disease (Bishop et al. 2010). A key regulator in autophagy is the signaling pathway of the kinase, target of rapamycin (TOR; Bishop et al. 2010). The TOR kinase is a major amino acid and nutrient sensor that controls cell growth and blocks autophagy when food is abundant. It activates an array of anabolic processes including protein synthesis, transcription, and ribosome biogenesis in response to nutrient presence (Bishop et al. 2010). Dysfunction or misregulation of this pathway may lead to altered cell volumes, which in turn may lead to developmental errors and subsequent pathological conditions. In contrast, when the TOR signal is inhibited, it increases the lifespan of many species, from yeast to mice (for full review, see Kenyon 2010). Further, administration of the TOR inhibitor, rapamycin, extends the lifespan in mice even when given in late life (Harrison 2009). As nutrition may be a driving factor in this age related pathway, caloric restriction may play a role in aging as well. It is interesting to note that caloric restriction and enhanced metabolic demands are a normal feature of the state of the pregnant-lactating female.

Caloric restriction, not to be confused with malnutrition, increases the lifespan of many species, including primates (Bishop et al. 2010; Colman et al. 2009). A reduced intake of calories improves verbal memory in humans (Witte et al. 2009), reduces amyloid-β deposition, and improves learning and memory in transgenic mouse models of Alzheimer's disease (Halagappa et al. 2007). This effect is believed to be mediated by gene expression changes—specifically in sirtuins. Sirtuins are NAD$^+$-dependant protein deacetylases whose expression is upregulated in animals that are calorie restricted (for a complete review, see Kenyon 2010). But their effects are subtle and possibly context-dependent. Sirtuins can increase or decrease lifespan, may be neuroprotective or harmful to neurons, under different conditions (for full review, see Bishop et al. 2010; Kenyon 2010). Although little work has examined AMP kinase in this paradigm, it has been identified as a nutrient and energy sensor similar to TOR kinase. Overexpression of AMP kinase signals have also been shown to lengthen the lifespan of worms (Apfeld et al. 2004) and is thought to extend the lifespan in response to dietary restriction. Additionally, the anti-diabetic drug, metformin, which activates AMP kinase, can extend life in mice (Anisimov et al. 2008). The plasticity of aging again comes to the fore, extending the gene-by-environment interaction to the latter stages of life.

To what extent might a plastic brain, sensitive to reproductive hormones and experiences, be a model for aging-related modifications? Until recently, the dogma was that at birth, we possess a finite number of brain cells which, only through cell death and damage, changes. This notion has been dismissed, because many data have shown that the adult brain produces new neurons in two areas of the brain, well into senescence. The subventricular zone of the lateral ventricle where the newly born cells migrate to the olfactory bulbs, and subgranular zone of the dentate gyrus of the hippocampus have been repeatedly shown to be rife with new neurons. Though most work has been done in rats, adult neurogenesis has been observed in humans as well (Kemperman and Gage 1998). Age is the number one negative regulator of adult hippocampal neurogenesis (Kemperman 2006), with numbers dropping significantly as age increases. In rats, however, the rate at which the new neurons are born is affected by such factors as stress (negatively; Gould et al. 1998), enriched environment (positively; Nilsson et al. 1999; Kemperman et al. 1997), hormones (like estrogen and prolactin: positively; Tanapat et al. 1999; Shingo et al. 2003; Kemperman et al. 1998, respectively), to name a few. Furthermore, in some paradigms, disruption of neurogenesis leads to behavioral deficits if neurogenesis is interrupted (Enwere et al. 2004; Mak and Weiss 2010).

Neurogenesis occurs throughout the adult life of primates, but it is susceptible to stress effects, as well, as alluded to above (Gould et al. 1998). High levels of glucocorticoids, which can bind to receptors in the hippocampus, significantly inhibit neurogenesis in adult trees shrews (Gould et al. 1997), marmoset monkeys (Gould et al. 1998), and rats (Cameron and Gould 1994; Gould et al. 1992). The effect is reversed when an antidepressant is administered to chronically stressed rats (Dagyte et al. 2010). In contrast to stressed animals, animals living in an enriched environment display heavier brain weights (Cummins et al. 1973, 1977); more dendritic branching (Volkmar and Greenough, 1972); a significantly greater number of glial cells (Altman and Das 1965). Enriched animals also perform better on tasks of learning and memory (Cummins et al. 1973), and show enhanced neurogenesis in their hippocampal dentate gyri (Nilsson et al. 1999; Kemperman et al. 1997; Kemperman, et al. 1998).

Some earlier work suggests parallels between parity and brain enrichment. Diamond and colleagues reported and discussed structural plasticity in the reproductive brain five decades earlier. For example, the width of the cortex in late-pregnant rats that were housed in impoverished conditions was equivalent to those of non-pregnant rats that were exposed to enriched conditions (Diamond et al. 1971; Diamond, Krech and Rosenzweig 1964). The "pregnant" cortex was stimulated in a manner akin to an environmentally "enriched" cortex. Other support for these older data indicating cortical plasticity in the maternal brain has also been reported (Stern 1996; Xerri et al. 1994) in the somatosensory cortex. Data have indicated that this important cortical region (which receives many stimuli from the rooting and suckling pups) undergoes neuronal reorganization and is significantly larger in lactating compared to virgin and non-lactating postpartum rats. According to Pascual-Leone et al. (2005), such changes in cortical "maps" denote plasticity of the type that fits the definition of, as they state, a "dynamic shift in the strength of preexisting connections across distributed neural networks". Further, such change clearly occurs "…in response to changes in afferent input or efferent demand" resulting from both pregnancy and offspring. That it supports reproduction fits both with Pascual-Leone's "evolution's invention" for the "value-added" superimposition of flexible learning mechanisms (onto genomic substrates), as well as with the requirement of marked neural responses to parity and enhancements in brain and behavior. Thus, the maternal brain is, by definition, a plastic organ, perhaps with effects that linger well into senescence.

Moreover, these enriching effects were observed in aged animals exposed to an enriched environment: they did not display the same rate of decline in neurogenesis compared to age matched, non-enriched controls (Kempermann and Gage 1998; Kempermann et al. 2002). These data are interesting as we learn that the brain (and the aging brain) has the potential for remarkable plasticity and change due to simple, natural changes in the environment. For example, exercise increases the endurance of cells and tissues in the brains of Alzheimer's patients, thereby enhancing the brain environment for further neurogenesis, memory improvement, and brain plasticity (Radak et al. 2010). Running has also been shown to increase adult hippocampal neurogenesis, learning, and long-term potentiation in the mouse, as well (Van Praag et al. 1999). It has been suggested that insulin-like growth factor-1(IGF-1) may be mediating the effect of physical activity on adult neurogenesis (Carro et al. 2001). Here, aged animals administered IGF-1 that mimicked the exogenous levels of younger animals showed a restoration of levels to those seen in younger adults (Lichtenwalner et al. 2001). These data demonstrate that plasticity is not based on external influences alone. Intrinsic factors play a role in rates of neurogenesis in adult and aging brains as well.

Enwere et al. (2004) have reported reduced epidermal growth factor receptor signaling in aged mice accompanied by diminished olfactory neurogenesis, which may be responsible for the mouse's impairment in fine olfactory discrimination.

It has been shown that adult-born neurons are physiologically functional in the olfactory bulb (Carlen et al. 2002; Carleton et al. 2003). Hence, their role in late-in-life olfactory function. Estrogen and prolactin have been shown to enhance neurogenesis in adult animals. Ovariectomized females show a decrease in neurogenesis, and animals that receive replacement estradiol show an enhancement (Tanapat et al. 1999). When cell death was examined in these animals, estrogen replacement not only stimulated cell proliferation but increased the survivability of the neurons as well, which suggest that estrogen may be neuroprotective for neurons (Suzuki et al. 2001; Wise et al. 2001). These data have many implications for hormone replacement therapy and the aging brain, as well as to suggest that natural reproductive experience, of the sort discussed here, may itself be neuroprotective. The fluctuations in the above mentioned hormones and changes in environment mimic another time of life, namely, that of pregnancy, birth, lactation, and maternal behavior. Could the changes we see in maternal female brains contribute to current research about the aging brain?

Shingo et al. (2003) reported that pregnancy and postpartum were accompanied by nearly double the number of new olfactory bulb interneurons. When they examined the factors that may contribute to this increase in neurogenesis, more closely; they found that prolactin, a chief pregnancy and lacational hormone, played a major role. Is the effect limited to the female? Could such a "female oriented hormone" such as prolactin play a role in neurogenesis in males? Few studies have examined the male paternal brain, but Mak and Weiss (2004) recently examined paternal recognition of young and the associated changes in the brain. They found that newly born olfactory interneurons in males were preferentially activated by their offspring's odors. The interruption of prolactin inhibited the production of new neurons, thus interfering with offspring recognition. Further, the recognition behavior was restored with the restoration of neurogenesis (Mak and Weiss 2010). So, could reproductive experience in this gender promote an "anti-aging" effect? Although there is a lack of thorough research pertaining to life expectancy and reproductive experience in males, one paper suggests that it may have a negative effect in males versus females. Rehm et al. (1984) revealed that Han:WIST female breeder rats live longer than virgins but that virgin males lived longer than their male breeder counterparts—an interesting trade-off, to be sure (sex versus longevity). These data would suggest that reproductive experience reduces lifespan in these animals, though it is important to keep in mind that this particular strain of rats does not often remain to rear the offspring, thus depriving themselves of the potential enriching benefits that may arise from the pup stimulation and care that we describe here. Interestingly, there are some data to suggest that "fatherhood" in primates results in similar brain changes as seen with animals living in an enriched environment (Kozorovitskiy et al. 2006). Specifically, primate fathers had a higher density of dendritic spines and vasopressin V1a receptors in the prefrontal cortex. Whereas we do not know if such changes would lead to a longer lifespan in males, parallel changes have been observed in the brains of maternal animals that appear to "age" better than their virgin counterparts. Thus, reproductive experience may improve the quality and not necessarily the quantity of lifespan. Further, paternal male mice have similar levels of circulating oxytocin compared to maternal mice, thereby making their brains "more maternal" when compared to non-parental males. As discussed above, the hormonal make-up of maternal animals is in part what contributes to their longevity or more successful aging. To the extent that males benefit from reproductive experience, those effects may be related more to care of offspring as opposed to reproductive experience, *per se*. In any event, the complexities of such endocrinological compromises certainly represent a fertile area for research. If such changes are occurring in maternal and paternal brains to promote and support cell birth and survival, would we expect to see positive persistent "anti-aging" effects in the maternal brain? As we will see in the next section, there are some interesting connections to suggest "yes".

3 Evidence for Reproductive Experiential Mediation of Nervous System Development

As we saw earlier in this chapter, and in the book overall, the complexities involved in the phenomenon of aging of the nervous system are astronomically complex, and the likely interactions seemingly infinite. There are, however, many strong and long-lasting modifications that occur in the female rat during and following pregnancy, changes that are directed at successful reproduction, and some of which that relate in interesting ways to the aging brain and nervous system. It stands to reason, therefore, that significant events in the life of the animal possess substantial neural consequences. Note that early development and sexual differentiation, as well as puberty, mark the animal for the remainder of its life. No one questions that these events are life-defining. Here, we argue, too, that reproduction may take its place alongside the aforementioned epochs in the animal's life.

For example, there are permanent modifications in the female's behavior toward her and other young, a facilitation of maternal behavior, which suggests a likely permanent alteration of cognition associated with reproduction and hence, underlying neural structures. Earlier work by Moltz et al. (1969) had shown that females who experience reproduction and subsequent young retain the memory for pups, and act maternal toward them, many months after their initial exposure. Bridges (1975) demonstrated the lengthy effects that both pregnancy and the experiences of parturition had on the maternal memory for pup responsiveness. A single early reproductive event and associated maternal experience is sufficient to mark the female for many months, the equivalent of decades in the human, in her maternal responsiveness.

Other evidence shows a form of cumulative change to the brain with repeated reproductive experiences. Svare and Gandelman (1976) reported that, in Rockland-Swiss albino mice, postpartum aggression continued to increase in intensity, through parity experiences five or six. The animals displayed more aggression with more pregnancies and lactations, as if the combination of pregnancy, hormone exposure, and pup sensory stimulation built a foundation upon which substrate, subsequent pregnancies acted, like stories in a high-rise building. Parity-induced changes in the endogenous opioid system likewise show a step-wise pattern of effects, with primiparous females being significantly different than virgins, and multiparous being significantly different than primiparous in their sensitivities to both endogenous and exogenous opiates (Kinsley et al. 1999; Mann and Bridges 1992; and in humans, Agaram et al. 2009). Their underlying opioid systems are similarly modified (Bridges and Hammer 1992). Further, Felicio and colleagues have shown a similar effect (parity-induced) on dopamine activity (Bridges et al. 1993; Felicio et al. 1996; Bridges and Grimm 1982). Together, these data suggest a blanket effect of parity on numerous neurochemical systems, a plasticity likely directed at ensuring successful reproduction.

These long term effects on the brain and learning-and-memory and emotional/affective consequences indicate a powerful effect of the most basic and natural of life's experiences: reproduction. Further, it begs the question: why should such manifest effects occur? We will explore these questions below.

Such neural effects suggest a significant maternal memory for young—for the many cues associated with pups (sights, smells, sounds, gustatory inputs, tactile stimulation, suckling stimulation, etc.)—that is resistant to the degradation of age or at least to interference. These data are suggestive of significant alterations of the substrate for learning, perhaps hippocampus, cortex, etc.

Recent data from our laboratory examined significant decreases in the number of degenerating neurons (as measured by silver-stained neurons) in the dorsal raphe, and frontal, parietal, and cingulate cortices (Love et al. 2005). The possible effects in the cortex are of interest because of recent reports which demonstrate that transient spatial memory information from the hippocampus is transferred to permanent storage in prefrontal and cingulate cortex (Maviel et al. 2004). Changes there include synaptogenesis and reorganization of the laminar layers, reminiscent of the cortical changes in somatosensory cortex in lactating females reported by Xerri et al. (1994). The data on the maternal brain, therefore, suggest an early, natural hormonal (estrogen, progesterone, oxytocin, etc.) mechanism that may eventually protect the aged, parous brain.

The same hormones shown to stimulate maternal behavior may also affect the foundation for such learning as presented above. For instance, in the hormonal profile characteristic of pregnancy, the powerful steroids progesterone (P) and estradiol (E_2) regulate the transition from non- to high responsiveness to neonates (Bridges 1984, 1990, 2009) through neuron-hormone interactions and alterations of neuronal activity. The temporal patterning of exposure to E_2 and P during pregnancy—in particular, the progressive alteration of the ratio of the two hormones over the 3 weeks of gestation, especially the rat's third trimester where E_2 is significantly elevated—are required for the eventual production and display of maternal behavior, and the requisite brain changes. Other substances, e.g., prolactin, placental lactogens, oxytocin, endogenous opioids, etc. also fluctuate temporally and are vital to both pregnancy and parturition, as well as to the onset of maternal behavior (see, e.g., Bridges et al. 1996, 1997; Kinsley 1994).

The striking influence exerted by natural levels of steroid and protein hormones extends into many areas. For example, the hormones of pregnancy exert significant changes in behavior and physiological regulation in the female, with the immediate onset of maternal behavior—characteristic of the postpartum, maternal female—being the most striking. These rapid and intense behavioral effects are preceded by significant actions of the hormones on the female's neural substrate. Several studies have reported that P and E_2 modify the structure of the neuron in the adult female brain, for example, increasing the concentration of apical dendritic spines in hippocampal neurons (McEwen and Woolley 1994; Woolley et al. 1990; Woolley and McEwen 1992, 1993; Yankova et al. 2001). These effects occur with relatively short exposure to the hormones, primarily E_2, during the female's estrous cycle; if the level or pattern of E_2 and P associated with the estrous cycle is prolonged or increased, as occurs during pregnancy, there may be even greater and/or long lasting effects on the morphology of the neuron.

A study using Golgi-Cox silver-staining revealed an example of neural plasticity due to pregnancy and its hormonal exposure (Keyser et al. 2001). Neurons in the medial preoptic area (mPOA), a region that strongly regulates maternal behavior (Numan and Insel 2003), were examined in a group of females from different hormonal groups (ovariectomized [OVX], diestrus, sequential progesterone and estradiol exposed [P and E2, respectively], late-pregnant, and lactating [d5] rats). There was no difference between OVX and diestrous females, and both had smaller somal areas compared to P and E2-treated and late-pregnant females. The area of the soma returned to diestrus/ovx levels in lactating females, suggesting a return to baseline that follows the female's pregnancy hormonal state. Further, there were similar hormone-induced effects on a number of dendritic branches and cumulative dendritic length in pregnant and hormone-treated groups compared to the OVX, diestrus, and lactating females. The increase in somal area denotes increased cellular activity (Miller and Erskine 1995), whereas the stimulatory effects on additional neuronal variables represent modifications in information processing capacity. Pregnancy and its attendant hormonal exposure, may stimulate neurons in the mPOA and, possibly other regions, which then contribute to the display of maternal behavior and its supporting activities.

For example, whereas the mPOA regulates pup-directed maternal behavior, ancillary sites, are also liable to undergo changes in their own right. Data show that the effects appear to extend to other behaviors and additional brain regions on which the female relies for carrying-out her maternal duties. Reproductive experience and exposure to offspring significantly modify the brain and behavior of the female, particularly those required for effective care and rearing of offspring. For example, age-matched female rats with multiple reproductive experiences (so-called multiparous females) exhibit reductions in opiate sensitivity relative to virgin or primiparous (one parous experience) females (Kinsley and Bridges 1988; Mann and Bridges 1992). Further, parous females performed significantly better in tests of spatial ability, compared to age-matched nulliparous (no reproductive experience) females (Kinsley et al. 1999). Together, the data suggest hormone-neuron interactions that have primarily maternal behavioral implications.

The potentiating experience of pregnancy, lactation, and pup exposure appear to alter female brain and behavior. In fact, it has been argued that these plastic changes are the hallmark of the female brain as it prepares to care for an expensive metabolic and genetic investment for a significant portion of its lifetime (Kinsley and Lambert 2006, 2008; Kinsley et al. 2008; Lambert et al. 2005; Love et al. 2005; Wartella et al. 2003). Our work here examines the extent to which the effects that are coming to light regarding RE are lengthy and widespread (Lambert and Kinsley 2009). For example, Gatewood et al. (2005) investigated the ramifications of reproductive experience for lifetime effects. They examined the spatial learning and memory of nulliparous, primparous, and multiparous rats out to 2 years post-reproduction. Their data suggest that the effects of RE are long-lived, persistent, and may contribute to late-in-life hippocampal integrity and function.

In the Gatewood et al. work, reproductive experience in females appears to regulate spatial learning and memory throughout life. In this study, the animals were behaviorally tested at six, twelve, eighteen, and twenty-four months of age on a spatial learning task, which required them to remember the location of a baited food well in the dry land maze (DLM; a Morris water maze analog; Kesner and Dakis 1995). Beginning at 12 months of age, and continuing until 24 months, the animals underwent additional testing (on a so-called reversal task), which required that they "un-learn" the original contingency, and learn a new location for the food reward, thereby assessing the flexibility of their learning.

At each age, and for both main and reversal tasks, the multiparous females remembered the baited location significantly faster than both the nulliparous and primiparous females; the primiparous females learned the mazes significantly faster than the nulliparous females at 12, 18, and 24 months (in the main task) and at 12 and 24 months (in the reversal task). These animals appeared to retain the enhanced learning and memory abilities that they developed around the experiences of reproduction. They function more efficiently well out to the latter stage of their lives. We will discuss other aspects of this persistent effect of reproductive experience.

With the behavioral data in hand, the authors sought to examine likely neural regulators—or, at least, associates—of the enhanced memory effect. Thus, when the brains of these animals were examined, the parous animals also had a neural substrate that appeared healthier, as well, at least with regard to a single measure: neural health. The parous rats generally had fewer hippocampal deposits of the deleterious substance, amyloid precursor protein (APP). Compared to nulliparous females, the parous animals' APP levels were negatively correlated with their performance in both the main and reversal tasks (i.e., more APP equaled poorer performance in the maze). This was a surprising find. In essence, all that these subjects had done to them was mating and pup exposure. Thus, the events surrounding reproductive experience (e.g., mating stimuli, elevations in pregnancy hormones such as estrogen, progesterone, oxytocin, etc., cues and/or stimulation from the young), may summate to produce a female brain that is both flexible and, perhaps, healthier in the aged female rat. The notion of pups as an enriched environment (see Kinsley et al. 1999; Vallée et al. 2001) is suggestive, and there is evidence that enriching environments have positive effects on Alzheimer's disease pathology (Bennett et al. 2003). Early life experiences (including reproductive?) may help to forestall some negative components of the aging process.

In a preliminary test of the neuroprotective effects of reproductive experience (Kinsley and Brown, unpublished), we examined transgenic mice that over-express the APP protein (Jackson Lab's B6.Cg-Tg[PDGFB-APP]5Lms/J). We mated half of these mice and allowed them to remain with their offspring through weaning. Then, ten days following weaning, the animals were trained to find a food reward in the DLM. These are preliminary data with a small N (3 three animals per group only.) Nonetheless, the data are provocative: the primiparous B6.Cg-Tg[PDGFB-APP]5Lms/J females significantly out-performed the un-mated females in a probe task which required the mice to spend time near the previously baited food well. Of special note is the fact that the behavioral improvements occurred at a time ($\sim$170d) when the age-progression of the APP effects begin to exert themselves in this animal model of age-related neurodegeneration. Therefore, a simple manipulation of these B6.Cg-Tg[PDGFB-APP]5Lms/J mice's parity status appear capable of marked positive effects on an otherwise baleful protein product associated with senescence and deterioration of cognitive ability. Follow-up experiments are ongoing.

Other recent work has examined some of the gene expression patterns of mothers versus non-mothers (Contino et al. 2007; Kinsley et al. 2008). As we have discussed, life for the single, non-reproductive animal is harsh enough; with a full litter of demanding and "costly" offspring, however, the parous female faces the even more daunting task of keeping herself and her young alive. It follows, therefore, that mechanisms, expressed transiently or otherwise, might have evolved to assist in the dual survival of mother and young. Reproductive experience significantly enhances spatial learning and memory in rats (Kinsley et al. 1999). Others (Tomizawa et al. 2003) have demonstrated the role of oxytocin and other mechanisms in such improvements, coupled to significant modifications of hippocampal CA1 dendritic spine density (Kinsley et al. 2006). How general are the effects of parity on learning and memory, related tasks, and the various neural substrates that either support it directly or rely on it for survival? In a DNA microarray study (Kinsley et al. 2008), we timed-mated females, producing parous, lactating and non-parous groups. At their respective and specific stages of reproduction, we killed the females and isolated the CA1 region from each brain. Using a modified pipette tip punch, we excised bilateral tissue (15–20 mg of tissue/brain). The results showed a significant set of differences between the lactating and NULL females, especially and interestingly, for the set of genes associated with the insulin family. In particular, we found that, overall, there were 91 genes expressed in lactating but not nulliparous females, and 49 genes expressed in the former but not the latter females (Kinsley et al. 2008). Focusing on genes with a two-fold difference or greater (in parentheses), where lactating females >nulliparous females, the following gene expression differences were observed for:

- Insulin-like growth factor (10.6)
- Sensory neuron synuclein (8.6)

- Synapotojanin (6.8)
- Proenkephalin (4.7)
- Calmodulin-dependent protein kinase (CaM kinase-GR) (4.6)
- Insulin-like growth factor binding protein (4.5)
- Insulin growth factor-binding protein (4.0)
- SNAP-25a (3.2)
- Huntington's disease mRNA (2.8)
- Potassium channel mRNA (2.7)
- Glutamate receptor (2.5)
- Interleukin-1 b-converting enzyme-related protease (2.2)

Where nulliparous females >lactating females, the following pattern was observed:

- 5-Hydroxytryptamine receptor (5HT5b) (11.2)
- Olfactory inositol 1,4,5-triphosphate receptor (3.4)
- Na–Ca Exchanger isoform NACA-1 (2.8)
- RET Ligand-2 (RETL2) (2.6)
- Neural receptor protein-tyrosine kinase (trkB) (2.4)
- Glycine receptor a (2.3)
- P2x (ATP) receptor (2.1)

These data (Kinsley et al. 2008) suggest some intriguing effects of reproductive experience. Among the genes showing expression pattern differences are those considered to be neuroplastic, in particular the insulin-family genes. The latter are interesting, too, in light of recent work by Toth et al. (2007) in which insulin growth factor (Igf) gene expression differences were implicated in the development of eusociality in insects. The authors here showed that in social wasps, females that engaged in brood care either through active reproduction or via foster care demonstrated an up-regulation of the same family of genes observed in mammalian mothers similarly engaged. It is intriguing to think that the act of parental care spreads across the phylogenetic scale a shared set of genetic changes uniting diverse species in the two core principles that have governed life on this planet since its inception: ingestion and reproduction. And it can be argued that feeding merely supports reproduction. Other effects on shared genes can be found in the next section. That feeding and aging well are related, has been shown in the earlier part of this chapter (and following) related to food restriction and insulin gene signaling in diverse species.

In other work, we time-mated females, producing parous (including MULT) and non-parous groups (Contino et al. 2007). At their respective and specific stages of reproduction, we killed the females and isolated the CA1 region from each brain. Using a modified pipette tip punch, we excised bilateral tissue (15–20 mg of tissue/brain). We then flash-froze the samples in –75°C acetone and stored them at –80°C until we determined relative levels of target mRNA expression via Quantitative, Reverse Transcriptase Polymerase Chain Reaction (qRT-PCR) analysis of the following neuroplasticity-related genes: brain derived neurotrophic factor (BDNF);

cAMP-response-element binding (CREB) protein; neurotrophic tyrosine kinase receptor, type 2 (NTrk2); spinophilin; and syntaxin. To date, the data indicate general increases in gene expression levels for the aforementioned genes. Interestingly, recent work from our laboratory has also shown a different parity effect: lactating females display a significant reduction in pre-frontal cortical APP compared to nulliparous females. It is worthwhile noting that for all the genes examined thus far, APP is the only one that showed a *decrease* compared to nulliparous females.

The data thus far indicate that, in general, reproductive experience (viz., females that are pregnant, or primiparous/multiparous), modifies the expression of these genes. For each gene, there were significant differences or parallel but non-significant trends among nulliparous, pregnant, and parous females. We are currently evaluating the differences between up-/down-regulation of these genes and their downstream products. In total, the data suggest that the parity-related changes in activity of these specific genes may be at least in part responsible for the observed augmentation in spatial memory (and, possibly other functions) in response to pregnancy and presence of young. Such alterations identify a robust and far-ranging modification of basic neuronal activity in service to the mother and her offspring. Further addressing these intriguing neuroplastic and neurodegeneration-affecting genes is likely to elucidate ways to positively modify the aging process. The insulin gene family is a fascinating thread running through the metabolic changes, of aging, and reproductive plasticity effects, we and others have seen. Together with the diversity of species that show parallel influences, these relationships mark the gene as a major instigator in the path that aging may take in an individual. In fact, the unanimity of aging-related effects across species suggests common mechanisms and present a fascinating basis for future study. In the next section, we examine some of these cross-species relationships.

4 Mechanisms and Likely Connections to Disparate Species and Systems

Facing one's own mortality may be unique to the human condition, but aging and death are not. In the continuing struggle against time, we have slowly begun to unlock the molecular mysteries of the aging phenomenon by studying the processes in other species. Among the goals in comparing humans to our distant (and not so distant; see Fig. 1) relatives is to find a mechanism—lifestyle, hormone treatment, or drug—that can not only to extend our existence, but to improve the time we do have. Among other benefits, improving aging could help prevent a wide range of age-related diseases and disorders, potentially including cancer and neurodegenerative conditions. Research on a wide array of species, including yeasts, nematodes, flies, rodents, monkeys, and humans has demonstrated conclusively that it is possible to increase lifespan (e.g., Fontana et al. 2010).

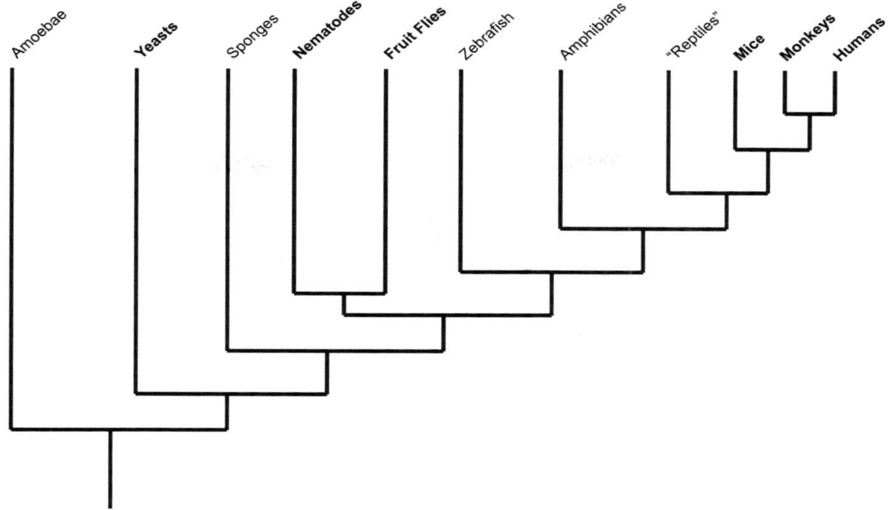

Fig. 1 Phylogenetic relationship among organisms in which aging and longevity have been studied (*bold*). Described here are studies involving mammals (humans, monkeys, mice), fruit flies (*D. melanogaster*), nematodes (*C. elegans*), and yeasts (*S. cerevisiae*). Despite the vast evolutionary distances between these species (approximately 1 billion years between yeasts and humans; Lucking et al. 2009), many of the same genetic pathways involved in regulation of aging are conserved. Tree modified from the Tree of Life Web Project

Pro-aging benefits have been achieved in several different ways and, as we discuss here, it appears that reproductive experience could be playing a role as well. To understand the ways in which aging, longevity, and reproductive experience are interrelated, it is important to have a general understanding of the underlying genetics pathways.

As we allude to above, perhaps the most thoroughly studied mechanism to increase longevity is caloric/dietary restriction (DR). Dietary restriction can increase the lifespan of the yeast *Sacchromyces cerevisiae* three-fold, has similar effects in the worm *Caenorhabditis elegans* and fruit fly *Drosophila melanogaster*, and can increase the lifespan of *Mus musculus* by up to 50% (Fontana et al. 2010). Lifespan-promoting benefits are primarily a result of decreases in gene expression in insulin or insulin-like pathways. Regulating insulin pathways result in decrease in age-related decline in function (e.g., errors in cell division) and disease (e.g., diabetes and cardiovascular disease; Fontana and Klein 2007; Anderson et al. 1999). Somewhat surprisingly, single-celled yeasts show these same age-related declines as found in rodents and humans, and improvements to longevity can be modulated through caloric restriction (Fabrizio et al. 2001).

In the presence of dietary restriction, yeasts are able to survive through modification of insulin pathways (Maroso 2005). Rather than dying off, *S. cerevisiae* actually experiences an extension lifespan. Studies have found that lifespan increases in yeast are possible through deletion of the serine-threonine kinase Sch9

(Fabrizio et al. 2001). Originally thought to be closely related to the human AKT—which has functions in insulin signaling, proliferation, and apoptosis—Sch9 is now considered to be homologous to the ribosomal protein kinase S6 K, another factor in the insulin signaling/aging pathway of a wide range of organisms (Kaeberlein et al. 2005). Under normal conditions, yeasts sense amino acids, activating the target of rapamycin kinase (TOR), whose functional properties include cell growth, proliferation, and cellular motility (Hay and Sonnenburg 2004; Beevers et al 2006), all of which drive the yeast to obtain the amino acids. TOR signaling then stimulates Sch9, eventually leading to cell growth, proliferation, etc. Disruption of this pathway, however, either by deletion of SCH9 (Fabrizio et al. 2001, 2003) or TOR (Kaeberlein et al. 2005; Pan and Shadel 2009) leads to a slowing of normal TOR function, but increases life expectancy by sending the yeasts into a state of reduced development and replication (Fig. 2). Further investigation of downstream targets of Sch9 has revealed how Sch9 mutants are able live up to three times as long as wild-type yeasts. Ge and collegues (2010) revealed that turning off Sch9 may trigger a longevity circuit that includes the regulation of rRNA processing and of a glucose response element. Evolutionarily, modification of the TOR-Sch9 pathway helps yeasts deal with stressful conditions, in particular, decreased amounts of glucose, their primary food source. At times of nutrient restriction, *S. cerevisiae* decreases cellular proliferation cycles, which in turn extends the time before replication-related dysfunction sets in (leading to aging and death). As an added bonus, modulation of the Sch9 and rRNA pathways confers additional stress-related buffers, including heat-shock resistance (Fabrizio et al. 2001, 2003; Ge et al. 2010). Given the multitude of advantages conferred to yeasts in this scenario, it is no surprise that pathways responding to food intake would be under significant evolutionary pressure and thus, be conserved by so-called "higher" organisms.

The TOR—Sch9 pathway is conserved as a mechanism in *Caenorhabditis elegans, Drosophila melanogaster,* and *Mus musculus* (in these organisms, Sch9 is S6 K; Kapahi et al. 2004; Doonan et al. 2008; Selman et al. 2009). The major difference between yeasts and these groups is that the food-response system is indirect in worms, flies, and mice, going through insulin/insulin-like growth factors (Ins/IGF-1-like; Henderson and Johnson 2001; Wijchers et al. 2006; Johnson 2008). Normal functioning of the Ins/IGF-1-F pathway results in cell functioning and replication, which increases the amount of cellular waste products (i.e., oxidants) and opportunities for cellular dysfunctions that lead to aging.

Dietary restriction (DR) reduces the amount of insulin needed to process nutrients, and therefore, lead to down-regulation of the Ins/IGF-1-like signally pathway. In *C. elegans*, disrupting insulin signaling leads to up-regulation of a forkhead FoxO transcription factor, *DAF-16*. The activity of FOXO proteins is similar in worms, fruit flies, and mammals—they lead to a variety of activities that promote longer lifespan including fat storage (Feige et al. 2008); modulation of the cellular stress response (Brunet et al. 2004); autophagy (Klionsky and Emr 2000); removal of free radicals (Doonan et al. 2008); and protecting against neurodegeneration (Liu et al. 2004; Mojsilovic-Petrovic et al. 2009). The effects of FOXO

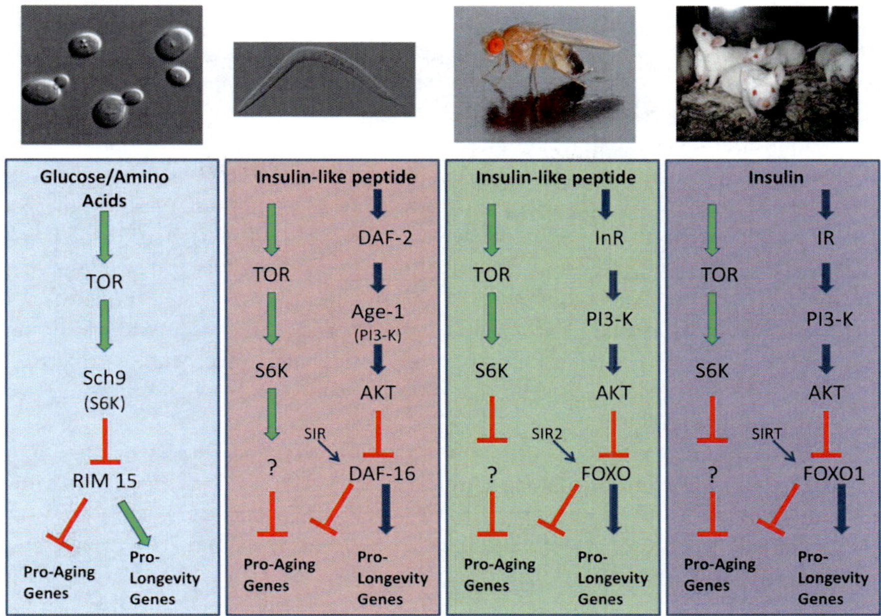

Fig. 2 A simplified model for the regulation of aging genes shared by yeasts, worms, flies, and mammals (including humans). In yeasts (*Sacchromyces cerevisiae*, far left), food intake triggers the TOR pathway, which triggers Sch9 and normal cellular activity (respiration, division, etc.). Inhibition of this pathway, either through dietary restriction or inactivation of TOR or Sch9 results in a decrease in normal cellular activity and an increase in longevity. The TOR/Sch9 pathway is conserved in other organisms as well. Inactivation of the pathway again inhibits aging genes through an as-yet unknown mechanism. Additionally, aging may be regulated through the activity of forkhead box transcription factors in *Caenorhabditis elegans*, *Drosophilia melanogaster*, and *Mus musculus* (DAF-16, FOXO, and FOXO1, respectively). In all three species, disruption of the insulin pathway allows functioning of the forhead box transcription factors leading to longer lifespan. Additionally, it is possible to increase functioning of DAF-16, FOXO and FOXO1 through the activity of sirtuin proteins. These proteins are notably found in the plant compound resveratrol, which has received much attention lately. (Modified from Russel and Kahn 2007; Fontana et al. 2010.)

are hardly modest—mutations in the Ins/IGF-1-like pathway that lead to FOXO expression can increase the lifespan of worms by 10 times and over-expression of the gene in *D. melanogaster* can yield a 52% increase (Hwangbo et al. 2004). In addition to decreasing the Ins/IGF-1-like signaling, dietary restriction can enhance the function of sirtuin deacetylase proteins (SIR/SIRT), which appear to help regulate FOXO in worms, flies, and mammals (Tissenbaum and Guarente 2001; Brunet et al. 2004; Frescas et al. 2005; Yang et al. 2005). Alteration to the amounts SIR/SIRT present in animals can lead to significant results. Loss of SIRT has been shown to increase aging in mice (Mostoslavsky et al. 2006) and increases in SIRT may increase lifespan in humans (Rose et al. 2003). Interestingly, the plant compound resveratrol—found in grapes, red wine, and peanuts—may target sirtuin proteins. Resveratrol has become a popular target in the popular media ("Drink wine to lose weight and live longer!") due largely to studies implicating its role in increasing the lifespan of *S. cerevisiae*, *C. elegans*, and *D. melanogaster*, fish, and even mice (Howitz et al. 2003; Wood et al. 2004; Viswanathan et al. 2005; Baur et al. 2006; Valenzano et al. 2006; Gruber et al. 2007). In addition to the genes and proteins discussed here, there are many more that are being discovered as players in the complex regulation of aging through the general insulin pathway (for an in-depth review, see Greer and Brunet, 2010). Although DR can have profound effects in increasing life-expectancy, some reviewers feel that there are too many complications with humans—decrease in immune function, individuals not wanting to fast—for it to be feasible as a life-extending technique in humans (Phelan and Rose 2005). Fortunately, the genetic underpinnings of aging may help clarify the role of endocrine hormones and reproductive experience in the aging process.

The discovery that hormones critical to sexual reproduction are associated with the regulation of insulin and aging mechanism is striking. Specifically, modification of pregnenolone (*C. elegans*; Broue et al. 2007) and ecdysone pathways (*D. melanogaster*; Simon et al. 2003) have both been shown to increase life expectancy in their respective species. Even more interestingly, the changes in steroid hormones are not thought to happen independently from the insulin pathway. Rather, studies suggest that several hormone pathways could be operating downstream of insulin receptors themselves. Importantly, these findings do not propose new functions for already important hormones, but the use of hormones rather than proteins makes possible interactions between multiple cell and tissue types; neurons, fat tissues, and other targets such as heart or liver tissue can all work with one another to regulate growth, cellular proliferation, and reproduction. Understanding the actions of hormones is crucial to understanding the positive effects of reproduction experience on aging.

The regulation of the insulin pathway by hormones mean that aging and anti-aging mechanisms are also affected by hormones in worms, flies, and mammals (Fig. 3). Surprisingly, these hormones are generated not only by the neuroendocrine and reproductive system, but also in fatty tissues. *Caenorhabditis elegans* utilizes cholesterol from the intestines to produce pregnenolone, a mammalian steroid precursor, which then interacts with DAF16 to trigger anti-aging genes/stress responses as discussed above (i.e., entering at the dauer hibernation stage; Hsin and Kenyon 1999). This interaction may be modulated by another factor in the intestines before sending signals—perhaps KLOTHO, a protein generated in the brain that has been shown to affect longevity in mice by blocking function of insulin receptors (Kuro-o et al. 1997, 2009; Kuroshu et al. 2005)—to target tissues, but it does seem clear that pregnenolone is interacting with both insulin-like factors in neurons and target tissues. In rats, insulin from the pancreas attaches to receptors on fat cells and target tissues to inhibit FOXO1, and thus, anti-aging genes. If, however, insulin signaling is blocked or reduced by either SIRT2 (in adipocytes) or KLOTHO, FOXO1 and anti-aging genes are activated allowing mice to live longer (Coschigano et al. 2003; Flurkey et al. 2001).

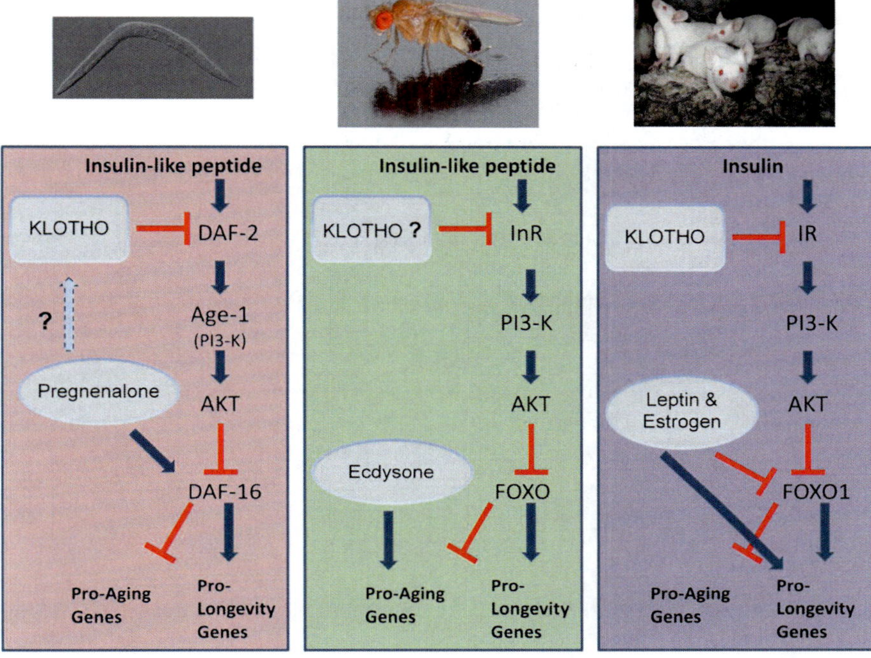

Fig. 3 A prospective model for the effect of hormones (*gray background*)—including those related to pregnancy (*oval background*)—on longevity. KLOTHO, generated in the brain, appears to block insulin receptors in mice and may be involved in signaling in *C. elegans*. In *C. elegans*, pregnenalone triggers DAF-16 activity which in turn promotes longevity. In flies, however, it is necessary to down-regulate the ovarian hormone ecdysone in order to see increases in longevity. The story is even more complex in mammals, as the reproductive hormones leptin and estrogen appear both to block FOXO1 signaling and to promote aging through another mediator. It appears that, in general, decreasing reproduction leads to an increase in aging. Pregnancy hormones do, however, appear to have regulatory and beneficial effects on the course of aging in a variety of organisms. Future focused research will enable a better understanding of these intriguing relationships

A link between fatty tissues and reproductive experience may be found in a study revealing that progesterone increases the number of insulin receptors (InR) in adipocytes, allowing the pregnant mother to store fat even though glucose is being preferentially shuttled to the fetus (Flint et al. 1979). This increase in InR would logically lead to decreased functioning of FOXO proteins and thus promote aging. More recent studies, however, have shown that a peptide hormone, leptin, is produced in adipocytes and usually correlates positively with overall body fat mass. Leptin function remains high, however, when insulin receptors are down regulated in fat-specific insulin-receptor (FIRKO) mice (Ahima and Flier 2000). FIRKO mice remain thin even when placed on a high-fat diet and outlive their littermates, who suffer from obesity-related diseases when fed the same diets, by

18% (Bluher et al. 2002, 2003). These data indicate that leptin may play a role in the anti-aging process by promoting insulin sensitivity and keeping animals thin (Russel and Kahn 2007). Lastly, Messines and collegues (2001) found that levels of leptin could be raised in non-pregnant women by adding a progesterone supplement. Thus, we might conclude that pregnancy-related progesterone not only induced adipocytes to create insulin receptors, but also to increase levels of leptin, potentially attenuating some of the negative side-effects of pregnancy on glucose insensitivity.

As previously noted, changes to the insulin pathway help yeasts and worms go into states of hibernation/developmental stasis in times of famine, a condition adaptive enough to have been preserved in metazoans such that mammals can now reap benefits from decreasing our caloric intake (conversely, increases in human obesity rates have recently lead to a decline in life expectancy (Olshansky et al. 2005). Whereas caloric restriction can increase life span, such decreases are also associated with negative side effects that could be considered, including decreased immune function (Fontana and Klein 2007) and infertility (Holliday 2005). The decrease in reproduction is a curious one; evolutionary theory would predict that organisms should optimize reproductive success by living longer. Although this can happen in certain species (Weladji et al. 2006), it is not the case for most organisms.

Several studies have looked for link between reproductive experience and longevity. Although one might suspect that living longer would allow for greater reproductive success, it appears that there is instead a trade-off between being long-lived and having reproductive success. The argument is essentially that there are costs associated with maintaining the mother and her offspring that a non-mother does not incur. In *C. elegans*, reproductive success does not increase with longevity during dietary restriction. Rather, reproduction is maximized when lifespan decreases to normal levels (Fontana et al. 2010). The same relationship is found in cheetahs (Pettorelli and Durant 2007) and humans. A study in 153 countries worldwide found that in every region there was a negative correlation between age and reproduction (Thomas et al. 2000). Indeed, pregnancy itself can have negative medical side effects such as gestational diabetes, (a disease associated with the insulin pathway; e.g., Kumangai et al. 2003) and work has suggested that knocking out reproduction entirely can have positive effects in mice (Conover and Bale 2007). What is likely happening, then, at least in the case of DR, is that disruption of the insulin pathway allows the organism to live longer and create *better* reproductive opportunities (i.e., in ideal environmental conditions rather than low-food conditions) rather than *more* opportunities. Aging better, it seems, is not just about the number of offspring produced.

There are a host of behavioral and biological modifications present in reproductive females that can lead to healthy aging, particularly enhancements in heart and brain function. Alterations to the insulin receptors have been shown to improve heart performance in fruit flies (Wessells et al. 2004). Estradiol, an ovarian sex steroid, helps mice survive oxidative stress (Behl et al. 1995), and pregnenolone, a steroid precursor hormone in humans (currently a popular, if unproven, anti-aging supplement), has been shown to improve neuronal survival (Gursoy et al. 2001) and modulate neurotransmitter systems by promoting microtubule assembly (Murakami et al. 2000). These protective features no doubt let the organism live a more productive (and perhaps happier) life, and may be related to the improvements we describe above.

Recent evidence from mice and rat approaches the question of aging from the opposite direction—discovering improvements in behavior and aging well and then attempting to discern the underlying hormonal basis for those changes. During pregnancy and into lactation, mother rats show superior performance at learning, memory, and physical challenges when compared to non-mothers (Kinsley et al. 1999) and mother rats show decreased stress and anxiety (Wartella et al. 2003). Our laboratory is even finding evidence that mother rats are better at "thinking" about the future (prospective memory) than virgins (Franssen, Rafferty et al. unpublished). What seems clear is that during pregnancy, steroid hormones—including pregnenolone progenitors—are reworking the mammalian brain. Oxytocin alters astrocyte morphology (Modney and Hatton 1994; Hatton and Zhao Yang 2002), neurogenesis increases (Furuta and Bridges 2005), and even short-term changes in estradiol and progesterone changes during pregnancy appear to be able to affect neuroplasticity in the brain (Kinsley et al. 2006; Kinsley 2008). Moreover, these changes are long-lasting. As we discuss above, mothers remain better at learning and memory, well beyond their reproductive years (Gatewood et al. 2005; Love et al. 2005; Kinsley et al. 2008), and are less likely to suffer from dementia (Victoria and Kinsley, unpublished).

5 Summary, Conclusions, and Future Directions

By studying the interactions among genes, proteins, and hormones, we can discover ways in which we can both increase life expectancy and improve our lifestyles in old age. Reproductive experience has been investigated regarding its relationship to life expectancy and in other sets of behavioral benefits. The next step is to incorporate the studies. It is important to discover what, exactly, is the molecular impact of pregnancy hormones and motherhood that lead to long-term memory, anxiety, and stress coping advantages over non-mothers. The work is being done. Progesterone receptors have been mapped in the hippocampus, the brain's major memory center (e.g., Guerra-Araiza et al. 2001; Auger and De Vries 2002), and neuroprotective capabilities such as increased myelination, and stimulation of mitochondria, are being elucidated (Koenig et al. 1995; Azcoitia et al. 2003). Perhaps most encouragingly, there is evidence that pregnenolone may act as a neurosteroid that can reverse age-related deficits in mammals (including non-mothers), suggesting a possible future treatment for neural diseases like dementia (Schumacher et al. 2003). Although these studies were not necessarily focused on aging, neuroprotection, and enhancement of brain activity are definitely important components of aging well. Understanding the genetic, hormonal, and reproductive data at our disposal, coupled to cross-species comparative approaches, we may soon be able to identify specific lifestyle choices (eat less and reproduce!) and/or drug options to improve the quality of life as life itself wanes.

References

Agaram R, Douglas MJ, McTaggart RA, Gunka V (2009) Inadequatepain relief with labor epidurals: a multivariate analysis of associated factors. Int J Obstet Anesth 18:10–14

Ahima RS, Flier JS (2000) Adipose tissue as an endocrine organ. Trends in Endocrinol Metlab 11:327–332

Altman J, Das GD (1965) Post-natal origin of microneurons in the rat brain. Nature 207:953–956

Anisimov VN et al (2008) Metformin slows down aging and extends life span of female SHR mice. Cell Cycle 7:2769–2773

Apfeld J, O'Connor G, McDonagh T, DiStefano PS, Curtis R (2004) The AMP-activated protein kinase AAK-2 links energy levels and insulin-like signals to lifespan in *C. elegans*. Genes Dev 18:3004–3009

Auger CJ, De Vries GJ (2002) Distribution and steroid responsiveness of progestin receptor immunoreactivity within vasopressin-immunoreactive cells in the bed nucleus of the stria terminalis and the centromedial amygdala of male and female rat brain. J Neuroendocrinol 14:161–167

Azcoitia I, DonCarlos LL, Garcia-Segura LM (2003) Are gonadal steroid hormones involved in disorders of brain aging? Aging Cell 2:31–37

Baur JA, Pearson KJ, Price NL, Jamieson HA, Lerin C, Kalra A, Prabhu VV, Allard JS, Lopez-Lluch G, Lewis K, Pistell PJ, Poosala S, Becker KG, Boss O, Gwinn D, Wang M, Ramaswamy S, Fishbein KW, Spencer RG, Lakatta EG, Le Couteur D, Shaw RJ, Navas P, Puigserver P, Ingram DK, de Cabo R, Sinclair DA (2006) Resveratrol improves health and survival of mice on a high-calorie diet. Nature 444:337–342

Beevers C, Li F, Liu L, Huang S (2006) Curcumin inhibits the mammalian target of rapamycin-mediated signaling pathways in cancer cells. Int J Cancer 119:757–764

Behl C, Widmann M, Trapp T, Holsboer F (1995) 17-β Estradiol Protects Neurons from Oxidative Stress-Induced Cell Death in Vitro. Biochem Biophys Res Comm 216:473–482

Bennett DA, Wilson RS, Schneider JA, Evans DA, Mendes De Leon CF, Arnold SE, Barnes LL, Bienias JL (2003) Education modifies the relation of AD pathology to level of cognitive function in older persons. Neurology 60:1909–1915

Bishop NA, Lu T, Yankner BA (2010) Neural mechanisms of ageing and cognitive decline. Nature 464:529–535

Bluher M, Peroni OD, Ueki K, Carter N, Kahn BB, Kahn CR (2002) Adipose tissue selective insulin receptor knockout protects against obesity and obesity-related glucose intolerance. Dev. Cell 3:25–38

Bluher M, Khan BP, Kahn CR (2003) Extended longevity in micelacking the insulin receptor in adipose tissue. Science 299:572–574

Bridges RS (1975) Long-term effects of pregnancy and parturition upon maternal responsiveness in the rat. Physiol Behav 14:245–249

Bridges RS, Grimm CT (1982) Reversal of morphine disruption of maternal behavior by concurrent treatment with the opiate antagonist naloxone. Science 218:166–168

Bridges RS (1984) A quantitative analysis of the roles of dosage, sequence, and duration of estradiol and progesterone exposure in the regulation of maternal behavior in the rat. Endocrinology 114:930–940

Bridges RS (1990) Endocrine regulation of parental behavior in rodents. In: Krasnegor NA, Bridges RS (eds) Mammalian Parenting: Biochemical, Neurobiological and Behavioral Determinants. Oxford University Press, New York, pp 93–117

Bridges RS, Hammer Jr RP (1992) Parity-associated alterations in medial preoptic opiate receptors in female rats. Brain Res 578:269–274

Bridges RS, Felicio LF, Pellerin LJ, Steuer AM, Mann PE (1993) Prior parity reduces post-coital diurnal and nocturnal prolactin surges in rats. Life Sci 53:439–445

Bridges RS, Robertson MC, Shiu RPC, Friesen HG, Stuer AM, Mann PE (1996) Endocrine communication between conceptus and mother: a role for placental lactogens in the induction of maternal behavior. Neuroendocrinology 64:57–64

Bridges RS, Robertson MC, Shiu RP, Sturgis JD, Henriquez BM, Mann PE (1997) Central lactogenic regulation of maternal behavior in rats: steroid dependence, hormone specificity, and behavioral potencies of rat prolactin and rat placental lactogen. Endocrinol 138:756–763

Bridges RS (2009) The Neurobiology of the Parental Brain. Academic Press, New York

Broue F, Liere P, Kenyon C, Baulieu EE (2007) A steroid hormone that extends the lifespan of *Caenorhabditis elegans*. Aging Cell 6:87–94

Brunet A, Bonni A, Zigmond MJ, Lin MZ, Juo P, Hu LS, Anderson MJ, Arden KC, Blenis J, Greenberg ME (1999) Akt promotes cell survival by phosphorylating and inhibiting a Forkhead transcription factor. Cell 96:857–868

Brunet A, Sweeney LB, Sturgill JF, Chua KF, Greer PL, Lin Y, Tran H, Ross SE, Mostoslavsky R, Cohen HY, Hu LS, Cheng H, Jedrychowski MP, Gygi SP, Sinclair DA, Alt FW, Greenberg ME (2004) Stress-Dependent Regulation of FOXO Transcription Factors by the SIRT1 Deacetylase. Science 303:2011–2015

Cameron HA, Gould E (1994) Adult neurogenesis is regulated by adrenal steroids in the rat dentate gyrus. Neuroscience 61:203–209

Carlen M, Cassidy RM, Brismar H, Smith GA, Enquist LW, Freisen J (2002) Functional integration of adult-born neurons. Curr Biol 12:606–608

Carleton A, Petreanu LT, Lansford R, Alvarez-Buylla A, Lledo PM (2003) Becoming a new neuron in the adult olfactory bulb. Nat Neurosci 6:507–518

Carro E, Nunez A, Busiguina S, Torres-Aleman I (2001) Circulating insulin-like growth factorI mediates effects of exercise on the brain. J Neurosci 20:2926–2933

Colman RJ et al (2009) Caloric restriction delays disease onset and mortality in rhesus monkeys. Science 325:201–204

Conover CA, Bale LK (2007) Loss of pregnancy-associated plasma protein A extends lifespan in mice. Aging Cell 6:727–729

Contino R, Friedenberg J, Christon L, Norkunas T, Worthington D, Jablow L, Drew M, Victoria L, Chipko C, Sirkin M, Ferguson T, Jones C, Bardi M, Lambert KG and Kinsley CH (2007). The expression of the maternal Brain: Specific genes involved in the neuroplasticity of motherhood. Paper presented at the Society for Neuroscience annual meeting, San Diego, CA, November

Coschigano KT, Holland AN, Riders ME, List EO, Flyvbjerg A, Kopchick JJ (2003) Deletion, but not antagonism, of the mouse growth hormone receptor results in severely decreased body weights, insulin, and insulin-like growth factor I levels and increased life span. Endocrinology 144:3799–3810

Cummins RA, Walsh RN, Budtz-Olsen OE, Konstantinos, Horsfall CR (1973) Environmentally induced changes in the brains of elderly rats. Nature 243:516–518

Cummins RA, Livesey PJ, Evans JG (1977) A developmental theory of environmental enrichment. Science 197:692–694

Dagyte G, Trentani A, Postema F, Luiten PG, Den Boaer JA, Gabriel C, Mocaer E, Meerlo P, Van de Zee EA (2010) The novel antidepressant agomelatine normalizes hippocampal neuronal activity and promotes neurogenesis in chronically stressed rats. CNS Neurosci Ther 16:195–207

Diamond MC, Krech D, Rosenzweig MR (1964) The effects of an enriched environment on the histology of the rat cerebral cortex. J Comp Neurol 123:111–120

Diamond MC, Johnson RE, Ingham C (1971) Brain plasticity induced by environment and pregnancy. Int J Neurosci 2:171–178

Doonan R, Mcelwee JJ, Matthijssens F, Walker GA, Houthood K, Back P, Matscheski A, Vanfleteren JR, Gems D (2008) Against the oxidative damage theory of aging: superoxide dismutases protect against oxidative stress but have little or no effect on life span in *Caenorhabditis elegans*. Genes Dev 22:3226–3241

Enwere E, Shingo T, Gregg C, Fujikawa H, Ohta O, Weiss S (2004) Aging results in reduced epidermal growth factor receptor signaling, diminished olfactory neurogenesis, and deficits in fine olfactory discrimination. J Neurosci 24:8354–8365

Fabrizio P, Pozza F, Pletcher SD, Gendron CM, Longo VD (2001) Regulation of longevity and stress resistance by Sch9 in yeast. Science 292:288–290

Fabrizio P, Liou LL, Moy VN, Diaspro A, Valentine JS, Gralla EB, Longo VD (2003) SOD2 functions downstream of Sch9 to extend longevity in yeast. Genetics 163:35–46

Feige JN, Lagouge M, Canto C, Strehle A, Houten SM, Milne JC, Lambert PD, Mataki PJ, Auwerx J (2008) Specific SIRT1 activation mimics low energy levels and protects against diet-induced metabolic disorders by enhancing fat oxidation. Cell Met 8:347–358

Felicio LF, Florio JC, Sider LH, Cruz-Casallas PE, Bridges RS (1996)Reproductive experience increases striatal and hypothalamic dopamine levels in pregnant rats. Brain Res Bull 40: 253–256

Fischer A, Sananbenesi F, Wang X, Dobbin M, Tsai LH (2007) Recovery of learning and memory is associated with chromatin remodeling. Nature 447:178–182

Flint DJ, Sinnett-Smith PA, Clegg RA, Vernon RG (1979) Role of insulin receptors in the changing metabolism of adipose tissue during pregnancy and lactation in the rat. Biochem J 182:421–427

Flurkey K, Papaconstantinou J, Miller RA, Harrison DE (2001) Lifespan extension and delayed immune and collagen aging in mutant mice with defects in growth hormone production. Proc Natl Acad Sci 98:6736–6741

Fontana L, Klein S (2007) Aging, adiposity and calorie restriction. JAMA 297:986–994

Fontana L, Partridge L, Longo VD (2010) Extending healthy life span from yeasts to humans. Science 328:321–326

Frescas D, Valenti L, Accili D (2005) Nuclear trapping of the forkhead transcription factor FoxO1 via sirt-dependent deacetylation promotes expression of glucogenetic genes. J Biol Chem 280:20589–20595

Furuta M, Bridges RS (2005) Gestation-induced cell proliferation in the rat brain. Brain Res Dev Brain Res 156:61–66

Gatewood JD, Morgan MD, Eaton M, McNamara IM, Stevens LF, Macbeth AH, Meyers EAA, Lomas LM, Kozub FJ, Lambert KG, Kinsley CH (2005) Motherhood mitigates aging-related decrements in learning and memory. Brain Res Bull 59:267–283

Ge H, Wei M, Fabrizio P, Hu J, Cheng C, Longo VD, Li LM (2010) Comparative analyses of time-course gene expression profiles of the long-lived *sch9Δ* mutant. Nucl Acids Res 38: 143–158

Gould E, Cameron HA, Daniels DC, Woolley CS, McEwen BS (1992) Adrenal hormones suppress cell division in the adult rat dentate gyrus. J Neurosci 12:3642–3650

Gould E, McEwan BS, Tanapat P, Galea LA, Fuchs E (1997) Neurogenesis in the dentate gyrus of the adult tree shrew is reguilated by psychosocial stress and NMDA receptor activation. J Neurosci 17:2492–2498

Gould E, Tanapat P, Hastings NB, Shors TJ (1998) Proliferation of granule cell precursors in the dentate gyrus of adult monkeys is diminished by stress. Proc Natl Acad Sci 95:3168–3171

Greer EL, Brunet A (2010) Signaling networks in aging. J Cell Sci 121:407–412

Gruber J, Tang SY, Halliwell B (2007) Evidence for a trade-off between survival and fitness caused by resveratrol treatment of *Caenorhabditis elegans*. Ann NY Acad Sci 1100:530–542

Guerra-Araiza C, Reyna-Neyra A, Salazar AM, Cerbon MA, Morimoto S, Camacho-Arroyo I (2001) Progesterone receptor isoforms expression in the prepuberal and adult male rat brain. Brain Res Bull 54:13–17

Gursoy E, Cardounel A, Kalimi M (2001) Pregnenolone protects mouse hippocampal (HT-22) cells against glutamate and amyloid beta protein toxicity. Neurochem Res 26:15–21

Halagappa VK et al (2007) Intermittent fasting and caloric restriction ameliorate age-related behavioral deficits in the triple-transgenic mouse model of Alzheimer's disease. Neurobiol Dis 26:212–220

Hara T et al (2006) Suppression of basal autophagy in neural cells causes neurodegenerative disease in mice. Nature 441:885–889

Harrison DE et al (2009) Rapamycin fed late in life extends lifespan in genetically heterogenous mice. Nature 460:392–395

Hatton GI, Zhao Yang Q (2002) Peripartum interneuronalcoupling in the supraoptic nucleus. Brain Res 932:120–123

Hay N, Sonenberg N (2004) Upstream and downstream of mTOR. Genes Dev 18:1926–1945

Henderson ST, Johnson TE (2001) *DAF-16* integrates developmental and environmental inputs to mediat aging in the nematode *Caenorhabditis elegans*. Curr Biol 11:1975–1980

Holliday R (2005) Food, reproduction and longevity: is the extended lifespan of calorie-restricted animals an evolutionary adaptation? Bioessays 10:125–127

Howitz KT, Bitterman KJ, Cohen HY, Lamming DW, Lavu S, Wood JG, Zipkin RE, Chung P, Kisielewski A, Zhang LL, Scherer B, Sinclair DA (2003) Small molecule activators of sirtuins extend *Saccharomyces cerevisiae* lifespan. Nature 425:191–196

Hsin H, Kenyon C (1999) Signals from the reproductive system regulate the lifespan of *C. elegans*. Nature 399:362–366

Hwangbo DS, Gersham B, Tu MP, Palmer M, Tatar M (2004) *Drosophila* dFOXO controls lifespan and regulates insulin signaling in brain and fat body. Nature 492:562–566

Johnson TE (2008) Caenorhabditis elegans. (2007). The premier model for the study of aging. Exp Gerontol 43:1–4

Kaeberlein M, Powers III RW, Steffen KK, Westman EA, Hu D,Dang N, Kerr EO, Kirkland KT, Fields S, Kennedy BK (2005) Regulation of yeastreplicative life span by TOR and Sch9 in response to nutrients. Science 310:1193–1196

Kapahi P, Zid BM, Harper T, Koslover D, Sapin V, Benzer S (2004) Regulation of lifespan in Drosophila by modulation of genes in the TOR signaling pathway. Curr Biol 14:885

Kempermann G (2006) Adult neurogenesis. Oxford, Oxford Press

Kempermann G, Gage F (1998) Closer to neurogenesis in adult humans. Nat Med 4:555–557

Kempermann G, Kuhn G, Gage F (1997) More hippocampal neurons in adult mice living in an enriched environment. Nature 386:493–495

Kempermann G, Brandon EP, Gage F (1998) Environmental stimulation of 129/SvJ mice causes increased cell proliferation and neurogenesis in the adult dentate gyrus. Curr Biol 8:939–942

Kempermann G, Gast D, Gage F (2002) Neuroplasticity in old age:sustained five-fold induction of hippocampal neurogenesis by long-term environmental enrichment. Ann Neurol 52:135–143

Kenyon CJ (2010) The genetics of ageing. Nature 464:504–512

Kesner R, Dakis M (1995) Phencyclidine injections into the dorsal hippocampus disrupt long-but not short-term memory within a spatial learning task. Psychopharmacology 120:203–208

Keyser L, Stafisso-Sandoz G, Gerecke K, Jasnow A, Nightingale L, Lambert KG, Gatewood J, Kinsley CH (2001) Alterations of medial preoptic area neurons following pregnancy and pregnancy-like steroidal treatment in the rat. Brain Res Bull 55:737–745

Kinsley CH, Bridges RS (1988) Parity associated reductions in behavioral sensitivity to opiates. Biol Reprod 39:270–278

Kinsley CH (1994) Developmental psychobiological influences on rodent parental behavior. Neurosci Biobehav Rev 18:269–280

Kinsley CH, Madonia L, Gifford GW, Tureski K, Griffin GR, Lowry C, Williams J, Collins J, McLearie H, Lambert KG (1999) Motherhood improves learning and memory: Neural activity in rats is enhanced by pregnancy and the demands of rearing offspring. Nature 402:137–138

Kinsley CH, Lambert KG, The Maternal Brain (2006) Pregnancy and motherhood change the structure of the female mammal's brain, making mothers attentive to their young and better at caring for them. Scientific American January 2006

Kinsley CH, Trainer R, Stafisso-Sandoz G, Quadros P, Keyser-Marcus L, Hearon C, Wightman N, Morgan MD, Kozub FJ, Lambert KG (2006) Motherhood and pregnancy hormones modify concentrations of hippocampal neuronal dendritic spines. Horm Behav 49:131–142

Kinsley CH (2008) The neuroplastic maternal brain. Horm Behav 54:1–4

Kinsley CH, Lambert KG (2008) Reproduction-induced neuroplasticity: natural behavioral and neuronal alterations associated with the production and care of offspring. J Neuroendocrinology 20:515–525

Kinsley CH, Bardi M, Karelina K, Rima B, Christon L, Friedenberg J, Griffin G (2008) Motherhood Induces and Maintains Behavioral and Neural Plasticity Across the Lifespan in the Rat. Arch of Sex Behav 37:43–56

Klionsky DJ, Emr SD (2000) Autophagy as a regulated pathway of cellular degradation. Science 290:1717–1721

Koenig HL, Schumacher M, Ferzaz B, Thi AN, Ressouches A, Guennoun R, Jung-Testas I, Robel P, Akwa Y, Baulieu EE (1995) Progesterone synthesis and myelin formation by Schwann cells. Science 268:1500–1503

Kozorovitskiy Y, Hughes M, Lee K, Gould E (2006) Fatherhood affects dendritic spines and vasopressin V1a receptors in the primate prefrontal cortex. Nat Neurosci 9:1094–1095

Kumangai S, Holmang A, Bjorntorp P (1993) The effects of oestrogen and progesterone on insulin sensitivity in female rats. Acta Phys Scand 149:91–97

Kuro-o M (2009) Klotho and aging. Biochem Biophys Acta 1790:1049–1058

Kuro-o M, Matsumura Y, Aizawa H, Kawaguchi H, Suga T, Utsugi T, Ohyama Y, Kurabayashi M, Kaname T, Kume E, Iwasaki H, Iida A, Shiraki-Iida T, Nishikawa S, Nagai R, Nabeshima YI (1997) Mutation of the mouse klotho gene leads to a syndrome resembling ageing. Nature 390:45–51

Kuroshu H, Yamamoto M, Clark JD, Pastor JV, Nandi A et al (2005) Suppression of aging in mice by the hormone Klotho. Science 309:1829–1833

Lambert KG, Kinsley CH (2009) The neuroeconomics of motherhood: the costs and benefits of maternal investment. In Bridges RS (ed.), The Neurobiology of the Parental Brain. 481–492, Academic Press

Lambert KG, Berry AE, Griffin G, Amory-Meyer EA, Madonia-Lomas LF, Love G, Kinsley CH (2005) Pup exposure differentially enhances foraging ability in primiparous and nulliparous rats. Physiol Behav 85:799–806

Lichtenwalner RJ, Forbes ME, Bennett SA, Lynch CD, Sonntag WE, Riddle DR (2001) Intracerebroventricular infusion of insulin-like growth factor-I ameliorates the age related decline in hippocampal neurogenesis. Neuroscience 107:603–613

Liu Y, Neve RL, Taylor P, Driscoll M, Clardy J, Merry D, Kalb RG (2004) FOXO3a is broadly neuroprotective in vitro and in vivo against insults implicated in motor neuron diseases. J Neurosci 29:8236–8247

Love G, Torrey N, McNamara I, Morgan M, Banks M, Wightman N, Glasper ER, DeVries AC, Kinsley CH, Lambert KG (2005) Maternal experience produces long lasting behavioral modifications in the rat. Behav Neurosci 119:1084–1096

Lucking R, Huhndorf S, Pfister D, Plata ER, Lumbsch H (2009) Fungi evolved right on track. Mycologia 101:810–822

Mak HY, Ruvkun G (2004) Intercellular signaling of reproductive development by the *C. elegans* DAF-9 cytochrome P450. Development 131:1777–1786

Mak GK, Weiss S (2010) Paternal recognition of adult offspring mediated by newly generated CNS neurons. Nat Neurosci 13:753–758

Mann PE, Bridges RS (1992) Neural and endocrine sensitivities decline as a function of multiparity in the rat. Brain Res 580:241–248

Maroso EJ (2005) Overview of caloric restriction and ageing. Mech Ageing Dev 126:913–922

Maviel T, Durkin TP, Menzaghi F, Bontempi B (2004) Sites of cortical reorganization critical for remote spatial memory. Science 305:96–99

McEwen BS, Woolley CS (1994) Estradiol and progesterone regulate neuronal structure and synaptic connectivity in adult as well as developing brain. Exp Gerontol 29:431–436

Melendez A et al (2003) Autophagy genes are essential to dauer development and life-span extension in *C elegans*. Science 301:1387–1391

Messines IE, Papageorgiou I, Milingo S, Asprodini E, Kollios G, Seferiadis K (2001) Oestradiol plus progesterone treatment increases serum leptin concentrations in normal women. Hum Repro 16:1827–1832

Miller S, Erskine MS (1995) Ultrastructural effects of estradiol and 5-alpha-androstane on neurons within the ventromedial nucleus of the hypothalamus. Neuroendocrinology 61:669–679

Modney BK, Hatton GI (1994) Maternal behaviors: evidence that they feed back to alter brain morphology and function. Acta Paediatr 83:29–32

Mojsilovic-Petrovic J, Nedelsky N, Boccitto M, Mano I, Georgiades SN, Zhou W, Liu Y, Neve RL, Taylor PJ, Driscoll M, Clardy J, Merry D, Kalb RG (2009) FOXO3a is broadly neuroprotective *in vitro* and *in vivo* against insults implicated in motor neuron diseases. J Neuro 29:8236–8247

Moltz H, Levin R, Leon M (1969) Differential effects of progesterone on the maternal behavior of primiparous and multiparous rats. J Comp Physiol Psychol 67:36–40

Mostoslavsky R, Chua KF, Lombard DB, Pang WW et al. (2006) Genomic instability and aging-like phenotype in the absence of mammalian SIRT6. Cell 124:315–329

Muller FL, Lustgarten MS, Jang Y, Richardson A, Van Remmen H (2007) Trends in oxidative aging theories. Free Radic Biol Med 43:477–503

Murakami K, Fellous A, Baulieu E, Robel P (2000) Pregnenolone binds to microtubule-associated protein 2 and stimulates microtubule assembly PNAS 97:3579–3584

Nilsson M, Perfilieva E, Johansson U, Orwar O, Erikccon PS (1999) Enriched environment increases neurogenesis in the adult rat and improves spatial memory. J Neurobiol 39:569–578

Numan M, Insel TR (2003) The neurobiology of parental behavior. Springer-Verlag, New York

Olshansky JS, Passaro DJ, Hershow RC, Layden J, Carnes BA, Brody J, Hayflick L, Butler RN, Allison DB, Ludwig DS (2005) A potential deline in life expectancy in the United States in the twenty-first century. N Engl J Med 352:1138–1145

Pan Y, Shadel GS (2009) Extension of chronological life span by reduced TOR signaling requires down-regulation of Sch9p and involves increased mitochondrial OXPHOS complex density. Aging 1:131–145

Pascual-Leone A, Amedi A, Fregni F, Merabet LB (2005) The plastic human brain cortex. Ann Rev Neuroscience 28:377–401

Pettorelli N, Durant SM (2007) Longevity in cheetahs: the key to success? Oikos 116:1879–1886

Phelan JP, Rose MR (2005) Why dietary restriction substantially increases longevity in animal models but won't work in humans. Age Res Rev 4:339–350

Radak Z, Hart N, Sarga L, Koltai E, Atalay M, Ohno H, Boldogh I (2010) Exercise plays a preventative role against Alzheimer's disease. J Alzheimers Dis 20:777–783

Rehm S, Deerberg F, Rapp KG (1984) A comparison of life-span and spontaneous tumor incidence of male and female Han: WIST Virginia and retired breeders. Lab Anim Sci 34:458–464

Rose G, Dato S, Altomare K, Bellizzi D, Garasto S, Greco V, Passarino G, Feraco E, Mari V, Barbi C, Bonafe M, Franceschi C, Tan Q, Boiko S, Yashin, AI, De Benedictis G (2003) Variability of the SIRT3 gene, human silent information regulator Sir2 homologue, and survivorship in the elderly. Exp Gerontol 38:1065–1070

Russel SJ, Kahn CR (2007) Endocrine regulation of ageing. Nat Rev Mol Cell Biol 8:681–691

Schumacher M, Weill-Engerer S, Liere P, Robert F, Franklin RJM, Garcia-Segura LM, Lambert JJ, Mayo W, Melcangi RC, Parducz A, Suter U, Carelli C, Baulieu EE, Akwa Y (2003) Steroid hormones and neurosteroids in normal and pathologicalaging of the nervous system. Prog Neurobiol 71:3–29

Selman C, Tullet JM, Wieser D, Irvine E, Lingard SJ, Choudhury AI, Claret M, Al-Qassab H, Carmignac D, Ramadani F, Woods A, Robinson IC, Schuster E, Batterham RL, Kozma SC, Thomas G, Carling D, Okkenhaug K, Thornton JM, Partridge L, Gems D, Withers DJ (2009) Ribosomal Protein S6 Kinase 1 Signaling Regulates Mammalian Life Span. Science 326:140–144

Shingo T, Gregg C, Enwere E, Fujikawa H, Hassam R, Geary C, Cross JC, Weiss S (2003) Pregnancy stimulated neurogenesis in the adult female forebrain is mediated by prolactin. Science 299:117–120

Simon AF, Shih C, Mack A, Benzer S (2003) Steroid control of longevity in *Drosophila melanogaster*. Science 299:1407–1410

Simonsen A et al (2008) Promoting basal levels of autophagy in the nervous system enhances longevity and oxidant resistence in adult *Drosophila*. Autophagy 4:176–184

Stern JM (1996) Somatosensation and maternal care in Norway Rats. In: Rosenblatt JS, Snowden CT (eds) Parental care: Evolution, mechanisms, and adaptive significance. Academic Press, New York, NY

Suzuki S, Gerhold LM, Böttner M, Rau SW, Dela Cruz C, Yang E, Zhu H, Yu J, Cashion AB, Kindy MS, Merchenthaler I, Gage FH, Wise PM (2001) Estradiol enhances neurogenesis following ischemic stroke through estrogen receptors alpha and beta. J Comp Neurol 6:1064–1075

Svare B, Gandelman R (1976) A longitudinal analysis of maternal aggression in Rockland-Swiss albino mice. Dev Psychobiol 9:437–446

Tanapat P, Hastings NB, Reeves AJ, Gould E (1999) Estrogen stimulates a transient increase in the number of new neurons in the dentate gyrus of the adult female rat. J Neurosci 19:5792–5801

Thomas F, Teriokhin AT, Budilova EV, Brown SP, Renaud F, Guegan JF (2000) Human longevity at the cost of reproductive success: evidence from global data. J Evo Biol 13:409–414

Tissenbaum HA, Guarente L (2001) Increased dosage of a sir-2 gene extends lifespan in Caenorhabditis elegans. Nature 410:227–230

Tomizawa K, Iga N, Lu Y-F, Moriwaki A, Matsushita M, Li S-T, Miyamato O, Itano T, Matsui H (2003) Oxytocin improves long-lasting spatial memory during motherhood through MAP kinase cascade. Nat Neurosci 6:384–390

Toth AL, Varala K, Newman TC, Miguez FE, Hutchison SK, Willoughby DA, Simons JF, Egholm M, Hunt JH, Hudson ME, Robinson GE (2007) Wasp gene expression supports an evolutionary link between maternal behavior and eusociality. Science 318:441–444

Valenzano DR, Terzibasi E, Genade T, Cattaneo A, Domenici L, Cellerino A (2006) Resveratrol prolongs lifespan and retards the onset of age-related markers in a short-lived vertebrate. Curr Biol 16:296–300

Vallée M, Mayo W, Le Moal M (2001) Role of pregnenolone, hydroepiandrosterone and their sulfate esters on learning and memory in cognitive aging. Brain Res Brain Res Rev 37:301–312

Van der Heide LP, Ramakers GM, Smidt MP (2006) Insulin signaling in the central nervous system: learning to survive. Prog Neurobiol 79:205–221

Van Praag H, Christie B, Sejnowski TJ, Gage F (1999) Running enhances neurogenesis, learning and long-term potentiation in mice. Proc Natl Acad Sci 96:13427–13431

Viswanathan M, Kim SK, Berdichevsky A, Guarente L (2005) A Role for SIR-2.1 regulation of ER stress response genes in determining *C. elegans* life span. Dev Cell 9:605–615

Volkmar FR, Greenough WT (1972) Rearing complexity affects branching of dendrites in the visual cortex of the rat. Science 176:1145–1147

Wartella J, Amory E, Madonia-Lomas L, Macbeth AH, McNamara I, Stevens L, Lambert KG, Kinsley CH (2003) Single or multiple reproductive experiences attenuate neurobehavioral stress and fear responses in the female rat. Physiol Behav 79:373–381

Weladji RB, Gaillard JM, Yoccoz NG, Holand O, Mysterud A, Loison A, Nieminen M, Stenseth NC (2006) Good reindeer mothers live longer and become better in raising offspring. Proc R Soc 273:15911239–124

Wessells RJ, Fitzgerald E, Cypser JR, Tatar M, Bodmer R (2004) Insulin regulation of heart function in aging fruit flies. Nat Gen 21:127–1281

Wijchers PJ, Burbach JP, Smidt MP (2006) In control of biology: of mice, men and Foxes. Biochem J 397:233–246

Wise PM, Dubal DB, Wilson ME, Rau SW, Böttner M, Rosewell KL (2001) Estradiol is a protective factor in the adult and aging brain: understanding of mechanisms derived from in vivo and in vitro studies. Brain Res Brain Res Rev 37:313–319

Witte AV, Fobker M, Gellner R, Knecht S, Floel A (2009) Caloric restriction improves memory in elderly human. Proc Natl Acad Sci 106:1255–1260

Wood JG, Rogina B, Lavu S, Howitz K, Helfand SL, Tatar M, Sinclair D (2004) Sirtuin activators mimic caloric restriction and delay aging in metazoans. Nature 430:686–689

Woolley CS, McEwen BS (1992) Estradiol mediates fluctuation in hippocampal synapse density during the estrous cycle in the adult. J Neurosci 12:2549–2554

Woolley CS, McEwen BS (1993) Roles of estradiol and progesterone in regulation of hippocampal dendritic spine density during the estrous cycle in the rat. J Comp Neurol 336:293–306

Woolley C.S, Gould E, Frankfurt M, McEwen B.S (1990) Long-term and short-term electrophysiological effects of estrogen on the synaptic properties of hippocampal CA1 pyramidal neurons. J Neurosci 10:4035–3225

Xerri C, Stern JM, Merzenich MM (1994) Alterations of the cortical representation of the rat ventrum induced by nursing behavior. J Neurosci 14:1710–1721

Yang Y, Hou H, Haller EM, Nicosia SV, Bai W (2005) Suppression of FOXO1 activity by FHL2 through SIRT1-mediated deacetylation. EMBO J 24:1021–1032

Yankner BA, Lu T, Loerch P (2008) The aging brain. Annu Rev Pathol 3:41–66

Yankova M, Hart SA, Woolley CS (2001) Estrogen increases synaptic connectivity between single presynaptic inputs and multiple postsynaptic CA1 pyramidal cells: a serial electron-microscopic study. Proc Natl Acad Sci 98:3525

Treatment Trials in Aging and Mild Cognitive Impairment

Jody Corey-Bloom

Abstract There are currently no FDA-approved therapies for mild cognitive impairment (MCI) as no treatment trial to date has convincingly demonstrated a significant effect on cognition or symptom progression. Whether the problem lies with the evaluated compounds, drugs previously shown to have therapeutic benefit in Alzheimer disease (AD), or the clinical trial designs themselves, remains unclear. However, future trials will likely need to use strategies to enrich for more homogeneous samples with appropriate biological characteristics at entry, define optimal treatment durations, and develop highly sensitive assessments and reliable outcomes with the power to detect change and treatment benefit in mildly impaired subjects.

Keywords Aging · Mild cognitive impairment · Treatment · Clinical trials

Contents

1 Introduction to the Problem 348
2 MCI Trial Designs 348
3 Treatment Trials in MCI 349
  3.1 Symptomatic Trials 349
  3.2 Treatment Trials to Delay Progression from MCI to AD 350
  3.3 Ginkgo Evaluation of Memory Study 352
4 Lessons Learned 353
5 Potential Strategies 354
6 Patient Management 354
References 355

J. Corey-Bloom (✉)
Shiley-Marcos Alzheimer Disease Research Center, University of California, San Diego, 8950 Villa La Jolla Drive (Suite C129), La Jolla, CA 92037, USA
e-mail: jcoreybloom@ucsd.edu

Curr Topics Behav Neurosci (2012) 10: 347–356
DOI: 10.1007/7854_2011_153

Published Online: 22 July 2011

1 Introduction to the Problem

One motivation to better understand normal aging, the heterogeneous concept of mild cognitive impairment (MCI), and the risk they impart for future development of dementia, is to provide early interventions that could halt or at least slow progression of symptoms. To date, unfortunately, there are no FDA-approved therapies for MCI. Further, amnestic MCI (aMCI) has received all of the attention with regard to treatment trials, with no trials investigating other distinct clinical subtypes of MCI. Of the existing treatment trials in MCI, most have used a "progression to Alzheimer disease (AD)" design with the focus on slowing cognitive decline and delaying conversion to AD. As a whole, the trials have been disappointing. This article summarizes the designs of clinical trials in MCI, the outcomes of completed MCI clinical trials, lessons learned, and potential strategies for overcoming these problems in future trials. It also briefly describes the Ginkgo Evaluation of Memory (GEM) study, a randomized double-blind, placebo-controlled trial of Ginkgo biloba, in the prevention of dementia (and especially AD) in normal elderly and those with MCI.

2 MCI Trial Designs

Two main research designs have been used in MCI trials to date: the symptomatic design, intended to detect an improvement in cognition, and the "progression to AD" design, aimed at slowing cognitive decline and delaying conversion to AD. The symptomatic design is usually of short duration, perhaps 6–12 months, double-blind, placebo-controlled, with the goal of demonstrating symptomatic improvement. The progression to AD design, on the other hand, is usually longer, perhaps 2–4 years in duration, double-blind, placebo-controlled, with the goal of delaying disease progression. As mentioned above, all studies to date have focused on aMCI to the exclusion of other subtypes. In addition, probably not surprisingly, the therapies that have been examined for MCI mirror the therapeutic approaches developed in AD (Table 1). Acetylcholinesterase inhibitors are one of the two classes of medications currently approved by the FDA for the treatment of AD. In clinical trials, they significantly improved cognitive performance and global status of patients with AD (Winblad and Jelic 2004). Antioxidants have been evaluated for the treatment of AD and other neurodegenerative diseases, but there is no clear consensus regarding their effect. A study by the Alzheimer's Disease Cooperative Study (ADCS) found that vitamin E and selegiline (l-deprenyl, a selective irreversible MAO-B inhibitor) slowed the rate of progression of AD (Sano et al. 1997). Although epidemiologic studies have suggested that anti-inflammatory agents might reduce the risk of AD, several clinical trials have failed to document beneficial effects for steroids or non-steroidal anti-inflammatory agents in AD (Aisen 2002; Aisen et al. 2003). Most studies of nootropic agents for AD have been

Table 1 Therapeutic approaches in MCI

• Acetylcholinesterase inhibitors
• Antioxidants
• Anti-inflammatory agents
• Nootropic agents
• Glutamate receptor modulators

negative, and a clinical trial of the Ampakine CX516, a glutamate receptor modulator, reportedly failed to show any significant benefit in elderly individuals with cognitive impairment.

3 Treatment Trials in MCI

3.1 Symptomatic Trials

Two symptomatic trials of donepezil in MCI have been reported to date. The first, a 24-week randomized, double-blind, placebo-controlled parallel group study, examined the efficacy and tolerability of donepezil in 269 aMCI subjects at 22 centers in the US (Salloway et al. 2004). This study was supported by Pfizer. To be enrolled, subjects had to have an MMSE score of $\geq$24 and clinical dementia rating (CDR) of 0.5 (memory box score 0.5–1.0). Subjects were randomized 1:1 to activate drug and placebo. Subjects receiving active drug took 5 mg for the first 6 weeks and had to tolerate escalation to 10 mg thereafter or be discontinued from the trial. Efficacy analyzes were performed on the intent-to-treat (ITT) and fully evaluable (FE) populations. The outcome measures followed an AD trial model with a primary cognitive (New York University (NYU) paragraph delayed recall) and a primary global (clinical global impression of change for MCI, CGIC-MCI) outcome. Secondary outcome measures included an 89 point modified ADAS-Cog (included a 12 word immediate/delayed recall and concentration/distractibility rating), WMS-R digit span backwards, symbol digit modalities, and patient global assessment. The donepezil-treated group did not show significant treatment benefits over placebo on the primary outcome measures, the NYU paragraph delayed recall or the CGIC-MCI. Although the primary endpoints were not met, some secondary measures, including digit span and symbol digit modalities, suggested a potential benefit of donepezil at 24 weeks. Interestingly, rather than decline, MCI subjects tended to show improvement on virtually all cognitive outcome measures, no matter which treatment group they were randomly assigned. Unfortunately, the CGIC-MCI showed limited sensitivity at 6 months, with no impairment or decline in some domains over the short-time frame. Adverse events, predominantly gastrointestinal, occurred at a higher frequency (88% in the donepezil group and 73% in the placebo group) than in AD trials. In addition, it should be noted that there was a significant discontinuation rate (32%) with dose escalation.

The 48-week symptomatic donepezil trial for aMCI involved 821 subjects at 74 sites in the US (Doody et al. 2009). This study was supported by Eisai Inc. and Pfizer Inc. To be enrolled in this study, subjects had to have an MMSE of 24–28, CDR of 0.5, and meet an education-adjusted cut-off on the logical memory II delayed paragraph recall subtest of the Wechsler Memory Scale-Revised (WMS-R). The study utilized a dual primary endpoint of significant improvement on the modified (89 point) ADAS-cog and CDR-SB at 48 weeks. Secondary efficacy measures evaluated cognition, behavior, and function. Subjects were assigned to treatment with donepezil (5 mg/day for 6 weeks followed by 10 mg/day) or placebo. The dual primary efficacy endpoint was not reached. There was a small, but statistically significant, benefit in favor of donepezil on the ADAS-Cog, but not on the CDR-SB, at study endpoint. Little change was observed for either group on any of the secondary efficacy measures. Adverse events and discontinuations due to adverse events were higher in the donepezil-treated group.

A third symptomatic trial in MCI investigated the efficacy and tolerability of piracetam in 675 subjects with MCI over 12 months. The principal outcome measure was a composite score of key outcomes from eight neuropsychological tests. By report, there were no significant differences between the Piracetam and placebo groups; however, the results have never been formally published.

3.2 Treatment Trials to Delay Progression from MCI to AD

The Alzheimer Disease Cooperative Study (ADCS)–sponsored study of vitamin E and donepezil for MCI involved 769 subjects at 69 centers in the US and Canada over 3 years (Petersen et al. 2005). Also called the Memory Impairment Study (MIS), it was the first reported large-scale clinical trial in MCI. The MIS was supported by the National Institute on Aging (NIA), Pfizer Inc. and Esai Inc. To be enrolled in this study, subjects had to have an MMSE of $\geq$24, CDR of 0.5, and meet an education-adjusted cut-off on the Logical Memory II Delayed Paragraph Recall subtest of the WMS-R. There were three treatment arms: vitamin E 2000 IU/day, donepezil 10 mg/day, and placebo. The primary trial endpoint was conversion to possible or probable AD as defined by NINCDS-ADRDA criteria. Secondary efficacy measures evaluated cognition, behavior, function, and quality of life. Although conversion to AD favored donepezil at 1 year, there were no differences among groups with regard to conversion to AD at 3 years. More than half of the subjects enrolled in this trial possessed at least one APOE ε4 allele. Possession of the ε4 allele was associated with a three fold greater risk of conversion from aMCI to dementia and, thus, clearly an important predictor of progression. When the authors looked at the progression to AD for APOE ε4 positive participants by treatment group, they found that the effect of donepezil was greater in ε4 positive individuals and persisted for 2 years. While neither of the two active arms reduced the risk of progressing to AD over the entire 36 months, donepezil reduced the risk of progression to AD for the first 12 months in all subjects and up to 24 months in those who were positive for the APOE ε4 allele. Therefore, a treatment effect for donepezil was noted for up to 12–24 months. No treatment effect was noted for vitamin E. The secondary cognitive and global measures essentially corroborated the primary outcomes. The dropout rate was about 12% per year. In this study, aMCI as defined by the Petersen criteria, APOE ε4 carrier status, and hippocampal volume at baseline strongly predicted progression to AD.

Other treatment trials have been less promising for halting conversion from MCI to dementia over time. A large trial of rivastigmine, called the Investigation into the Delay to Diagnosis of AD with Exelon (InDDEx), was a double-blind, placebo-controlled study of 1018 aMCI patients supported by Novartis. The InDDEx trial had many of the same features as the ADCS study but was conducted in 14 countries using multiple languages and translations of the neuropsychological instruments (Feldman et al. 2007). The trial was initially intended as a 3-year study but was extended to 4 years because of slow enrollment. The primary efficacy measures were time to clinical diagnosis of AD in addition to change from baseline on cognitive function as measured by z-score on a cognitive test battery covering multiple domains (working memory, immediate and delayed recall, cued recall, attention/concentration, language, executive functioning, and praxis). There was a high-dropout rate in this study, with only 51% of rivastigmine-treated and 63% of placebo-treated subjects completing the trial. At baseline, arms were not well matched with regard to frequency of APOE ε4 genotype, which was 46% in the placebo arm but only 37% in the rivastigmine arm. The study also had a lower conversion rate to AD than expected, with only 21.4% of placebo-treated and only 17.3% of rivastigmine-treated subjects progressing to AD over 4 years. Although rivastigmine was favored, the results were not statistically significant, and secondary assessments were also not significant.

Investigation of the efficacy of another acetylcholinesterase inhibitor, galantamine, also failed to reveal a significant effect of galantamine on conversion to dementia in those with aMCI in either of two trials supported by Johnson & Johnson (Winblad et al. 2008). The Gal-Int-11 and Gal-Int-18 trials were both 2-year, double-blind, placebo-controlled international studies of 995 and 1062 aMCI patients, respectively. The trial designs were identical except for inclusion of magnetic resonance imaging (MRI) in the Gal-Int-11 study. Patients took 16 or 24 mg/day of galantamine or placebo. At baseline, the frequency of the APOE ε4 genotype was only 26% in the galantamine-treated arm and 30% in the placebo-treated arm in the Gal-Int-11 study; but only 24% in both the galantamine- and placebo-treated arms in the Gal-Int-18 study. The primary efficacy measure was progression to dementia (CDR1) at 24 months. Secondary efficacy measures included a 90-point modified ADAS-Cog, CDR-SB, DSST, and a modified ADCS-ADL scale. For the Gal-Int-11 study, 13% of galantamine-treated and 18% of placebo-treated subjects converted to dementia over 2 years. For the Gal-Int-18 study, 17% of galantamine-treated and 21% of placebo-treated subjects converted to dementia over 2 years. Thus, there was no effect on conversion to dementia for galantamine at 24 months. In Gal-Int-11, galantamine was superior on CDR-SB at

12 and 24 months, but there were no differences on the mADAS-cog or mADCS-ADL. Dropout rates are not available for either study. Greater mortality was observed in the galantamine group in both studies but the significance remains unclear.

And finally, a large randomized, placebo-controlled, double-blind study examined the ability of the COX 2 inhibitor, rofecoxib, to delay disease progression in 1457 aMCI subjects at 46 sites in the US (Thal et al. 2005). This study was sponsored by Merck. Subjects received either 25 mg/day of rofecoxib or placebo. The trial was initially intended as a 2-year study but was extended to 4 years because of low rates of conversion to AD. In addition, the memory inclusion criteria were modified at 6 months to enhance recruitment. The primary endpoint was conversion to CDR ≥ 1 and incident AD. At baseline, only about 35% of subjects possessed an APOE ε4 allele. There was an extraordinarily high-dropout rate in this study with only 55% of subjects completing the study (only 40% on drug). There was a lower than expected annual rate of conversion to AD (approximately 5–6%) and it is unclear if the modified memory inclusion criteria contributed to that lower conversion rate. It is worth noting however that the placebo subjects declined by less than one point on the ADAS-Cog over 4 years. Conversion to AD actually favored placebo in this trial but the authors dismissed the significance of this finding because the secondary cognitive measures did not corroborate the primary outcome.

3.3 Ginkgo Evaluation of Memory Study

The Ginkgo Evaluation of Memory (GEM) Study was a randomized, double-blind, placebo-controlled trial of the effectiveness of *G biloba* versus placebo in reducing the incidence of all-cause dementia and AD in elderly individuals with normal cognition and those with MCI (DeKosky et al. 2008). It was conducted at five academic medical centers in the US between 2000 and 2008 and encompassed 3069 community-dwelling individuals aged 75 years or older with normal cognition (n = 2587) or MCI (n = 482) at baseline. Subjects were randomized to twice-daily *G biloba* 120 mg or placebo and assessed every 6 months for incident dementia. The primary outcome measure was progression to dementia and AD, as determined by expert panel consensus. Five hundred and twenty three individuals developed dementia with 92% of the dementia cases classified as possible or probable AD, or AD with evidence of vascular disease of the brain. Rates of dropout and loss to follow-up were low (6.3%), and the adverse effect profiles were similar for both groups. The overall dementia rate was 3.3 per 100 person-years in participants assigned to G. biloba and 2.9 per 100 person-years in the placebo group. The hazard ratio (HR) for G. biloba compared with placebo for all-cause dementia was 1.12 (95% confidence interval (CI), 0.94–1.33; P = 0.21) and for AD, 1.16 (95% CI, 0.97–1.39; P = 0.11). G. biloba also had no effect on the rate of progression to dementia in participants with MCI (HR, 1.13; 95% CI,

0.85–1.50; P = 0.39). Thus, G. biloba was not effective in reducing either the overall incidence rate of dementia or AD incidence in elderly individuals with normal cognition or those with MCI.

4 Lessons Learned

In hindsight, several important factors likely influenced the results of the MCI studies, which are summarized in Table 1. First and foremost, there was a very variable rate of progression from aMCI to AD seen in these trials. Annual rates of conversion (Kaplan–Meier annualized conversion rate, adjusted for dropouts) from MCI to AD ranged from 5 to 6% in the Rofecoxib trial to 13–16% in the ADCS vitamin E/Donepezil study. Sources of this variability likely include subject heterogeneity, with regard to impairment level, culture, language, and APOE ε4 carrier status, in addition to likely simple differences in implementation of enrollment criteria. The current recruitment strategy is probably too heterogeneous with regard to subject populations who, although variable in their level of impairment, are likely closer to normal subjects than to demented ones. Clearly, subjects with MCI progress at a slower annual rate than patients with AD. It is striking that, rather than decline, MCI subjects showed improvement on cognitive outcome measures, regardless of treatment assignment, in the 24-week symptomatic donepezil trial and that the CGIC-MCI showed limited sensitivity over the short 6-month time frame. Higher APOE ε4 carrier rates are also associated with greater likelihood of progression and it is significant, in this regard, that the trial with the highest ε4 carrier rate, the ADCS vitamin E/Donepezil trial, showed the highest rates of conversion to AD. That small differences in implementation of enrollment criteria can have a significant effect has been suggested by the Rofecoxib trial where modification of the memory criteria at 6 months has been blamed for the inclusion of milder, possibly normal, subjects who were less likely to convert to AD. Another significant lesson from these trials is that MCI patients may show increased awareness of, and lower tolerability for, adverse events, as evidenced by the significantly higher discontinuation rates in MCI trials as compared to AD studies (Table 2).

These trials highlight the fact that outcome measures currently used for trials may be insensitive to the real impairments of MCI. The "conversion to AD" trial design essentially dichotomizes a continuous variable. The outcome measures used to date in MCI trials have essentially followed an AD model. They have been chosen to comply with prior FDA guidelines for AD trials that required demonstration of improvement on performance-based cognitive and global outcome measures. What is needed are cognitive, global, and ADL instruments sensitive to mild impairment. In the 24-week symptomatic donepezil trial, MCI subjects showed improvement and a capacity to learn, rather than decline, even in the placebo group. There was no impairment or decline in some domains of the CGIC-

Table 2 Clinical trials in aMCI

Agent	Mechanism	*N*	Duration	Endpoint	Outcome
Donepezil	AChEI	269	24 weeks	Symptoms	Negative
Donepezil	AChEI	821	48 weeks	Symptoms	Partially Positive
Donepezil/vitamin E	AChEI/vitamin	769	3 years	AD	Partially Positive
Rofecoxib	NSAID	1200	2–3 years	AD	Negative
Galantamine	AChEI	995	2 years	CDR 1	Negative
Galantamine	AChEI	1062	2 years	CDR 1	Negative
Rivastigmine	AChEI	1018	3–4 years	AD	Negative

aMCI amnestic MCI, *AD* Alzheimer disease, *AChEI* Acetylcholinesterase inhibitor, *NSAID* Non-steroidal anti-inflammatory drug, *CDR* clinical dementia rating

MCI even in the placebo group. The placebo groups declined by less than 1 point per year on the ADAS-Cog in most of the MCI trials, and by less than one point over 4 years on the ADAS-cog in the Rofecoxib trial. In fact, Winblad et al. (2008) noted a significant ceiling effect in MCI subjects participating in the Gal-Int-11 study, with more that 80% of subjects having perfect scores at baseline on multiple ADAS-cog subscales.

5 Potential Strategies

To date, no drug has convincingly demonstrated symptomatic or disease delaying effects in aging or MCI. Future trials may benefit from less heterogeneous recruitment, better outcome measures, novel imaging outcomes, larger sample sizes, and longer trials.

With regard to recruitment, there will need to be more stringent entry criteria with enriched populations for sensitive predictors of progression such as APOE $\varepsilon 4$ carrier status, positivity on in vivo amyloid imaging with Pittsburgh compound B (PiB) (or newer ligands labeled with fluorine, which has a longer half-life), hippocampal atrophy, and CSF (low Aβ and high τ levels) biomarkers.

There will need to be better outcome measures including continuous rather than dichotomous variables; more sensitive cognitive assessments; global and ADL instruments that reflect subtle impairments in complex activities of daily living; and novel imaging (structural, FDG-PET, amyloid) outcomes.

6 Patient Management

Despite the disappointing results from many of the MCI treatment trials to date, most clinicians recommend a careful discussion with the patient and family regarding the evolving nature of MCI and the fact that approximately 10–15% of patients with MCI progress to AD per year. Patients should be encouraged to use the time to plan for the future. Clinicians should have a frank discussion with the patient and family regarding the fact that, although considered "off-label", some would indeed proceed with the use of a cholinesterase inhibitor. Patients should be encouraged to work closely with their primary care physician to control vascular risk factors (such as blood pressure, hyperlipidemia, diabetes) and treat concomitant conditions such as depression or thyroid disease. They should be urged to make dietary modifications; engage in moderate exercise or physical activity; and lead an active and socially integrated lifestyle. Many clinicians recommend cognitive training for patients with mild cognitive impairment. Although there is some support from the literature (Tsolaki et al. 2011), a recent Cochrane review (Martin et al. 2011) reported that, although cognitive interventions can lead to performance gains, often the effects cannot be attributed specifically to cognitive training, as the benefits observed did not exceed the improvement in active control conditions. Clearly there is a need for a well-controlled randomized trial to assess the efficacy of cognitive training in MCI.

References

Aisen PS (2002) The potential of anti-inflammatory drugs for the treatment of Alzheimer's disease. Lancet Neurol 1(5):279–284

Aisen PS, Schafer KA, Grundman M, Pfeiffer E, Sano M, Davis KL, Farlow MR, Jin S, Thomas RG, Thal LJ, Alzheimer's Disease Cooperative Study (2003) Effects of rofecoxib or naproxen vs placebo on Alzheimer disease progression: a randomized controlled trial. JAMA 289(21):2819–2826

DeKosky ST, Williamson JD, Fitzpatrick AL, Kronmal RA, Ives DG, Saxton JA, Lopez OL, Burke G, Carlson MC, Fried LP, Kuller LH, Robbins JA, Tracy RP, Woolard NF, Dunn L, Snitz BE, Nahin RL, Furberg CD, Ginkgo Evaluation of Memory (GEM) Study Investigators (2008) Ginkgo biloba for prevention of dementia: a randomized controlled trial. JAMA 300(19):2253–2262

Doody RS, Ferris SH, Salloway S, Sun Y, Goldman R, Watkins WE, Xu Y, Murthy AK (2009) Donepezil treatment of patients with MCI: a 48-week randomized, placebo-controlled trial. Neurology 72(18):1555–1561

Feldman HH, Ferris S, Winblad B, Sfikas N, Mancione L, He Y, Tekin S, Burns A, Cummings J, del Ser T, Inzitari D, Orgogozo JM, Sauer H, Scheltens P, Scarpini E, Herrmann N, Farlow M, Potkin S, Charles HC, Fox NC, Lane R (2007) Effect of rivastigmine on delay to diagnosis of Alzheimer's disease from mild cognitive impairment: the InDDEx study. Lancet Neurol 6(6):501–512 Erratum in: Lancet Neurol. 2007 Oct; 6(10):849

Martin M, Clare L, Altgassen AM, Cameron MH, Zehnder F (2011) Cognition-based interventions for healthy older people and people with mild cognitive impairment. Cochrane Database Syst Rev 1:CD006220

Petersen RC, Thomas RG, Grundman M, Bennett D, Doody R, Ferris S, Galasko D, Jin S, Kaye J, Levey A, Pfeiffer E, Sano M, van Dyck CH, Thal LJ, Alzheimer's Disease Cooperative Study Group (2005) Vitamin E and donepezil for the treatment of mild cognitive impairment. N Engl J Med 352(23):2379–2388

Salloway S, Ferris S, Kluger A, Goldman R, Griesing T, Kumar D, Richardson S, Donepezil 401 Study Group (2004) Efficacy of donepezil in mild cognitive impairment: a randomized placebo-controlled trial. Neurology 63(4):651–657

Sano M, Ernesto C, Thomas RG, Klauber MR, Schafer K, Grundman M, Woodbury P, Growdon JCW, Pfeifer E, Schneider LS, Thal LJ, for the Members of the Alzheimer's Disease Cooperative Study (1997) A two-year, double-blind, randomized multicenter trial of Selegiline and α-tocopherol in the treatment of Alzheimer's disease. New Eng J Med 336(17):1245–1247

Thal LJ, Ferris SH, Kirby L, Block GA, Lines CR, Yuen E, Assaid C, Nessly ML, Norman BA, Baranak CC, Reines SA, Rofecoxib Protocol 078 study group (2005) A randomized, double-blind, study of rofecoxib in patients with mild cognitive impairment. Neuropsychopharmacology 30(6):1204–1215

Tsolaki M, Kounti F, Agogiatou C, Poptsi E, Bakoglidou E, Zafeiropoulou M, Soumbourou A, Nikolaidou E, Batsila G, Siambani A, Nakou S, Mouzakidis C, Tsiakiri A, Zafeiropoulos S, Karagiozi K, Messini C, Diamantidou A, Vasiloglou M (2011) Effectiveness of nonpharmacological approaches in patients with mild cognitive impairment. Neurodegener Dis 8(3): 138–145

Winblad B, Jelic V (2004) Long-term treatment of Alzheimer disease: efficacy and safety of acetylcholinesterase inhibitors. Alzheimer Dis Assoc Disord 18(Suppl 1):S2–S8

Winblad B, Gauthier S, Scinto L, Feldman H, Wilcock GK, Truyen L, Mayorga AJ, Wang D, Brashear HR, Nye JS, GAL-INT-11/18 Study Group (2008) Safety and efficacy of galantamine in subjects with mild cognitive impairment. Neurology 70(22):2024–2035

Erratum to: Behavioral Neurobiology of Aging

Marie-Christine Pardon and Mark W. Bondi

Erratum to: Springer, Current Topics in Behavioral Neurosciences, Vol. 10, DOI: 10.1007/978-3-642-23875-8

In front matter the Series Editor Information should read:
Series Editors
Mark A. Geyer, La Jolla, CA, USA
Bart A. Ellenbroek, Wellington, New Zealand
Charles A. Marsden, Nottingham, UK

Mark W. Bondi's address should read:
Dr. Mark W. Bondi
Director, VA Neuropsychological Assessment Unit
Professor, Dept of Psychiatry, University of California San Diego
VA San Diego Healthcare System (116B)
3350 La Jolla Village Drive
San Diego, CA 92161
USA

At the end of the preface:
Nottingham is the affiliation of Marie-Christine Pardon
La Jolla is the affiliation of Mark W. Bondi

The online version of the original front matter can be found under 10.1007/978-3-642-23875-8.

M.-C. Pardon (✉)
School of Biomedical Sciences, Queen's Medical Centre, University of Nottingham Medical School, Nottingham, NG7 2UH , UK
e-mail: pardon@nottingham.ac.uk

M. W. Bondi
Department of Psychiatry, University of California San Diego, 3350 La Jolla Village Drive, San Diego, CA 92161, USA
e-mail: mbondi@ucsd.edu

Curr Topics Behav Neurosci (2012) 10:E1-E1
DOI: 10.1007/978-3-642-23875-8_154

Index

A
Acetylcholine, 149, 153
Affective auditory verbal learning test, 278
Age-associated memory impairment (AAMI), 15
Age-related behavioural decline, 293, 296, 302, 308
Aging, 113–114, 116–118, 124–126, 128–130, 165–166, 170–176
Aging, 294, 296, 305, 310, 348, 350, 351, 354
Alzheimer's disease (AD), 37, 40, 42, 187–195, 197–205, 250, 254, 256, 273–274, 278, 282, 284
Alzheimer's disease, 114, 128, 303, 319
Amygdala, 60
Amyloid precursor protein (APP), 317
ANS reactivity, 59
Antecedent-focused strategies, 55
Anterior cingulate cortex (ACC), 43, 189
Anxiety, 244–247, 249, 253, 256–259, 261
ApoE, 40, 189, 191–192, 200, 204–205, 275, 278
Apolipoprotein E (APOE), 18, 99
Appraisal, 53
Arousal, 56
Autophagy, 320

B
Basolateral amygdala, 146
Biomarker, 35, 193, 205
Brain derived neurotrophic factor (BDNF), 329

C
Cache county study: CCS, 5
Caenorhabditis elegans, 331
Caloric restriction, 331
cAMP-response-element binding (CREB), 330
Categorization of emotionally-relevant stimuli, 55
Cerebrovascular disease, 214, 223, 236
Cingulate cortices, 325
Clinical assessments, 8
Clinical trail, 347–351, 354
Cognition, 214, 219, 225, 227, 229–231, 233
Cognitive decline, 113–130
Cognitive engagement, 24
Cognitive impairment not dementia (CIND), 15
Cognitive reserve, 300
Corpus callosum, 144–145

D
Declarative memory, 68–70, 73, 81, 83
Default-mode network, 97, 98
Dementia (AACD), 15
Dementia, 219, 221, 223–236, 337
Dendrites, 140, 145–147
Dendritic spines, 323, 326
Diagnostic and statistical manual (DSM), 244–245, 256–259
Digit symbol substitution, 191
Dopamine, 137–153
Dorsal raphe, 325
Dry land maze (DLM), 327

Curr Topics Behav Neurosci (2012) 10: 357–359
DOI: 10.1007/978-3-642-23875-8

E
Ecdysone, 334
Emotion regulation, 54, 61
Emotional distress, 52
Emotional experience, 57
Emotional expression, 54
Emotional valence, 56
Endogenous opioid, 325
Entorhinal cortex, 188, 190, 194–195
Enviromimetics, 294
Environmental enrichment, 293, 297, 304
Epigenetic, 40
Episodic memory, 187–194, 197–198, 200, 202, 226–227, 229–232, 235
Estradiol, 167, 174
Estrogen, 138, 150–153, 323
Event-related potentials, 60
Everyday action, 226–227, 233–236
Executive function, 229–232, 234, 236
Experience or feeling of emotion, 53

F
Facial emotion expression, 58
Facial expressions of emotions, 55
fMRI, 60, 113–129
Fos imaging, 67
Frontal, 325
Fruit fly *Drosophila melanogaster*, 331
Fuld Object Memory Test, 191

G
Gender, 293, 294, 301–302
Glucose utilization, 319
Glutamate, 149–150

H
Hippocampus, 41, 137, 139, 143, 145, 147, 150, 165, 167–168, 171–172, 174–175, 188, 190, 194–195, 197–198, 275, 283, 321
Hormone treatment, 139, 149–153
Human, 68–70, 73–75, 77–79, 81, 85

I
Individual differences, 294, 296, 309
Insulin/IGF-1, 319
International affective picture system, 56

K
Klotho, 334

L
Language, 224–226, 228, 231–232
Lifestyle, 297, 299, 301–303
Logical memory subtest, 191

M
Magnetic resonance imaging (MRI), 195, 198, 201, 206, 247, 280
Medial prefrontal cortex (MPFC), 43
Medroxyprogesterone acetate, 138, 151
Memory, 17, 165–166, 168–172, 174
Menopause, 150, 165–166, 168–172, 174–175
Mild cognitive impairment (MCI), 187–191, 193–203, 205, 274, 276, 279, 281, 283–284
Mild cognitive impairment, 15, 347, 355, 358
Mini-mental state exam (MMSE), 188, 278, 281–282, 284, 286
Mitochondrion, 319
Mood, 57
Mouse, 69, 70, 73–74, 76, 83
Myelination, 144–145

N
Nested-case-control, 7, 9
Neurogenesis, 293, 296, 304, 307
Neurogenesis, 321
Neuron number, 142–143
Neurotrophic tyrosine kinase receptor, type 2 (NTrk2), 330

O
Older adults, 220–221, 223, 229, 233
Olfactory bulb, 323
Orbito-frontal cortex (OFC), 43
Oxidative stress, 336
Oxytocin, 325

P
Paced auditory serial addition test (PASAT), 278
Parietal, 325
Parkinson's disease (PD), 248, 250, 253

Parturition, 317
Perceptual discrimination, 55
Physical exercise, 260, 274, 277, 285
Physiologic change, 54
Placental lactogens, 325
Plasticity, 293, 295, 298, 303, 305, 319
Population-based studies, 10
Preclinical models, 293, 305
Prefrontal cortex (PFC), 40, 280
Prefrontal cortex, 19, 60, 137–139, 142, 145–146, 148–149, 151, 165, 167, 172–173
Pregnancy, 317
Prevalence, 4
Prodromal AD, 15
Progesterone, 138, 151–152, 325
Prolactin, 323
Protein, 330
Psychophysiological measures, 54

Q
Quantitative, reverse transcriptase polymerase chain reaction (qRT-PCR), 329

R
Reactive oxygen species (ROS), 319
Reproductive experience, 317
Reproductive success, 336
Response-focused strategies, 55
Resveratrol, 333
Rodents, 300, 306

S
Schizophrenia, 47, 244–247, 249, 254–256, 261
Screening, 7
Selective reminding test, 191
Self-assessment manikin, 56
Sirtuin deacetylase proteins (SIR/SIRT), 333
Small-vessel disease, 220, 223–224, 226–227, 236
Socioemotional selectivity theory, 61
Spinophilin, 330
Stereology, 140
Stress, 294, 296, 307
Study, 5
Synapses, 137, 140, 145–147, 151–153
Synaptogenesis, 325, 146–147, 151–152, 152
Syntaxin, 330

T
Target of rapamycin (TOR), 320, 332
Telomere, 40
Treatment, 347–351, 353–354

V
Vascular cognitive impairment (vascular), 13, 15, 224, 236
Vascular depression, 247, 251
Vasopressin, 323
Ventromedial prefrontal cortex, 60

W
White matter, 140, 144
Working memory, 68, 74–75, 77, 81, 83

Y
Yeasts, 331

Printed by Publishers' Graphics LLC
SRO20130111.19.20.6